EARTH LAB
Exploring the Earth Sciences

Claudia Owen
Florida International University

Diane Pirie
Florida International University

Grenville Draper
Florida International University

BROOKS/COLE

™

THOMSON LEARNING

Australia • Canada • Mexico • Singapore • Spain • United Kingdom • United States

BROOKS/COLE

THOMSON LEARNING

Sponsoring Editor: *Nina Horne*
Developmental Editor: *Marie Carigma-Sambilay*
Marketing Manager: *Rachael Alvelais*
Assistant Editor: *Samuel Subity*
Editorial Assistant: *Rebecca Eisenman*
Production Editor: *Janet Hill*
Production Service: *Martha Emry Production Services*

Cover Design: *Roger Knox*
Cover Photo: *Laurie Schmidt*
Text Illustration: *Jim Atherton*
Composition: *Carlisle Communications*
Photo Research: *Kathleen Olson*
Text Printing and Binding: *Transcontinental Printing*
Cover Printing: *Transcontinental Printing*

For more information about this or any other Brooks/Cole product, contact:
BROOKS/COLE
511 Forest Lodge Road
Pacific Grove, CA 93950 USA
www.brookscole.com
1-800-423-0563 (Thomson Learning Academic Resource Center)

Printed in Canada

10 9 8 7 6 5 4 3 2 1

Library of Congress Cataloging-in-Publication Data
Owen, Claudia.
    Earth lab : exploring the earth sciences / Claudia Owen, Diane Pirie, Grenville Draper.
        p. cm.
    ISBN 0-534-37953-2 (paper)
    1. Earth sciences. I. Pirie, Diane. II. Draper, Grenville. III. Title.
    QE28 .O94 2001
    550'.78—dc21   00-049399

Contents

SECTION 3

Earth's Surface and the Fluid Earth

SECTION 4

Resources

Contents

Preface

Science and technology currently provide major stimuli of change in our society. Science courses prepare students to participate in these developments and, for this reason, are commonly part of college general education (core) requirements. In addition, a scientific understanding of our natural environment may be critical to our survival as a species on our fragile planet. Awareness of science and our natural environment comes more deeply and fully when students participate actively in data gathering, performing experiments, handling natural materials, and observing the behavior of natural phenomena. This experiential learning is essential in a laboratory environment and refines a student's ability to reason critically and to solve problems creatively.

The intent of *Earth Lab: Exploring the Earth Sciences* is to guide the student by way of active participation to a better understanding and appreciation of the fundamental concepts presented in an introductory Earth science or physical geology course. We have written this manual in an informal style, with the aim of making the substance and methods of Earth science more accessible to the students. We avoid the use of technical jargon except where our design is to teach terminology.

We, along with scores of other instructors, have used most of the labs in this book at Florida International University (FIU) for over ten years. Such experience has allowed us to incorporate past improvements to the material for this edition. Thousands of students have tested this material. Their feedback has helped us shape and hone this book to ensure that the material and activities are effective and sound and help students focus on the important concepts.

Message to Instructors

Content is directly related to the exercises and lab experiences included in this book. We encourage instructors to coordinate the use of this lab book with a lecture and text, such as *Earth Science Today* by Murphy and Nance (Brooks/Cole) that can complete the theoretical explanations necessary for an in-depth understanding of the material. We have also designed the book for flexibility. Most of the 17 labs have more activities than can be completed in a single lab session, so instruc-

tors can pick and choose between activities or, in some cases, expand specific topics to cover more than one lab period. Those instructors inclined to emphasize mineral and rock identification have a rich body of instructional materials to choose from, ranging from high-quality photographs to reference materials and charts. In addition, the rock masses, geologic time, oceans, atmosphere, and resources chapters are particularly extensive, so that instructors can expand any of these subjects to two labs if that is their desired emphasis.

Instructor's Manual

Accompanying this book is an instructor's manual that provides answers to all questions, has sample data for experiments, gives example graphs, and answers computer exercises. Many of the exercises and activities require particular materials to be present in the laboratory for use by the students. The instructor's manual provides a list of materials needed to do each lab activity and a source guide for the resources needed. We provide helpful hints for instructors to enhance lab activities and avoid common pitfalls. Also included are suggestions of methods to shorten or expand selected activities to emphasize certain themes.

Hands-on Activities

Initially at FIU, we started writing this manual for our own classes because we found that most manuals, although excellently produced, did not have a strong enough focus on handling specimens, doing activities, and using the various tools of science. We strongly believe that a lab should be the most pragmatic of teaching venues. Students need practice doing science and practical experience with the phenomena that they are studying. The more the students are engaged, both physically and mentally, the more the students' depth of understanding increases. Wherever we could, we have tried to include activities involving experiments, and handling samples, working with materials in ways that are fun and informative and facilitate a deeper understanding and appreciation of the subject.

Computer Activities

Many students find exploring a visual program on a computer more compelling and interesting than working with the same material from a text. For those who prefer the interactive nature of computer programs, we have incorporated activities using commercially available CD-ROM disks throughout the manual. Other lab activities and text cover the same material, so labs that do not have computers will not be disadvantaged, but those that have such facilities have another opportunity to enhance students' learning.

Acknowledgments

Numerous lab instructors and countless students at FIU have tested many of the activities in this lab manual over more than ten years. We thank them for their patience and feedback in producing this manual. We particularly thank Lois Geier, who not only taught many of the labs but also coordinated their scheduling, ordered the materials, and brought many innovations to our attention. Lois Geier, Tom Beasley, Florentin Maurrasse, Laurel Collins, and Mike Gross have helped provide materials for images in this book, for which we thank them. The Geology Department of FIU provided photographic equipment and samples for the mineral and rock photographs. We would also like to express appreciation to our spouses and family members who have endured our lack of attention to them during the many months of propelling this project to completion.

We thank our editor, Nina Horne, for all her encouragement and her support of this project. We also thank others of the Brooks/Cole book team for their contributions: Development Editor Marie Carigma-Sambilay, Marketing Manager Rachel Alvelais, Assistant Editor Samuel Subity, interior design John Edeen, cover designer Roger Knox, and Project Editor Janet Hill. Especially, many thanks to Martha Emry of Martha Emry Production Services for her expert guidance and professional expertise in moving us and our book materials through all the stages of production.

We would especially like to thank John Degenhardt of Texas A&M University, Nathan L. Green of the University of Alabama, and Carrie E. Schweitzer of Kent State University for their time and energy reviewing this manual. We thank the following organizations for providing maps, photos, or data: U.S. Geological Survey, Pennsylvania Topographic and Geologic Survey, Grand Canyon Natural History Association, Nevada Bureau of Mines, Washington Division of Mines and Geology, National Oceanic and Atmospheric Administration, Weather.com, and the Agriculture Stabilization and Conservation Services of the U.S. Department of Agriculture.

Future success of new editions of this manual relies on comments and recommendations from those who use it. We encourage you to contact us at EarthLab2001@yahoo.com with your suggestions.

Claudia Owen
Diane Pirie
Grenville Draper

The Scientific Method

> **OBJECTIVES**
>
> 1. To learn the purpose and structure of this lab book
> 2. To understand the scientific method

As the first generations to look back at Earth from space, we realize, as Carl Sagan does in his book *Pale Blue Dot,* that Earth is our planet, full of our friends and loved ones, where "every human being who ever was, lived out their lives." There is no other planet we can call home, and for the foreseeable future this is the only planet we will live on. The importance of the Earth sciences is obvious: we need to know our home. Ignorance of the workings of our home planet denies us the wonder of its intricate grandeur. The workings of this planet—some grand, dramatic and dangerous, others delicately balanced and fragile—combine to create a home that periodically overwhelms us but that we now may overwhelm if we do not understand it.

This book guides you toward a deeper understanding of Earth's processes and materials, through hands-on laboratory activities and experiments. The reward of a laboratory class is the opportunity to work with the materials and methods of the Earth sciences. The knowledge and skills you gain through laboratory experience are superior to mere lecture and book learning, because, in a lab, you learn by doing.

Organization of This Book

We have organized our lab manual into four sections.

The first section consists of this introduction about the book and the scientific method and the first lab on the reading and making of maps (cartography), skills that are fundamental in the study of the Earth and its systems.

The second section concerns the materials and processes that occur in the solid Earth. Here we show how to understand and recognize the minerals and rocks that make up the crust of the Earth; then how to reconstruct the events of the past from the information contained in the solid Earth; and, finally, how pieces of our planet move, causing earthquakes and creating our landscape.

Section 3 deals with the surface of the Earth and its fluids. It illustrates how we represent the Earth's surface and how oceans cover it, running water sculpts it, and atmosphere envelops it. We will explore the workings of the oceans and atmosphere, and the actions of water beneath and at the Earth's surface.

The final section explores geologic maps and resources. Geologic maps provide a method to document the location and occurrence of rocks including those of economic value. Materials from the Earth, such as oil, coal, iron, that are major resources sustain our global civilization. These materials occur in everyday life and are commonly included in our clothing, accessories, and buildings. Some of these power our civilization, heat or cool our homes, and fuel our cars.

The Scientific Method

The nonscientist may be tempted to think that science is merely the accumulation of "facts." Indeed, facts are the bricks in the structure of science, but science also consists of hypotheses, models, and theories that connect facts and make them meaningful. This larger structure is called the *scientific method.* The purpose of your laboratory course is not only to introduce you to the materials and the processes active on the Earth but also to give you insight into the workings of the scientific method. Here we outline how the scientific method operates.

The **scientific method** has evolved into a sophisticated system to investigate the natural world. The system uses a rational and logical framework that does not make appeals to the supernatural to explain nature's phenomena. The structure of scientific methodology has several components: gathering data; formulating hypotheses, models, and theories; and testing these. Communication continues throughout these processes.

Data Gathering

A preliminary stage in scientific investigation is to gather information through observations about the particular natural circumstances or processes being researched. Investigators strive to be as objective as possible. Facts must be distinguished from fiction and rumor, guesses and preconceived notions. The best observations are those that can be verified by independent investigators and are, therefore, also carefully documented.

A second aspect of data gathering is to collate and classify the information to see whether any patterns are present in the data. At this stage it may be possible to formulate various **laws,** which are generalizations derived from looking at many particular observations. For example, Newton's law of gravity does not explain gravity, but summarizes its action.

Hypotheses and Hypothesis Testing

Hypotheses are attempted explanations of observed natural occurrences. They try to connect diverse circumstances by stating a common underlying principle. A hypothesis should be more than a wild guess; it should explain all the observed data and also make predictions about features and occurrences not yet seen. These predictions provide a means of testing the hypothesis with further observation and/or experimentation. It is important to note that confirmation does not prove the hypothesis, but refutation or **falsification** disproves it. Correct predictions may support the hypothesis, but some other explanation may also account for these new observations. Hypotheses are either unsuccessful, in which case they are discarded, or they continue to be upheld by tests and observations.

At the early stages of an investigation several equally reasonable hypotheses may seem to fit the data. Scientists often use the method of **multiple working hypotheses** until they can eliminate the hypotheses that do not work. This is good scientific practice as it helps scientists to remain objective. After making further observations, scientists should throw out a hypothesis that has been proven incorrect or modify a hypothesis to fit new information, but they should *never* change the data or recorded observations to fit the hypothesis. Scientists also need to avoid only noticing phenomena or data that fit the hypothesis and ignoring information that refutes it.

1. Let's take a classroom example. In an experiment on cloud seeding students observe whether large or small droplets freeze first. Suppose a technical difficulty causes you to start your experiment after the rest of the class. Quite a few students have already posted their data on the blackboard, and all of them reported that larger droplets froze first. You run your experiment but notice that smaller droplets froze first. What should you do?
   a. Do not report your results because they are obviously wrong.
   b. Report your data regardless of what others have reported.
   c. Do the experiment over until you get the same results as others.

2. In another situation, you are measuring the amount of water draining from a tube during a certain period of time. You accidentally let two additional drops fall after the time for measuring is up. What should you do?
   a. Do not report your results because they are obviously wrong.
   b. Report your data regardless of the problem.
   c. Do the experiment over until you do it correctly.

In the first case, you have done everything right but got an unexpected result. These are real data and should be reported. In the second situation, you know you made an error in your measurements, so you need to do them over. The second case is not one of changing data but correcting the error. Most scientists in this situation would write down the result in their notebook with a note about what went wrong, rather than completely ignoring it.

Models

A *model* in science is a simplification or representation of some aspect of nature that is used as an analogy (or likeness). Models aid in the visualization and understanding of natural phenomena. Occasionally scientific models are actual physical replicas of some location or feature analogous to a model airplane, but unlike physical models, many scientific models are "constructed" mathematically or of ideas that imitate the aspect of nature being studied. Models are usually constructed to test the theory or hypothesis.

Modelers often make certain simplified assumptions. If the model is unsuccessful at explaining a natural occurrence, the scientist reformulates it using more complicated assumptions. If a model, in its simplified form, imitates nature fairly well, then scientists know that factors left out of the model are less important than those included. For particular investigations, models help researchers determine which factors of a system are significant and which are not.

Theories

A scientific **theory** is what a hypothesis or group of hypotheses becomes if their predictions are repeatedly successful. A theory also has a wide scope: it links together many diverse phenomena and has wide and profound explanatory power. The common disparaging comment "It's only a theory" is a contradiction; rather, a theory has been repeatedly tested and should not be casually disregarded.

Nonetheless, theories are not carved in stone and may be temporary, but less so than hypotheses. As new data are gathered that are not explained by the theory, it may have to be modified; theories are rarely simply abandoned. The new theory has broader scope than the old theory; in fact, the old theory may just be viewed as an approximation, applicable in certain restricted circumstances, of the new theory.

When many related hypotheses have withstood the test of time and can be integrated into a single theory that explains a vast range of data and observations within a field of science, a **unifying theory** is born. For the solid Earth sciences (geology), the theory of plate tectonics is such a unifying theory.

The concepts you learn in this introduction form an important foundation for Earth science; you will use them again at various times in later labs. You will apply the scientific method to develop and test hypotheses. However, it takes many years to develop a theory.

Communication

Besides conducting research, scientists need to disseminate and publish their results. Throughout their work, they share their data, ideas, hypotheses, and models with others; otherwise, they might as well not do the work in the first place. They submit the results for **review** as an article or **paper** to specialized magazines called **journals.** The journal editor sends copies of the paper to experts, called **referees,** who decide whether the work is new and valid science and therefore publishable. Editors reject "recycled" and bad data or hypotheses. Students are not the only ones who are graded!

Once the article is published, other scientists can examine the research and the observations or new hypotheses and incorporate them into new research. This process spreads science and introduces it to the community. Conferences are also important for transmitting results, and the Internet is being used more and more for this process. In science, books are less important in transmitting knowledge than they are in the humanities, for example.

Cyclic Aspect of the Scientific Method

As implied earlier, science is a dynamic process. Observations lead to laws and hypotheses, hypotheses and models to theories, and theories to newer theories or unifying theories. Hypotheses, models, and theories are constantly being evaluated by comparing their predictions with new observations (Fig. 1). In the experimental sciences (physics, chemistry, aspects of biology), scientists design and observe experiments specifically to test hypotheses. In the Earth sciences, astronomy and certain fields of biology we often rely on direct observation of the natural world. As a result of the comparison of hypotheses with the "real world," we adjust or reformulate them to provide a better explanation of the workings of nature, and then we test again. There is no room, therefore, for dogma in science. Scientists know that even their most treasured theories will eventually be replaced by other (and better) ones. Scientists by training are critical people. They continue testing hypotheses by looking for weak points and new observations. Scientists are also critical when examining each other's data because a hypothesis can only be examined properly if the data are accurate and truly reflect nature.

Special Features of the Earth Sciences

Practicing scientists, whether employed in industry, government, or universities, conduct research, which is the backbone of science. Sometimes the research is applied, involved

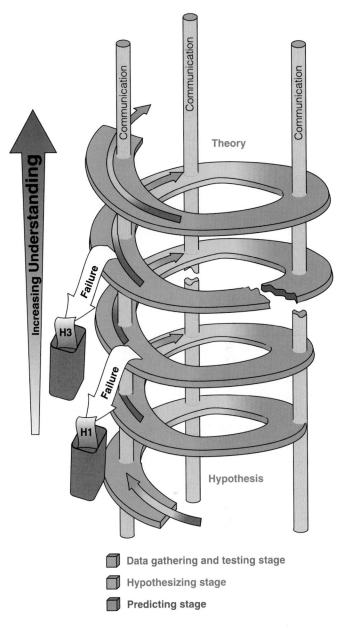

Data gathering and testing stage

Hypothesizing stage

Predicting stage

Figure 1 Scientific method cycle. Our understanding of natural phenomena is increased as we gather data, formulate hypotheses, use the hypotheses to make predictions, and then test the hypotheses by gathering more data related to the predictions. Our knowledge increases in a spiral fashion as we continue to hypothesize, predict, and test. After many cycles, represented by the break in the diagram, hypotheses increase in scope and reliability to become theories. Communication continues throughout the process.

in solving a particular practical problem; at other times the investigations are simply to understand the way nature works, regardless of the social usefulness of the results. The standards and rigor in both types of research are the same.

Earth scientists, in general, conduct research in a similar way to other scientists, but the Earth sciences have several features that distinguish them from the other sciences. The main difference is that Earth scientists cannot rely on experiments alone to provide observations. They must seek "case studies" as preserved in rocks in the field or as observed in the oceans or atmosphere as their primary source of information. This is why fieldwork and field trips are important in the Earth sciences. In the Earth sciences, particularly in geological fields, this means that the scientists conduct their research in the manner of detectives or forensic scientists, who must try to infer events from what is left over at the "scene of the crime." Earth scientists often have to deduce events logically from clues left within rocks, sediments, ice cores, or water and air samples. They must, therefore, be astute observers and have Sherlock Holmes-type minds. While this is a demanding intellectual exercise, it also makes the practice of the Earth sciences great fun.

Apply the Scientific Method to Your Own Life

You can use the scientific method in everyday life to find out whether your assumptions are correct. Let's take an example. Imagine you are looking for something that you have lost. Maybe it is a valuable ruby ring or your car key. You say to yourself, "I know it is somewhere in my room." You turn your room upside down and keep telling yourself it has to be here somewhere. After the fifth time you have gone through everything in your room and it just isn't

there, you go into the living room and find it on the end table, just where you left it.

How could the scientific method help you here? By saying the object is in your room, you form a hypothesis. You should then test that hypothesis by thoroughly searching your room. Once should be enough to make you reevaluate your hypothesis and start searching elsewhere, such as the living room. The stubbornness of holding onto your one hypothesis leads to an excessive waste of time. It would be better to compose multiple working hypotheses: "It is in my room or the living room." You could do a quick search of both rooms first, then a more thorough search if it doesn't turn up right away. This approach would be the most effective method for this situation.

3. What is another situation in which the scientific method could be useful to you personally? If necessary, have a discussion with your lab partners or with the whole class to answer this question.

_____

_____

_____

_____

_____

_____

Introduction to Maps

OBJECTIVES

1. To understand the differences between some basic types of maps
2. To be able to apply the information in legends and keys in map reading
3. To understand the concept of map scale
4. To understand map grid systems such as latitude and longitude and the Township and Range System
5. To understand different map projections and their distortions
6. To understand the basis for the Global Positioning System

You have probably used maps many times, whether studying in a geography class, planning a vacation, or visiting friends in unfamiliar towns, but you could also use them in many other ways. You may want to consult a flood zone map while building a house, look over a topographic map before hiking in an unfamiliar area, or even consult a geologic map before investing in a mining company. Maps are very useful in Earth sciences to display observations and data by location—that is, in a geographic context. Simply put, we need to know *where* rocks or sediments are sampled, *where* wind or ocean velocities are measured, or *where* plate boundaries are located. We can do this with maps.

Types of Maps

Maps can show the location and distribution of almost anything, but in the Earth sciences certain kinds of maps are frequently used:

Planimetric maps are the simplest type of map. They depict the location of major cultural and geographic features such as towns, rivers, roads, and railroads. Most people are familiar with this kind of map (Fig. 1.1). Gas station road maps, city plans, and the campus map in your college handbook are examples.

Topographic maps are more complicated but also more useful types of maps, which, in addition to showing the things planimetric maps show, also attempt to indicate the shape of the land surface. Features such as valleys, steep slopes, and mountain peaks—which may only be vaguely indicated on a planimetric map by the positions of rivers, winding roads, and elevation points—are clearly represented using various types of colors or symbols on a topographic map (Fig. 11.5). Lab 11 covers topographic maps in more detail.

Bathymetric maps do for the ocean floor what topographic maps do for the land (Fig. 1.2). They show the trenches, ridges, and plains that make up the ocean floor; in coastal areas bathymetric maps are particularly useful for navigation. We will look at additional bathymetric maps in Lab 12.

Geologic maps are used to show the distribution of various rock bodies (Fig. 1.3). Colors or patterns symbolize the different rock bodies, which are usually superimposed on a topographic map. Such maps are essential for mineral exploration and construction projects. Lab 16 involves understanding geologic maps.

Weather maps show the distribution of aspects of the weather, such as storm systems, precipitation, and temperature, including warm and cold fronts (Fig. 1.4 and Fig. 15.13). Some include variations in atmospheric pressure, wind direction, and speed. You will learn more about weather maps in Lab 15.

Other maps: There are many more types of maps than any one person can imagine, but some examples include maps of ocean currents (Fig. 12.15), maps showing heat flow from the Earth (Fig. 1.5), vegetation maps, land use maps, and maps showing plate boundaries (Fig. 9.7).

1. Draw a map in the space on page 8 showing directions from your house or apartment to school, or between any two nearby locations you know. Do not worry about being extremely precise at this stage; just draw a sketch as you might if giving someone directions. What type of map

Figure 1.1 Map of North America

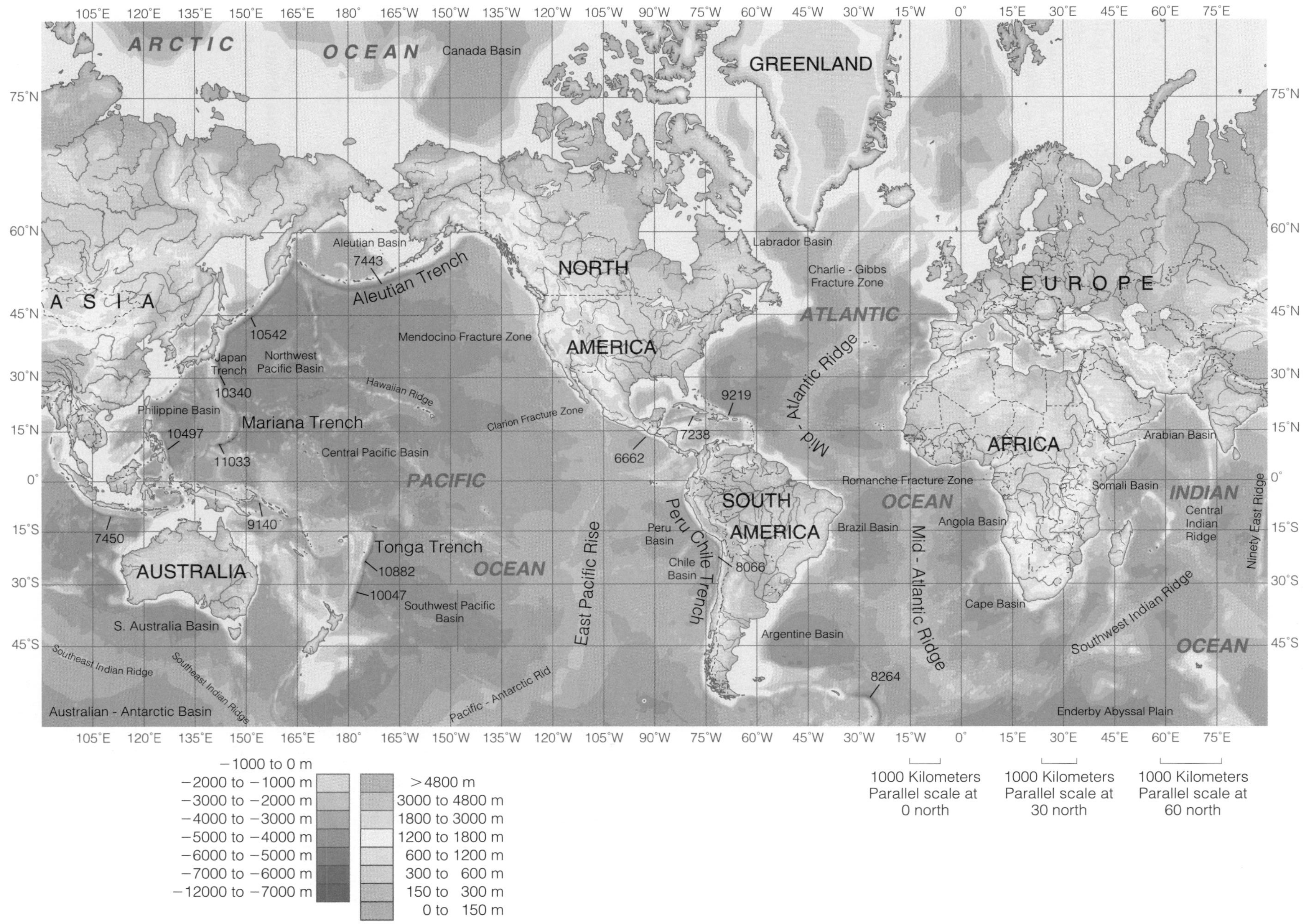

Figure 1.2 Bathymetric map of the world's oceans

have you drawn? _____ We will come back to this map later.

Map Legend

All maps, whatever their type, should be accompanied by a **legend** (also known as a **key** or **explanation**). Without a legend, it is difficult to understand what the various colors and symbols on the map mean. Although the information contained in the legend may vary, good map legends will includes the following:

North arrow: All maps should contain some indication of their orientation, so most maps usually have at least a north arrow. The north arrow customarily indicates the direction of geographic north, which is the direction toward the northern end of Earth's axis of rotation, called **true north.** You may think that north is just north, but there are other types of north arrows besides true north, which include magnetic north and grid north. **Magnetic north** is the direction a compass or magnetized needle points. **Grid north** is the orientation of the north-south set of grid[1] lines of a regional coordinate system used on the map, which for various reasons may not be exactly north-south. These three types of north are not always coincident. On maps that show a large area (small-scale maps), north may vary from one part of the map to another. On such maps, lines of longitude (north-south lines) each show a slightly different orientation for north (see, e.g., Figs. 1.15–1.17a and 1.19).

Scale: Features on a map represent features on the Earth. The scale indicates the size reduction needed to convert the actual feature into its representative on the map. A scale tells us how much smaller the items on the map are compared to the actual size of those features on the Earth. More about scale appears later in this chapter.

....

[1]A *grid* is a system of lines intersecting at right angles to form rectangles.

Location: Maps usually contain information indicating where they are on the Earth's surface. Showing and labeling lines of latitude and longitude on the map can uniquely establish the location of the map. If drawing the complete lines clutter the map, tick marks can show the position and orientation of the lines at the boundary of the map (Figs. 12.2 and 12.4). A small inset map of a larger, easily recognized area may show the location of the map's borders.

Symbols: Most maps will have additional symbols that need to be defined in the legend. Examples are the different color coded lines for different kinds of roads on a highway map; different colors representing ranges of elevation on some kinds of topographic maps; and symbols for swamps, springs, mountain peaks, cities or populated areas, city size, political boundaries, cold fronts, wind direction markers, and ocean currents. The possibilities are infinite. Some publishers use standardized legends and do not key all features on every map; instead, they provide a separate legend pamphlet available or, in an atlas, a separate legend page. A good example is topographic contour maps (which we will cover in more detail in Lab 11), such as those published by the U.S. Geological Survey.

2. Examine Figure 1.1. What symbol on the map shows a major city? Draw the symbol. _____

   What symbol is used for political boundaries?

   _____

   What type of map is this? _____

3. On the map in Figure 1.2, locate the deepest part of the Pacific Ocean. What is this place called? _____

   _____ Locate an undersea mountain range in the Atlantic Ocean. What is it called? _____

   _____

North Arrow Computer Exercise

- Start **TASA Topographic Maps.**
- Click stop(◼) to get to the main menu, then click **2. General Topographic Map Information** click advance (▶) twice and **skip** twice. Advance through this section answering the following questions:

4. How many different kinds of north are shown on the topographic map? _____

5. What are the first two mentioned and what do they show? _____

   _____

   _____

   _____

6. Which one of these moves slightly from year to year? _____

7. Where was the magnetic north pole at the time this program was written? Show this on Figure 1.1.

8. Where in North America is magnetic north west of true north? _____ Where is it east of true north? _____

9. What is the angle and direction of magnetic north shown? _____
What is it called? _____

10. Where on the map does magnetic north coincide with true north? Name two states or provinces:
_____ _____

11. According to the map, what is the magnetic declination for your location (skip this question if your location is not on the map)? _____ What is the magnetic declination of San Francisco, California (the SF Bay is visible on the map)? _____

12. What is GN, and what does it show? _____ _____

■ Click stop (■) then **Exit.**

■

13. What is the significance of the colors in the map in Figure 1.5? _____
How do you know this? _____
_____
Where is there high heat flow? Name one location. (Refer to Fig. 1.2 if needed.) _____
Where is the heat flow low? _____

14. Use the map legends in Figure 1.3 to locate granite and a fault. What color is used for granitic rocks? _____
Draw the symbol used for one type of fault on the map.
_____ On a map such as this, with no north arrow, how would you determine what direction is north? _____
_____

Where is this map area located? _____
_____

15. Look at Figures 1.2 and 1.3. What other sorts of things or features can be shown on a map? For each map, list two additional features shown and their symbols. Try to find items that are very different from each other.

Figure 1.2 Item 1 _____ Symbol _____
Figure 1.2 Item 2 _____ Symbol _____
Figure 1.3 Item 1 _____ Symbol _____
Figure 1.3 Item 2 _____ Symbol _____

Scale

A map is a representation of part of the Earth's surface, and to be of practical use it must be smaller than the area it depicts. Therefore, all maps are drawn to a specific scale, and all maps, including sketch maps, should have some indication of the scale. Even the map that declares "not drawn to scale" has a relative scale, although it is not accurately drawn. Scale on maps is quantitatively expressed in the following ways:

Statement (or **verbal scale**): On many simple maps the scale may be expressed as an ordinary statement. Examples are "One inch equals one mile" or "1 cm = 100 m."

Representative fraction (or **fractional scale**): A more common way of expressing a map scale is to give the fraction or ratio of a given distance between two points on the map to the distance between the two points on the Earth's surface. For this fraction to be meaningful, the same units of measurements must be used. Thus, if two towns are separated by 2 cm on the map, and by 1 km on the Earth's surface, the representative fraction is 2 cm:1 km, or 2 cm:100,000 cm, which in turn reduces to 1:50,000. Similarly, a map 1 inch to the mile has a representative fraction of 1:63,360 (because there are 63,360 inches in 1 mile).

Scale bar (or **graphic scale**): A simple and very effective way of representing the scale of a map (or any diagram) is to draw a line representing distances on the Earth's surface. Thus, the scale on a 1:50,000 map can be shown as a line, or bar, 2 cm long and labeled "1 km." Figure 1.6 gives an example of two scale bars, one English and one metric, but at the same scale or ratio. The advantage of this visual method of representation is that the scale is instantly grasped without the need of mental calculations. A further advantage is that the scale bar remains an accurate indicator of scale even if the map is reduced or enlarged in a copy machine or photograph. This is not true for verbal or fractional scales.

Figure 1.3 (*see next three pages*) Geologic Map of Part of Washington State

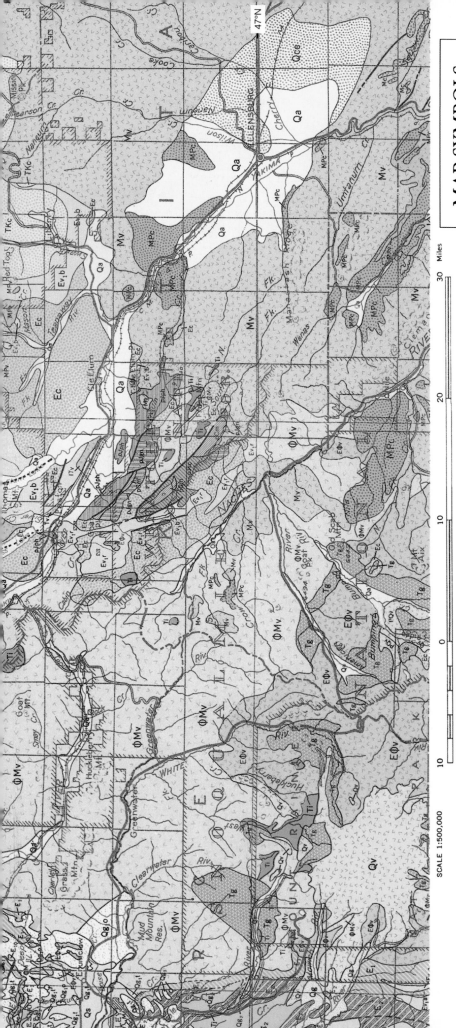

MAP SYMBOLS

Contact

Known, approximate and inferred are undifferentiated.

Scratch boundary

Contact uncertain, or edge of specific mapped area.

Fault

Character not designated; dashed where approximately located.

Concealed fault

Thrust fault

Saw-teeth on upper plate.

SCALE 1:500,000

10   0   10   20   30   Miles

GEOLOGIC MAP OF WASHINGTON

Geology by Marshal T. Hutting, W. A. G. Bennett,
Vaughn E. Livingston, Jr., and
Wayne S. Moen

Base from U. S. Department of Agriculture
Forest Service
Mark D. Haun, Cartographer

From: Washington Division of Mines and Geology

LOCATION OF AREA

EXPLANATION
SEDIMENTARY AND METASEDIMENTARY ROCKS

QUATERNARY

Recent

Qa — Alluvium

Qs — Landslide and mudflow deposits

Pleistocene

Qg₁, Qg₁t, Qg₁o — Younger glacial drift

Qg — Glacial drift, undivided

Qg₂ — Older glacial drift

Qc, Qce, Qcl — Quaternary nonmarine deposits

TERTIARY

Miocene

MP — Miocene-Pliocene marine rocks

MPc — Miocene-Pliocene nonmarine rocks

Oligocene

Φ — Oligocene marine rocks

E₁ — Upper upper Eocene nonmarine and marine rocks

E₂ — Lower upper Eocene marine and nonmarine rocks

Eocene

Ec — Eocene nonmarine rocks

Paleocene

TKc — Paleocene-Cretaceous nonmarine rocks

CRETACEOUS

pT — Pre-Tertiary sedimentary and metasedimentary rocks, undivided

JK, JKs — Upper Jurassic-Lower Cretaceous sedimentary and volcanic rocks

JURASSIC

pJ, pJs — Pre-Middle Jurassic sedimentary and volcanic rocks

PERMIAN

B — Permian rocks

CB, CBs — Carboniferous-Permian sedimentary and volcanic rocks

EXTRUSIVE IGNEOUS ROCKS

Qv — Pleistocene-Recent volcanic rocks

MPv — Miocene-Pliocene volcanic rocks

Mv — Miocene volcanic rocks

ΦMv — Oligocene-Miocene volcanic rocks

Tv, uTv, lTv — Tertiary volcanic rocks, undivided

Φv — Oligocene volcanic rocks

Ev₁, Ev₁a, Ev₁b, Ev₁r — Upper Eocene volcanic rocks

EΦv — Eocene-Oligocene volcanic rocks

Ev — Eocene volcanic rocks

Ev₂ — Middle and lower Eocene volcanic rocks

METAMORPHIC AND INTRUSIVE IGNEOUS ROCKS

Ti — Tertiary dikes, sills, and small intrusive bodies

Tg — Tertiary granitic rocks

Td — Tertiary dunite intrusive rocks

bi — Basic intrusive rocks, undivided

TKg — Tertiary-Cretaceous granitic intrusive rocks

TKbi — Tertiary-Cretaceous basic intrusive rocks

pTm — Pre-Tertiary metamorphic rocks, undivided

pTb — Pre-Tertiary ultrabasic intrusive rocks

Mzg — Mesozoic granitic rocks, undivided

pJgn — Pre-Upper Jurassic gneiss

pJsc — Pre-Upper Jurassic metamorphic rocks of the medium and high-grade zones

pJgs, pJph — Pre-Upper Jurassic metamorphic rocks of the low-grade zone

pCm — Pre-Carboniferous crystalline complex

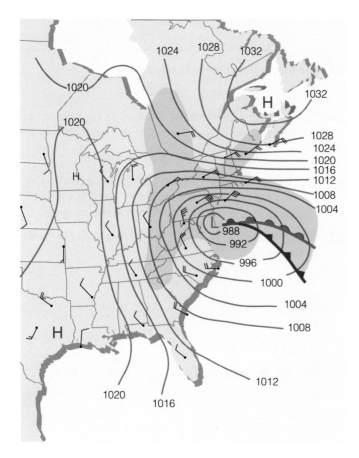

Figure 1.4 Surface weather map for 7 A.M. Eastern Standard Time, December 11, 1992, showing extremely low pressure (L), strong winds ( ), and heavy precipitation (green). This storm caused hundreds of millions of dollars of damage.

16. Assume the line just below is the length of a road on a map at the scale in Figure 1.6. Determine the length of the road four ways as described.

————————————

a. Use the scale bar in Figure 1.6 to determine how long the road is in miles. _____

Hint: Read the scale carefully. On a scale bar it is common to place smaller divisions to the left of 0. This aids measuring distances off a map to the smaller divisions, but beware, some students assume 0 starts at the left side.

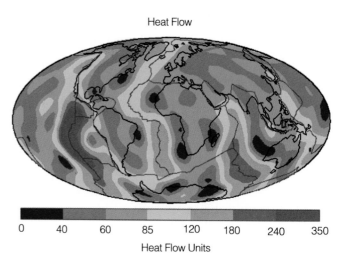

Figure 1.5 Heat flow map. Heat from the Earth's interior escapes at different rates in different areas. The colors on this map show areas of higher and lower heat flow.

b. Use the scale bar to determine how long the road is in kilometers. _____

c. Use the verbal scale and an appropriate ruler to measure the road in miles. _____ Compare your answer to that in **a**.

d. Use the fractional scale and an appropriate ruler to measure the road in kilometers. _____ Compare your answer to that in **b**.

In general terms, a map may be referred to as **small scale** or **large scale**. *Large* and *small* here refer to the representative fraction. Thus, a map with a scale of 1:1,000,000 is considered small scale, whereas one of 1:5,000 is considered large scale (because 1/5,000 is larger than 1/1,000,000). A small-scale map on the same size sheet as a large-scale map covers a larger area because the represented area is shrunk relative to the larger-scale map. Objects on a small-scale map appear smaller than the same objects on a large-scale map.

17. Compare your sketch map in Exercise 1 with a local street map of the same area. See whether you can determine the scale of your map. It is quite likely that different parts of your map have different scales. Draw scale bars on your sketch in at least two places showing the scale for that part

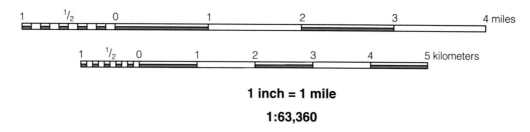

1 inch = 1 mile

1:63,360

Figure 1.6 Example scales. The first two scales are both scale bars, one in miles and the other in kilometers. The next scale is a verbal scale, and the last one is a fractional scale. All of these represent the same scale.

of the map. Notice that your map does have scale; it just isn't likely that the scale is consistent throughout your map. This is what we call "not drawn to scale."

18. Calculate the representative fraction of the following verbal scales showing how the calculations are done for each:
   a. 5 cm equals 10 km

   b. 1 in. equals 4 mi

   c. 4 mm equals 100 m

   d. 2 5/8 in. equals 4 mi

19. Draw a scale bar below, for both kilometers and miles, for a 1:12,000 scale map and also for a 1:100,000 scale map. Label which is which.

20. Which scale above is for a larger-scale map?

   _____

21. In Figure 14.8 determine the distance between two pluses in Henrys River, one plus west of the word *Henrys* and the other plus labeled "Mile 5." _____
   Check your answer in Table 14.1.

22. Determine the fractional scale for the map in Figure 10.23. Show your work.

 ## Scale Computer Exercise

- Start **TASA Topographic Maps.**
- In the program, click (■) to get to the main menu, then click **3. Map Scale.**
- Click the advance (▶) button six times to move through the program.

23. What are two types of scales shown? _____

   _____

24. The first type shows the relationship between

   _____.

25. The numerator represents one unit on

   _____.

26. This one unit represents the distance on the ground (actual earth) equal to the number of the same units in the

   _____ of the fraction.

- Click the advance (▶) button and continue reading this section until you reach the multiple-choice questions.
- Click **Skip** for the first two questions in the program.

27. Which of the maps covers the largest area (i.e. is the smallest scale)?

   1) Map A—1:24,000

   2) Map B—1:62,500

   3) Map C—1:250,000

   4) Map D—1:50,000

- Click (■) to get to the main menu, then click **Exit.**

■

Global Location System: Latitude and Longitude

A major property of a map is that it compares the locations of features within the map area. The fundamental system for locating features in relation to the entire Earth's surface is latitude and longitude. The poles of Earth's axis of rotation, otherwise known as the *geographic* or *North* (N) and *South* (S) *Poles* are the basis of this system. Visualize a plane intersecting the center of the Earth. The circle defined by the intersection of this plane with Earth's surface is the largest circle possible on the surface and hence is known as a **great circle.** A family of great circles exists that pass through the N and S Poles. These divide up the Earth into segments a little like segments of an orange as shown in Figure 1.7a. The half great circle between the N and S Poles can be used to specify the east-west position of a location and are called **meridians,** or **lines of longitude.** One of these meridians, the **prime meridian** or **Greenwich meridian,** has the value 0° and for historical reasons passes through Greenwich, United Kingdom. The other meridians are identified by their angle from the prime meridian measured in degrees in the plane of the equator as shown in Figure 1.7b. The measurements are toward the east or west, making the highest numbered meridian 180° from the prime meridian.

A second set of circles used for location is placed perpendicular to the meridians, with centers on the Earth's rotational axis. The circles in this set are called **parallels,** or **lines of latitude.** Only one of these, the equator, is a great circle; the rest are small circles (Fig. 1.7c). **Small circles** are

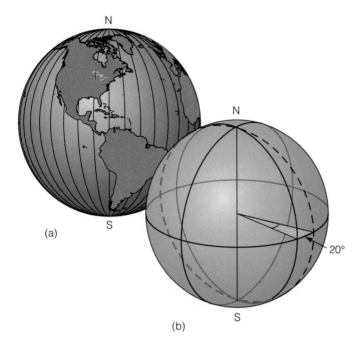

(a)

(b)

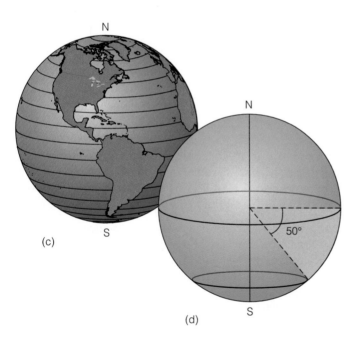

(c)

(d)

Figure 1.7 Latitude and longitude. (a) Lines of longitude are great circles through the North and South Poles. (b) Angle of a line of longitude measured from the Prime Meridian (dashed). (c) Lines of latitude are small circles with their centers lining up along the axis of rotation. (d) Angle of a line of latitude measured from the equator. In the Northern Hemisphere, this can also be measured as in Figure 1.8, between the North Star and the horizon.

defined as any circle on the globe whose radius is less than that of a great circle and therefore not centered at the Earth's center. The equator defines the origin (0°) of the system, and the angle, north or south, from the equator specifies each parallel (Fig. 1.7d; Fig. 1.8). The distance between lines of latitude is constant, with a spacing of 1° corresponding to about 110 km. However, between lines of longitude the spacing diminishes to 0 at the poles and is widest at the equator.

The criss-crossing lines of latitude and longitude form a reference grid on the surface of the Earth called a **graticule,** which can be used to specify the location of any point. Thus, a point on the Earth's surface is specified by its angular measurements of longitude and latitude. Angles can be specified either as degrees/minutes/seconds[2] or as decimal divisions of degrees. The hemisphere of either must be specified. In longitude this means E or W; east is sometimes characterized as positive (+) and west as negative (−). In latitude, the N or S must be indicated; north is often specified as positive (+) south as negative (−).

Longitude and Global Timekeeping

The prime meridian is also the basis of global timekeeping. The Earth rotates from west to east, 360° in 24 h, which means that it rotates 15° every hour. Twelve noon at 15°W, therefore, occurs 1 h later than at the prime meridian; equally, 12 noon at 15°E occurs 1 h before it does at the prime meridian. The world's time zones are thus 15° of longitude wide and differ by 1 h compared to adjacent time zones. The local time in these zones is compared to the number of hours that they are ahead or behind the time at the prime (or Greenwich) meridian. The time at the prime meridian is called **Universal Time** (UT, or previously *Greenwich Mean Time* or GMT). The dividing lines between time zones are generally at 7.5°E and 7.5°W and then every additional 15° from there. In practice, individual political units decide on their own local time but most are based on this system. Approximately at 180° longitude, the day and date change; the actual boundary is called the **International Date Line** and is not exactly at 180° everywhere.

28. Using a globe or map, find the latitude and longitude for each location in Table 1.1 and enter them in the table. If it is 9 A.M. on May 15, 2003, UT, what time and date is it in each location in Table 1.1? Enter these answers in the table also.

Distance Between Two Points on the Globe

29. Using the map in Figure 1.1, determine what appears to be the shortest route between Anchorage and Guatemala City. Write down the latitude and longitude of the

···········
[2]A degree has 60 minutes (60′), and a minute has 60 seconds (60″).

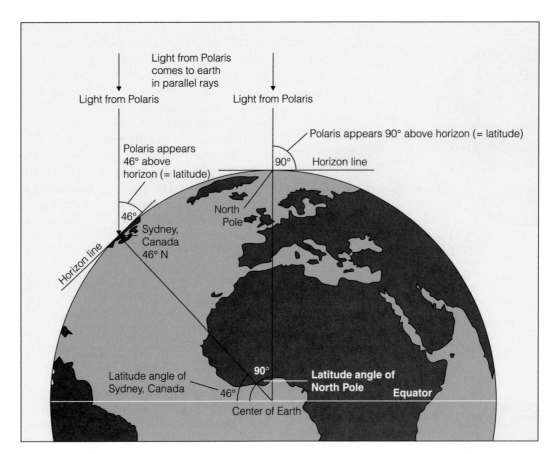

Figure 1.8 Latitude can be determined in the Northern Hemisphere by the angle that the North Star, Polaris, makes with the horizon.

Table 1.1

Location	Latitude	Longitude	Time	Date
Kansas City				
London				
Beijing				
Nairobi				
Rio de Janeiro				
Honolulu				

midpoint. _____ _____ Use the scale on the map to determine the distance between the cities; neatly record your calculations here. We will come back to this.

The shortest distance between two points X and Y on a globe is the segment of a great circle that passes through them. If you ever fly to Europe from North America, you will fly close to a great circle route, which takes you somewhat north then back south rather than traveling mostly east. This is because great circle routes are shorter distances than flying east from North America to Europe. Check this on the globe.

30. With a lab partner, determine the distance between San Francisco and Paris, France, using a globe, a string, a washable marker, and a metric ruler.
    a. First, use the string to measure the circumference of the globe. The circumference of the Earth is 40,000 km. Calculate the scale of the globe.

    b. Next, extend the string between the cities. Using a marker, place a dot on the string at each of the two cities. Take the string off the globe, and measure the length between the two marks. Use the globe's scale to calculate this great-circle distance between the cities. Write down your measurements and calculations below.

31. Do the same thing again to estimate the distance between Anchorage and Guatemala City. What are the latitude and longitude of the midpoint along this great-circle route? _____ _____ What is the distance from Anchorage to Guatemala City? _____ Now compare the locations of the two midpoints on the route from Anchorage to Guatemala City on the map (Exercise 29) and on the globe. Do the two locations coincide? Explain your results.

_____

_____

_____

_____

32. Imagine you work for a world disaster relief organization stationed in New York City and a major earthquake strikes Kushiro, Japan, at 43°N, 144°25′E. You need to know how far Kushiro is so you can charter a plane to carry relief supplies. How far is it from New York to Kushiro? _____ How much farther would the flight be if the plane flew no farther north than Kushiro? _____

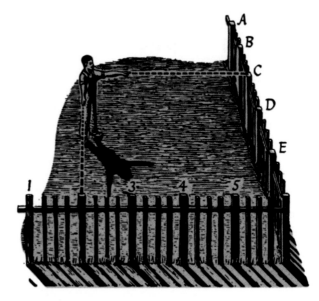

Figure 1.9 For small areas, a rectangular grid is more convenient than latitude and longitude to designate location. Where is the man in this picture standing? Giving the answer in latitude and longitude is much more cumbersome than using a rectangular grid set up by parallel lines from the fence posts.

Map Grids and Orientation

Grids and UTM

Specifying the location of points on a map using latitude and longitude is useful for a large area but is often clumsy and inconvenient in small areas. On maps of a scale of 1:250,000 or larger, a local rectangular grid is often more convenient (Fig. 1.9). The grid is prepared by placing two sets of straight, parallel, equally spaced lines on the map. One set of lines is oriented north-south, the other set east-west (or at least approximately so). The spacing between the lines depends on the territory or country concerned. The most widely used spacing (including in the United States) is 1000 m, but others, such as 1000 yd, may occasionally be encountered.

The N-S trending lines are numbered increasing eastward and are called **eastings;** the E-W trending lines are numbered increasing northward and are called **northings** (Fig. 1.10). The origin of this coordinate system is chosen so that all points in the country or region of interest will have positive northing and easting values. This often means that the origin lies outside the territory under consideration. You can easily specify points within the map area using their coordinates, much the same way as you would indicate the position of a point on a graph (Fig. 1.10). Specify eastings (*x* coordinates) first and northings (*y* coordinates) second. You can indicate points between grid lines using a dec-

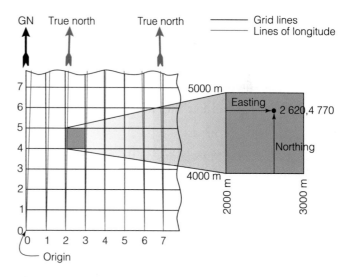

Figure 1.10 Example grid system based on 1000-m spacing.

imal subdivision of the grid squares. Note, however, that although this decimal subdivision is included in the grid reference, the decimal points themselves are not written (Fig. 1.10). One such system is the **Universal Transverse Mercator** (or UTM) coordinates, which are commonly included on topographic maps.

Universal Transverse Mercator (UTM) Computer Exercise

- Start **TASA Topographic Maps.**
- Click stop (■) to get to the main menu, then click **6. Location: Universal Transverse Mercator.** Click advance (▶). Advance through this section answering the following questions:

33. Each UTM grid zone is _____ degrees of longitude wide, numbered from the _____ meridian consecutively _____.

34. Grid zones are _____ degrees of latitude north-south, labeled with _____ starting with _____ between 72° and 80°S.

35. Since meridians of longitude _____ at the poles, the _____ lines of the UTM grid do not follow the lines of longitude on a map. The direction of north lines of the UTM grid is called _____ (_____).

36. Locations using UTM are in what units?

_____

What value is the equator given for measurements in the Northern Hemisphere? _____

For the Southern Hemisphere? _____

37. Within a UTM grid area, the central meridian is assigned _____ m east.

38. What are the UTM locations on the map at the end of this section for each of the following?

A: _____

B: _____

C: _____

D: _____

E: _____

- Click stop (■) then exit.

■

Remember grid north? The north-south lines of the grid are often not quite true north and true south. This may result from surveying errors when the original grid was set up, or it may be because a rectangular grid does not fit perfectly on the nearly spherical surface of the Earth. Whatever the reason, the north arrow on the map may show grid north, as discussed earlier, as being slightly different than true north (Fig. 1.10).

In parts of the United States, another type of grid system is in common use, known as the **Township and Range** System.[3] An east-west line through the origin or reference point of this system is known as the **base line,** and a north-south line through the point is the **principle meridian,** not to be confused with the prime meridian. In the Township and Range System, the reference point may be in the middle of the area of interest, and points are measured east or west and north or south of the point. The grid has a spacing of 6 mi, with the resulting squares known as **townships.** The townships are numbered as shown in Figure 1.11, where *T* and *R* stand for "township" and "range" and N, S, E, and W are the points of the compass. For example, T3S, R2E—or Township 3 South, Range 2 East—is shown enlarged at the right in Figure 1.11. Thirty-six squares called **sections** divide each township, where each section is 1 square mile. The numbering of sections starts in the NE corner, as shown in Figure 1.11. Each section can be further subdivided into quarter sections and each quarter section into quarters again. Figure 1.11 shows Section 20 subdivided into 16 quarter-quarter sections. The tiny point on the small-scale map is a quarter-quarter section (1/4-mile square area, 1/16th of a square mile) designated as SW¼, NW¼, Sec. 20, T3S, R2E. You would read this "the southwest one quarter of the northwest one quarter of Section 20 in Township 3 South, Range 2 East."

39. Place a dot on Figure 1.11 and label it with the appropriate letter for each of the following township and range locations:
    a. Sec. 11, T4S, R5E
    b. Sec. 26, T2N, R1W
    c. SE¼, NW¼, Sec. 5, T4S, R2W
    d. SW¼, SW¼, Sec. 17, T4S, R2W

40. For the dots labeled **x, y,** and **z** in Figure 1.11, write out their locations using the Township and Range System. For **x,** specify the section and township; for **y** and **z,** also specify the quarter-quarter sections.

    x. _____

    y. _____

    z. _____

Direction on the Globe

At a particular point, you can specify directions on the globe with reference to the direction of true north. In practice, we determine directions using a magnetic compass. The magnetic North Pole is centered near Baffin Island, not at the geographic pole (Fig. 1.12). This means that the magnetic north direction rarely coincides with geographic north. The

••••••••••
[3]Also called the Land Office Grid System or Public Land Survey.

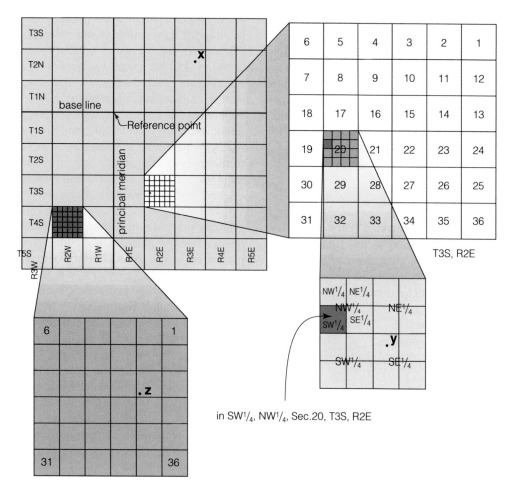

Figure 1.11   Township and Range System

angle between the two at any particular point is called the **magnetic declination.** Magnetic north lines up with true north along a curving line that passes through western Ontario and near Springfield, Illinois, and Mobile, Alabama. Magnetic north is west of true north for any place east of this line and east for any place west of the line. The magnetic field is not fixed but migrates slowly so that the magnetic declination changes from year to year by a few minutes. Also, the declination values and rate of change become greater near the poles. There is a useful declination calculator available on the Internet at the Canadian Geological Survey site at **www.geolab.nrcan.gc.ca/geomag/e_cgrf.html.**

41. Visit the Web site and find the magnetic declination in your hometown. You will need its latitude _____

and longitude _____.

What is the magnetic declination? _____

Also find the magnetic declination of your school location, if it is different. _____

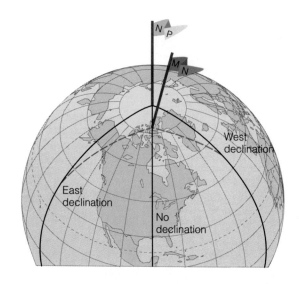

Figure 1.12   The magnetic north pole (M N) is a small angular distance from Earth's rotational pole, the North Pole (N P).

42. Draw a simple map of your lab room or another room, as directed by your instructor, on a separate piece of paper. You will need to decide on a scale and make a few measurements. Use a compass to determine the direction of magnetic north.[4] Draw a magnetic north arrow on your map. Use a protractor to determine and draw in true north, using the magnetic declination you obtained in Exercise 41. Don't forget to include a scale and legend on your map.

Projections

A fundamental problem of map making (cartography) is how to represent the features of a spherical Earth and the latitude-longitude graticule onto a flat sheet of paper. Can you peel an orange or cut a rubber ball and lay out the peel or rubber flat? The peel or rubber either has to be stretched and distorted or has strips and gaps (Fig. 1.13).

For maps, visualize a transparent globe with the latitude-longitude lines drawn on it. A small light source placed at, say, the center of the globe will then cast shadows of the lines

....

[4]Be careful to stay away from objects that can deflect the compass needle, including walls. If the compass points different directions from place to place, it is being deflected. Mapping an outside object should avoid this problem.

on any surface touching the globe. This is called **projection.** Moving the light source changes the projection of the lines. The surface does not have to be flat—it could be a cone or a cylinder—but for a useful projection, it has to be capable of opening out into a flat surface. The process of projection results in certain distortions of angles, areas, and scales, which are predictable for any given projection. There are many different kinds of projections, each with different properties that serve particular purposes. Here we will restrict ourselves to those projections most commonly used for regional scale maps—namely, the Mercator and transverse Mercator, conical, and stereographic projections.

The surface onto which the projection is made is usually cylindrical, conical or planar as these surfaces can be readily opened out and placed flat. Figures 1.14a, b, and c show the projection of the Earth's surface onto a cylinder, cone, and plane, respectively.

In the **Mercator projection,** perhaps the most familiar, the global graticule is projected onto a cylinder aligned with the geographic axis, with the cylinder touching the equator (Figs. 1.14a and 1.15a). The lines of latitude thus project as straight, parallel lines. The straight lines of longitude plot perpendicular to straight lines of latitude. While the distortion associated with this projection is minimal at the equator, it is considerable at higher latitudes and is extreme (infinite) at the poles.

In the **transverse Mercator projection,** the projection cylinder touches a particular meridian of interest (Fig. 1.15b).

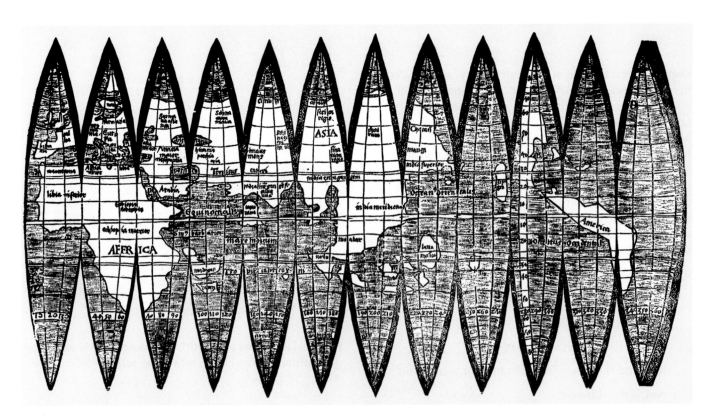

Figure 1.13 This map, published in 1507, is the first to show America as separate from Asia. The segments were designed to fit together to make a globe.

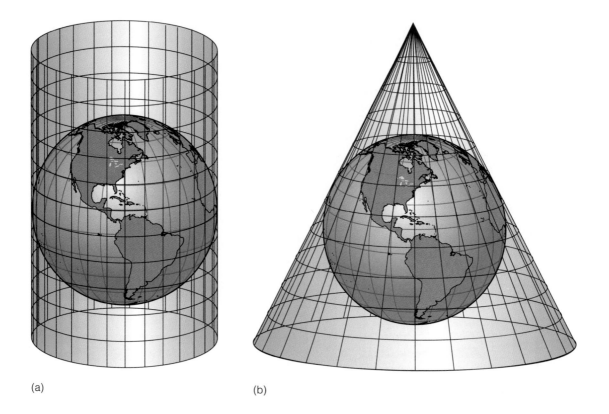

(a)

(b)

Figure 1.14 Map projections. (a) Projection onto a cylinder. (b) Projection onto a cone. (c) Projection onto a plane.

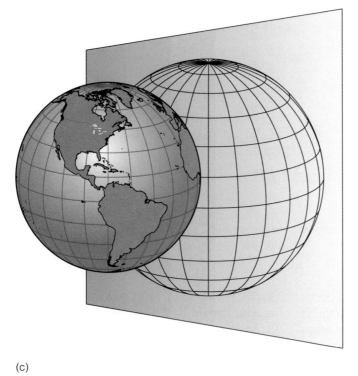

(c)

Meridians and parallels, except the central ones, are both curved in this projection, but there is minimal distortion around the central meridian. This projection has great utility as an almost universally applicable projection for large-scale maps and as the basis for a global map grid system (UTM), but it is rarely seen in small-scale maps. We are so used to seeing a regular Mercator projection that we may not notice the

distortion, but in a small-scale transverse Mercator map, such as Figure 1.15b, the distortion is readily evident.

In **conical projections,** the projection of the graticule is made onto a cone whose axis is coincident with the geographic axis. The cone "rests" on a particular line of latitude, the **standard parallel** (Fig. 1.14b). This and other parallels project as a series of concentric circles on the map. Meridians

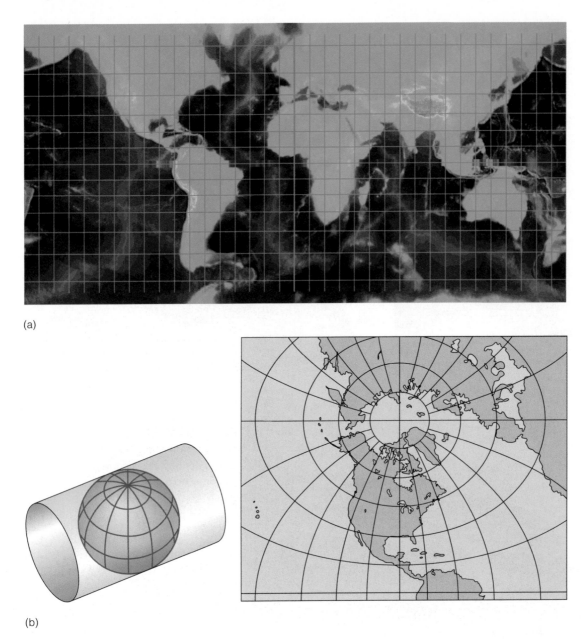

(a)

(b)

Figure 1.15 (a) Mercator projection of world topography and bathymetry. (b) Transverse Mercator projection with 40°W as the central meridian.

project as radiating straight lines perpendicular to the circles (Fig. 1.16). The distortion adjacent to the standard parallel is minimal but increases at parallels away from it. Many maps of midlatitude regions use the polyconic projection, including many U.S. Geological Survey maps. Visualize a **polyconic projection** as a series of strips pieced together where each strip has a slightly different standard parallel and only the least distorted part is used. If these strips are infinitesimally small, a continuous map results. In the polyconic projection, both meridians and parallels are curved.

Polar projections are often **projections onto a tangential plane** (Fig. 1.14c), in which a plane is tangent (touching at one point) to the Earth's surface, at the poles in the case of Figure 1.17b and c. The projection may be made from inside the Earth, from the opposite surface of the Earth (stereo-

graphic projections), or from a perspective outside the Earth. Imagine that the light placed at these different locations casts the shadows that produce the different variations of maps. For **stereographic projections,** the projection point (light) is on Earth's surface exactly opposite the tangential plane. Usually, the plane is tangent at 0° longitude and 0° latitude (equatorial projection, Fig. 1.17a) or at the N or S Poles (polar projection, Fig. 1.17b and c), but it is possible to choose any latitude or longitude (oblique projection).

Distortions of Shape, Scale, and Area

Showing the round Earth on a flat map involves some distortion of the appearance of the Earth. Some projections roughly preserve the shape of a region on the globe but distort the rel-

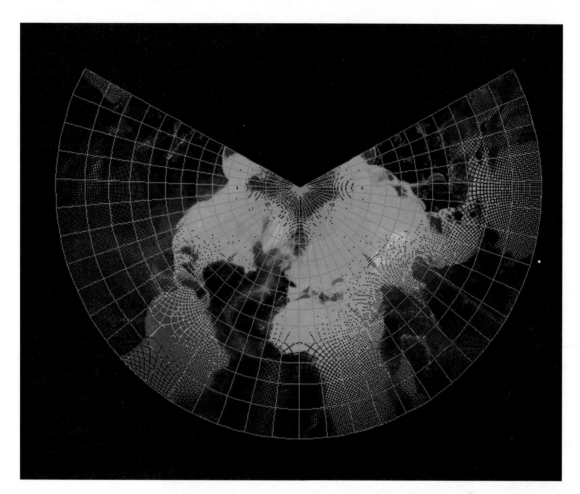

Figure 1.16 Conic projection

ative areas covered by those regions. Such projections, where different directions at a point have the same scale, are said to be **conformal.** Other projections can be devised that preserve the relative areas such as Figure 1.18; these are called **equal area projections,** but they usually distort the shape of the region. In Figure 1.19, distortion is especially noticeable in areas near the map's edges. Mercator projections are conformal but not equal area. Different projections also distort the graticule in different ways. Some, such as the Mercator projection, make the lines of longitude into parallel lines instead of converging lines. Others cause the lines of latitude (parallels) to become curved and not parallel (Fig. 1.18). Many projections cause the points at the poles to become lines or curves.

43. Use a globe and a rectangular piece of plastic and washable markers. Attach the edges of the rectangular plastic to make a cylinder, and put it over the globe so that the cylinder is in contact with the equator. Sketch in a rough outline of the continents (or just North America) including Greenland and Australia, while looking perpendicular to the globe's surface. Move your head to keep your lines of sight straight out from the globe area you are drawing. Notice that you cannot map all the way to the poles. Disconnect the edges and flatten your plastic cylinder so you can trace your plastic map onto a piece of paper. What type of map projection is this? _____

_____ Look at Greenland and Australia on your map (or in Fig. 1.2). Which appears to be larger? _____ Now examine these on the globe. Which is actually larger? _____ What is the most distorted part of the Mercator projection? _____ Where is the distortion the least? _____

44. Australia is the smallest geographical continent. Is Greenland a geographical continent? _____

45. Make a piece of plastic that is part of a circle into a cone by joining the straight edges together. Place this on the globe so that the point of the cone is over the North Pole. Follow the same procedure described in Exercise 43 to trace a map onto the plastic and then onto a piece of paper.

What type of projection is this? _____

Look at the conic projection you made and compare it to the globe. What is the most distorted part of this projection? _____ Where is the distortion the least? _____

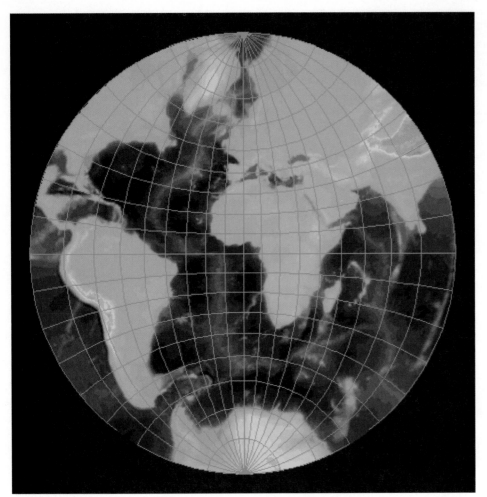

(a)

Figure 1.17 Stereographic projection. The projection point (or point of perspective) is on the other side of the globe from the point of tangency (map center) on the sphere. (a) Center at 0°, 0°, equatorial projection; (b) center at North Pole; (c) center at South Pole. Both b and c are polar projections.

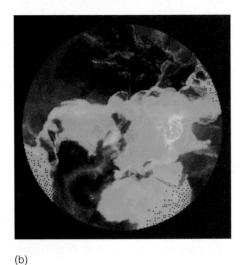

(b)

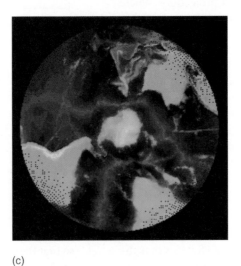

(c)

46. Attach a flat piece of plastic to the North Pole of the globe and follow the same procedure as before. What type of projection is this? _____ Look at the projection you made and compare it to the globe. What is the most distorted part of this projection? _____

Where is the distortion the least? _____

47. Look at Australia and Greenland in the two polar projections in Figure 1.17. How do their sizes compare in these maps? _____

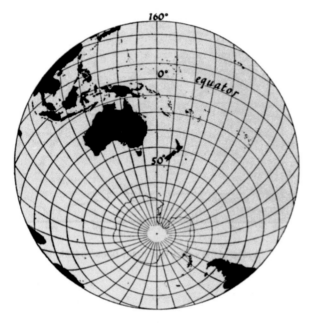

Figure 1.18 This Lambert equal-area projection has a point of contact at 50°S, 160°E, and is an example of a projection that causes parallels to become curves.

48. What types of distortion predominate in a Mercator map? Circle all that apply.
a. Shape distortion—objects do not maintain their same shape on the map as they have on Earth.
b. Scale distortion—equal distances on Earth are not equal on the map.
c. Area distortion—equally sized areas on Earth are not equal areas on the map.
d. Parallels are not parallel.
e. Parallels are not straight.
f. Meridians are parallel, not converging at the poles.
g. Meridians have unusual curvature.

49. What types of distortion predominate in a map made by conic projection (Fig. 1.16)? Using choices **a**–**g** from Exercise 48, list all that apply. _____

50. What types of distortion predominate in a map made by stereographic projection (Fig. 1.17)? Using choices **a**–**g** from Exercise 48, select all that apply. _____

U.S. Geological Survey Maps

The U.S. Geological Survey is the government agency responsible for creating official maps of the United States. The most popular varieties of maps are their topographic quadrangle series, which are identified by the number of degrees of latitude and longitude that they cover. Table 1.2 gives the latitude × longitude covered and corresponding fractional scales, and Figure 1.20 shows their relative size and scale. Topographic maps and how to read them are covered in Lab 11.

Table 1.2 USGS Topographic Quadrangles

Quadrangle Name	Quadrangle (lat. × long.)	Scale
2 degree sheet	1° degree × 2°	1:250,000
1 degree sheet	30′ × 1°	1:100,000
30 minute quad	30′ × 30′	1:125,000
15 minute quad	15′ × 30′	1:62,500
7 1/2 minute quad	7 1/2′ × 7 1/2′	1:24,000

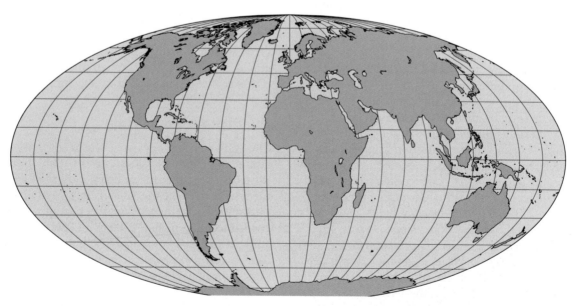

Figure 1.19 Equal area projection (Mollweide projection)

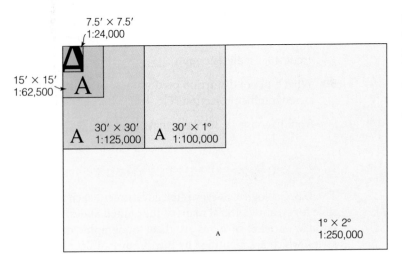

Figure 1.20 Comparison of the scales and relative sizes of standard U.S. Geological Survey Topographic maps. The letter A on each map represents the 10-mi tall letter A lying on the ground.

Global Positioning System (GPS)

The *Global Positioning System* (GPS) is a satellite-based system of radio location originally developed by the U.S. Department of Defense for military purposes. Civilians now use the system widely for various purposes, many in Earth sciences, both for general location and for more accurate surveying necessary in the construction of large-scale maps. GPS is especially useful in areas of flat or subdued topography; in fact, they may be the only reliable way of location, outside of surveying, in such areas.

How the Global Positioning System Works

The system consists of a constellation of 24 orbiting satellites deployed in six orbits with four satellites per orbit. The orbits are circular, about 20,200 km above the Earth's surface, and are inclined at 55° to the equator. This arrangement ensures that a GPS receiver can gather signals at any location on the Earth at any particular time. In general, a receiver can detect four to eight satellites at a given time.

A GPS receiver (Fig. 1.21) is a multichannel microwave receiver with a computer that converts the transmissions from the GPS satellites to a geographic position. The satellites transmit several codes as microwaves with their position, the precise GPS time, satellite identification, and various correctional factors. They receive their position from globally distributed satellite tracking stations. The computer part of the receiver uses the information to calculate the position of the unit. It determines the range (distance) from the receiver to a given satellite by knowing the velocity of the satellite relative to the receiver and the satellite's position.

The receiver's computer needs information from a minimum of four satellites to calculate its position. It can determine latitude, longitude, and elevation, but because of the configuration of the satellite orbits, the determination of elevation is less precise than that of latitude and longitude. Out-

Figure 1.21 Receiver for Global Positioning System

put from a receiver is usually in the form of latitude/longitude or Universal Transverse Mercator (UTM) coordinates. As the UTM system is used as the basis of the grid of many topographic maps, the UTM output of the receiver presents a grid reference that can be plotted directly on a topographic map.

Precision of GPS Locations

The GPS technology is capable of determining the latitude/longitude of a location to within 10 m. Originally, for security reasons, the military authorities who developed the system arranged for a deliberate scrambling of the signal transmitted from the satellites that degraded the precision on civilian receivers to about 100 m. In May 2000, however, U.S. authorities ended this practice. Now a commercially available receiver can give a location to about 20 m (i.e., the indicated position may be up to 20 m away from the actual location). Determination of altitude is less precise, however, due to the geometry of the global positioning system.

51. Determine how far off the 20-m level of precision is on a 1:62,500 map. That is, how many millimeters is equivalent to 20 m on the ground at this scale?

_____ In your opinion, is this pretty good or

pretty bad precision? _____ How far off is an

error of 20 m on a 1:24,000 map? _____

The precision of location using the GPS is inherent in the system itself. In other words, a more elaborate and expensive receiver will *not* give you a more accurate location than a cheaper one. The more elaborate receivers will give you a position faster and will detect weaker signals than cheaper ones and will have more software options.

Reception Problems

Use a GPS receiver in as open a location as is possible, because GPS satellites broadcast microwave band frequencies, which do not transmit through the ground, concrete, or heavy vegetation. Thus, in a steep-sided valley, a GPS may not be able to "see" enough satellites to determine a position. Equally, GPS receivers will not work well, if at all, under a heavy forest canopy. Large trees may also block part of a receiver's horizon. Many GPS receivers have a display that shows the approximate position in the sky of the transmitting satellites, and in the field this may be used to select a better place from which to determine the site's position.

Using a GPS Receiver

52. If a GPS receiver is available, take it outside and determine your location. Your instructor will give directions on using the receiver or will do this as a demonstration. Notice the numbers on the receiver's screen.

a. What are the numbers? _____

What do they mean? _____

_____

_____

_____

b. Plot the location given by the receiver on a map.

53. Try the GPS receiver near a building or under trees. What happens?

_____

_____

54. How would a GPS receiver be useful to an Earth scientist? Try to think of a specific example.

_____

_____

_____

_____

_____

_____

Physical Properties of Minerals

OBJECTIVES

1. To learn the definition of a mineral
2. To become familiar with the physical properties shown by minerals
3. To determine these properties for unknown minerals

Definition of a Mineral

Many of you may think of a mineral as something contained in a multivitamin capsule. *Mineral* in this sense is really an abbreviation of "mineral salt," something that is derived from a mineral, in the geoscientific sense of the word. Geoscientists have a very precise definition of what a mineral is. A **mineral** is a naturally occurring, usually inorganic, chemically homogeneous crystalline solid with a strictly defined chemical composition and characteristic physical properties.

Let's look at the idea that a mineral is **crystalline.** This means that the atoms composing it are arranged in an orderly way with a distinct structure (your lab instructor may have some models to show you). A **crystal** is a single grain of a mineral in which the structural planes of atoms extend in the same directions throughout the grain. The orderly arrangement of atoms controls many of the properties of the mineral: the external shape of the mineral if it was free to grow, or its **habit;** and the way the mineral breaks, or its *cleavage* or *fracture.* The crystalline structure can even influence the *hardness* and *density* of a mineral. For example, diamonds and graphite are both pure carbon, but their crystalline structures are very different. These differences give rise to the hardest substance known, diamonds, and a substance, graphite (used as pencil lead), which is so soft we can write with it.

Crystallization of Minerals

Minerals may crystallize in any one of four geological environments:

1. *From molten rock.* This is the typical mode of formation of minerals in igneous rocks (see Lab 4).
2. *From solution.* This typically happens in various stages of the formation of sedimentary rocks (see Lab 5).

3. *Within living cells.* Many creatures precipitate crystalline substances within their cells to build a skeleton or shell. Shellfish, for example, precipitate *calcite* or *aragonite* in their shells. We humans precipitate a calcium phosphate, called apatite, in our bones and teeth. Although these materials are generally not considered minerals while part of the living organism, they are an important source of minerals in certain kinds of sedimentary rocks once the organism dies.
4. *By recrystallization of preexisting minerals into the same or new minerals.* As rocks undergo changing conditions of pressure and temperature in the Earth, minerals may break down and re-form into the same or new minerals while in the solid state. This process is typical in the formation of metamorphic rocks (see Lab 6). Recrystallization of the remains of organisms (see Lab 8) or minerals precipitated from solution may recur during the formation of sedimentary rocks.

 ## Mineral Computer Exercise

This first computer exercise gives a quick overview of material covered later in this lab. Starting with this computer exercise, you will find that the later discussion of color, luster, hardness, and cleavage will reinforce and expand on what you learn here.

- Start **In-TERRA-Active.**
- Click **Materials,** then click **Minerals.**
- Click **Intro,** then click the play button (▶) twice.
- Click COLOR:

   1. Why is color not always a useful mineral property?

   _____

   _____

   Another reason is that more than one mineral can have the same color.

- Click LUSTER:

2. What are two types of luster mentioned?

_____ _____

■ Click HARDNESS:

3. a. How hard is a fingernail? _____

  b. How hard is glass? _____

  c. How hard are diamonds? _____

■ Click CLEAVAGE and FRACTURE:

4. A mineral cannot have _____ different directions of cleavage. (Choose one of the following: 0, 1, 2, 3, 4, 5, 6.)

5. Finish the following sentence:

Whatever cleavage a specific mineral has, it's important to realize that that mineral _____ _____.

■

Types of Physical Properties Shown by Minerals

Crystal Habit

Crystal habit is the external shape a mineral exhibits given favorable conditions when a crystal is well formed and unbroken. Habit is determined by the crystal's internal structure; the external shape and symmetry of a crystal are reflections of its internal atomic order. We can analyze the symmetry of these shapes and use this analysis to help identify a mineral. Some common terms used to describe crystal habits are illustrated in Figure 2.1. If you wish to describe the habit of a crystal you first look at the individual grains and determine their external shape.

Crystals are often in groups or clusters with their edges touching one another. We then say that these groups are **aggregates.** The crystal boundaries of minerals in rocks interlock in this way. Some typical aggregate types are shown in Figure 2.2.

6. Look at an aggregate of several pyrite crystals:
  a. Describe the shape of the individual crystals. Some illustrations and shapes are shown in Fig. 2.1.

  _____

  _____

  b. Describe the aggregate type (refer to Fig. 2.2).

  _____

  _____

7. Your instructor will provide a Crystal Habit Display. Select 4 samples from the display, some with individual crystals and some aggregates. In Table 2.1, list each sample and one or more terms from Figures 2.1 and 2.2.

Luster

Luster describes how _light reflects_ from a fresh surface. You can observe luster on samples provided and in the photographs on pages 38–42. There are two broad classifications of luster: **metallic** and **nonmetallic.** Metallic lusters include **metallic** (galena) and **submetallic** (magnetite, p. 41; and graphite, p. 42). Nonmetallic lusters are many and varied. The most common nonmetallic lusters are **adamantine** (garnet), **vitreous** (quartz), **waxy, resinous** (biotite), **greasy, pearly** (talc), **silky** (satin spar gypsum), and **dull/earthy** (kaolin). You will understand these terms better after you do Exercise 8 in which you associate the luster terms with simple English descriptions of the terms by looking at actual examples of each luster.

8. Examine the samples in the Luster Display, which have labeled lusters, and identify which lusters correspond to

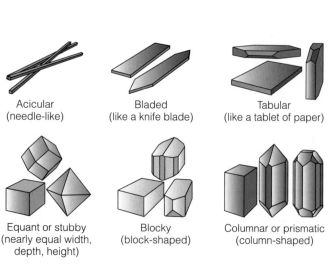

Acicular
(needle-like)

Bladed
(like a knife blade)

Tabular
(like a tablet of paper)

Equant or stubby
(nearly equal width,
depth, height)

Blocky
(block-shaped)

Columnar or prismatic
(column-shaped)

Figure 2.1   Crystal habits of individual minerals

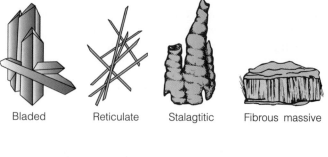

Bladed   Reticulate   Stalagtitic   Fibrous massive

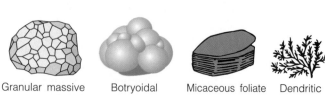

Granular massive   Botryoidal   Micaceous foliate   Dendritic

Figure 2.2   Mineral aggregate habits

the descriptions in Table 2.2. Remember that you are looking for *surface* shine or appearance, not color, transparency, or opaqueness.

Color

On a fresh, unaltered surface, the color of a mineral may help in identification, but beware: Color is a very unreliable property to use in identifying minerals. Impurities within a mineral may give rise to a variety of colors. Quartz, for instance, has many varieties: amethyst (purple), rock crystal (colorless), rose quartz (pink), smoky quartz (gray), citrine (yellow), and milky quartz (white). (See quartz on display and on pp. 39–40.) In addition, more than one mineral may have the same color. For example, amethyst and some varieties of fluorite are purple (p. 42), and quartz, K feldspar, calcite, and gypsum all have pinkish varieties.

Hardness

The resistance of a mineral to abrasion (scratching) is termed **hardness.** Mohs scale of hardness (H), established by mineralogist F. Mohs, has ten minerals arranged in an order of increasing relative hardness, each of which will scratch the mineral of lower hardness on the scale (Table 2.3). To determine the relative hardness of a specimen, you scratch an unweathered surface of the unknown with an object of known hardness. If, for example, glass of hardness 5.5 scratches the mineral, then try to scratch the mineral with a

penny (H = 3) and so on until you no longer mark the surface or until you have no softer object with which to test it. You can also check to see whether the unknown sample can scratch the object of known hardness. *When you think you see a scratch,* check to make sure you have not simply left powder of the scratching object behind; wipe away any powder, and look at the surface closely. Also, the physical nature of a mineral specimen may prevent correct determination of hardness if a mineral is, for example, splintery or granular and falls apart when tested.

Imagine that you are recording the properties of a sample you think is fluorite. You might have noticed that fluorite is on the Mohs scale of hardness and has a hardness of 4.

Table 2.3 Mohs Scale of Hardness

Mohs Number	Mineral	Common Object
1	Talc	
2	Gypsum	
		Fingernail ($\cong 2.5$)
3	Calcite	Copper penny ($\cong 3$)
4	Fluorite	
5	Apatite	
		Pocket knife ($\cong 5.5$)
		Glass ($\cong 5.5$)
6	Orthoclase (a feldspar)	Streak plate ($\cong 6.5$)
7	Quartz	
8	Topaz	
9	Corundum	
10	Diamond	

Note: Hardness may vary in some minerals (1/2 to 2 points) from crystal face to crystal face as seen in kyanite, which has a hardness of 5 parallel to its length and 7 across the length.

Table 2.1 Mineral Habits (Exercise 7)

Sample	Habit

Table 2.2 Examples of Luster (Exercise 8)

Luster Description	Luster	Sample
Like a bright shiny metal		
Shines like glass		
Like a tarnished metal		
Like wax		
Looks like it is coated with grease		
Not shiny, like chalk or dirt		
Very bright and gemlike shine		
Like pearls		
Shines like plastic		
Shimmers like silk fibers		

However, this does not mean you should write down 4 for its hardness. Instead, you should use the evidence obtained by the scratching tests to provide a range of possible hardness. You should not jump to conclusions when testing minerals but carefully record your observations. What if you thought the mineral was a purple variety of quartz instead of fluorite and decided, knowing the hardness of quartz is 7, that you would just write down 7? This conclusion would mislead you in the mineral identification. In fact, it would be an example of formulating a hypothesis and then changing the data to fit the hypothesis—a definite scientific no-no.

Streak

Streak is the color of a mineral when powdered. The color of the powder is less variable than the color of a mineral, so streak is a more reliable property than color. A streak plate made of porcelain is used to obtain a small amount of powder from a specimen. Look at the example of the streak for hematite shown on page 42. Notice that although one sample of hematite is silvery gray, both samples have a brown streak. Since the hardness of porcelain is about 6 to 7 on Mohs scale, the streak of a mineral with hardness greater than 6 cannot be easily determined and can be said to have **no streak.** Softer minerals should each have streak, but keep in mind that it may be hard to see a white or colorless streak on a white streak plate. Look closely.

Broken Surfaces of Minerals

The broken surfaces of a mineral often have characteristics that can help in mineral identification. See examples in the Fracture and Cleavage Display.

Fracture   If the broken surfaces of a mineral are irregular and nonplanar, the mineral is said to have **fracture.** When fracture is determined, it should be described with one of the following terms:

Conchoidal: A smooth, curved surface that looks like the inside of a clam shell. See the displays and photographs of obsidian (a volcanic glass; Fig. 4.13) and conchoidal fracture of quartz (p. 40). Well-crystallized quartz displays conchoidal fracture on its broken surfaces.

Fibrous: Fracture surface has a threadlike appearance, similar to the coarser, splintery appearance of wood. A good example is asbestos, another is satin spar gypsum.

Hackly: A sharp, irregular surface, the same as *jagged.* Both garnet and wollastonite may produce a hackly fracture.

Uneven/Irregular: A general term that can be applied to many different minerals that otherwise defy definition

Cleavage   If one or more of the broken surfaces of a mineral are planar surfaces, the mineral has **cleavage.** The way that a mineral breaks depends on the crystal structure. A cleavage plane is a plane of weakness in the structure of the

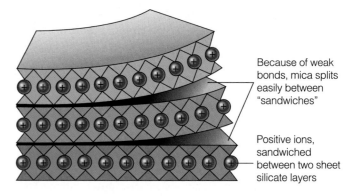

Because of weak bonds, mica splits easily between "sandwiches"

Positive ions, sandwiched between two sheet silicate layers

Figure 2.3   Cleavage in minerals occurs where weak bonds in planes of atoms allow the mineral to split apart easily, such as the mica shown here.

mineral (Fig. 2.3). A mineral may have many or no cleavage planes, and these planes may be perfectly planar, as in muscovite, or slightly irregular, as in pyroxene or hornblende. A mineral with no planes of weakness (e.g., quartz) will fracture. This does not necessarily mean that the mineral is exceptionally tenacious or hard, only that inherent weaknesses are not planar. A complete description of cleavage includes the relationship of the cleavage planes. For instance, galena and halite both have cubic cleavage (pp. 41–42), which is equivalent to three cleavage planes at 90° to each other (Fig 2.4). In some cases, the angle between the cleavages helps to distinguish two minerals. Augite pyroxene and hornblende amphibole both have two cleavages, but the angle between them is near 90° for augite and much more oblique for hornblende (Fig. 2.5).

Parting   Not a common characteristic generally speaking, **parting** is a roughly planar break in a mineral that is not a true cleavage plane. In a sense, parting is an intermediate condition between cleavage and fracture. A clear example of parting can most often be seen in hand specimens of corundum. Another mineral with parting is garnet; this parting leads to garnet's hackly fracture.

9. Examine the samples with unknown fracture or cleavage. List the minerals that show cleavage in the appropriate place in Table 2.4; then list those with fracture. Describe the cleavage and fracture in more detail referring to earlier information and Figure 2.4 if necessary.

Distinguishing Cleavage from Crystal Faces

Many students look at a well-formed crystal of quartz with flat crystal faces and say the sample has cleavage. This is wrong. By now you realize that quartz has conchoidal fracture, not cleavage. The flat surfaces on quartz did not break along planes of weakness but grew that way when the mineral formed.

How can you tell the difference between cleavage and crystal faces? Since cleavage is an inherent planar weakness in the mineral, you will almost always see multiple examples of

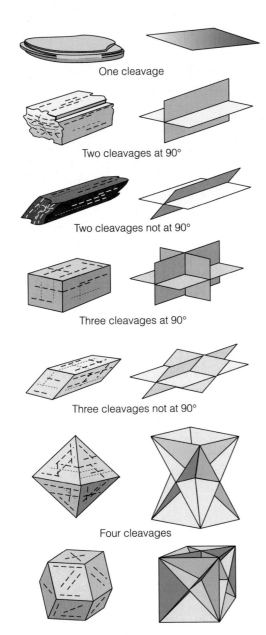

Figure 2.4 Types of cleavage

One cleavage

Two cleavages at 90°

Two cleavages not at 90°

Three cleavages at 90°

Three cleavages not at 90°

Four cleavages

Six cleavages

Table 2.4 Cleavage and Fracture (Exercise 9)

Samples with Cleavage	Describe the Cleavage
Samples with Fracture	**Describe the Fracture**

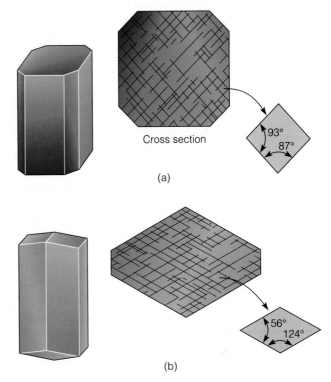

Cross section

93°
87°

(a)

56°
124°

(b)

Figure 2.5 Distinctive cleavage angles. The two cleavage planes in (a) augite pyroxene form a nearly 90° angle, but in (b) hornblende amphibole, the planes meet at more oblique angles (56° and 124°).

a particular cleavage plane exhibited. These may show up as a steplike surface feature (hornblende, galena and halite). Look closely at samples for these steps—you may even want to use a hand lens.

10. Examine a quartz crystal. Notice the flat planes on this sample. Since all quartz has fracture, not cleavage, how would you account for the flat planes in this natural sample of quartz? _____

_____

11. Examine the two samples provided. For each sample, sketch the planar features and show any steps in your sketch.

   a. Which sample has fracture and crystal faces?

   _____

   b. Which sample has cleavage? _____

Density (Specific Gravity)

Density is defined as the mass per unit volume of a substance. Another way to express this property is **specific gravity,** which is the ratio of the weight of a mineral specimen to an equal volume of water. In hand specimen, we are usually concerned more with relative density, which is crudely comparing the density of one mineral specimen to another. You can make general comparisons by using the "heft test," in which you make a comparison of two specimens of about the same size to say which is heavier. This is apparent when you compare, for example, quartz, which has a fairly normal or average density, with a metal sulfide such as pyrite, which has a high density, or with graphite, which has a low density. It turns out that humans are quite good at judging density by feel. With some practice, you could probably make fairly accurate estimates of mineral density this way. The reason for this is that your brain has to tell your arm how much force to apply to pick up an object. When you pick it up, you don't want it to go flying through the air because you applied too much force, nor do you want it to fail to budge. So when you see an object, from past experience, your brain decides what density it is likely to have, figures the weight and tells your arm just the right amount of force to apply. Sometimes you misjudge the density of a new substance and are surprised to find how light or how heavy it is. An example would be a fake rock made of foam. You look at it and think it is a rock, your brain processes that information, but when you pick it up—surprise—you lift it much too fast because it is much lighter than you thought.

12. Examine the samples provided. Heft the samples. While looking at them, judge their relative density. The reason you should look at the samples is to allow your eyes to judge the size while your hands judge the weight; together the two give you an estimate of the density. Now rank the samples in order of their density from least dense to most dense.

    least dense _____ _____ _____ _____ **most dense**

13. Your instructor will provide the actual densities of these minerals.
    a. How accurate were you in distinguishing the densities when the samples were close to the same size?

    _____

    b. When the samples were quite different in size?

    _____

    c. What was the closest difference in density you were able to distinguish? _____

Special Properties

Some minerals have special properties such as effervescence in acid, magnetism, unusual visual properties, or a distinctive smell, taste, or feel.

Effervescence Seen in carbonates (primarily calcite; p. 41), **effervescence** is a fizzing or bubbling that takes place when a dilute solution of hydrochloric acid (HCl) is applied (please use sparingly, and clean off the sample with a tissue or paper towel afterward). This property is especially useful in distinguishing calcite from other common rock-forming minerals. Dolomite will effervesce when it is powdered; aragonite will do so easily but is not so often seen in hand specimen. The chemical reaction that occurs when HCl is applied to calcite or aragonite ($CaCO_3$) is

$$2H^+ \quad + \quad CO_3^{--} \quad \Rightarrow \quad H_2O \quad + \quad CO_2$$
$$\text{(from HCl)} \quad \text{(from } CaCO_3\text{)} \quad\quad \text{(water)} \quad \text{(escaping gas in fizzing)}$$

Magnetism The mineral is attracted to a magnet. Magnetite (Fe_3O_4) is strongly magnetic (p. 41). Magnetism is especially useful in distinguishing magnetite from other common rock-forming minerals. Hematite (Fe_2O_3) and magnetite often occur together in specular hematite samples, making what looks like a pure hematite sample magnetic.

Feel Some minerals have a diagnostic feel (e.g., the greasy feel—not greasy luster—of graphite).

Fluorescence Minerals may fluoresce when they are put under ultraviolet (UV) light. The short wave radiation of the UV is absorbed by the mineral and radiated back as longer-wave visible radiation. A mineral that often shows fluorescence is calcite.

Double Refraction Diagnostic of Iceland spar calcite, **double refraction** is when the light, upon entering the crystal, is broken into two rays (Fig. 2.6). If the crystal is placed over some writing, for example, what you see is a double image. If you rotate the crystal, one image will stay fixed and the other will move around it.

Smell Some minerals, such as sulfur, possess a distinctive odor. (However, students in a lab setting have been putting acid on samples for years, which tends to give all of the samples a smell.)

Taste A few minerals have a characteristic taste. Halite tastes salty. Sylvite tastes bitter. *Please do **not** taste the sam-*

Figure 2.6 Double refraction. Calcite (*Iceland spar*) has double refraction.

ples since students have been putting acid on them. However, taste is a valid mineral property and can be quite helpful in identifying some minerals. Ask your lab instructor if you want to know how a mineral tastes. In the field, geologists do sometimes taste samples, but they must be aware that some minerals are poisonous.

Tenacity

Tenacity is the cohesiveness of a specimen, a description of a mineral's resistance to mechanical deformation (breaking, bending, crushing, etc.). These tests are destructive, so please do *not* perform them on the lab mineral samples unless your lab instructor supplies tenacity test samples. The following are examples of this property:

Brittle: This is the most common type of tenacity; a brittle mineral breaks as easily as a hard candy does.

Ductile: A specimen can be pulled out into an elongated shape.

Elastic: A sample bends when force is applied and resumes its previous shape when force is released.

Flexible: A specimen bends without breaking but will not spring back to its original shape.

Sectile: A mineral can be shaved with a sharp blade as if it were wax or soap.

Determining the Properties of Minerals

14. Complete Table 2.5 by testing the mineral properties of known samples (ones with the names labeled) and filling in the blanks in the table. Figure 2.7 shows examples of silicate minerals—those with chemical compositions containing silicon and oxygen. Figure 2.8 has photographs of nonsilicate minerals.

Materials needed:

- Known minerals (see also Figs. 2.7–2.8): biotite mica, calcite, microcline feldspar, plagioclase feldspar, fluorite, galena, almandine to pyrope garnet, gypsum, halite, oolitic or red ocher hematite, specular hematite, hornblende, kaolin, magnetite, muscovite mica, pyrite, augite pyroxene, quartz.
- Mineral Testing Kit
- Squeeze bottle of 10% hydrochloric acid solution. (Use as little as possible—one drop is enough—and clean off the sample with a tissue when you finish.)

Table 2.5 Mineral Identification Chart of Known Samples

Name	Chemical Formula	Color	Streak	Luster	Hardness	Density (g/cm³)	Cleavage (CL) or Fracture (F)	Crystal Habit and Form	Distinguishing Characteristic or Special Property
Biotite mica	$K(Mg,Fe)_3AlSi_3O_{10}{}^-(OH)_2$		Light brown		2 1/2–3	2.8–3.2			Color, luster and cleavage, thin sheets, elastic
Calcite	$CaCO_3$	Colorless to white	White	Vitreous to waxy		2.7		Rhombohedral	
Feldspar (microcline or orthoclase)	$KAlSi_3O_8$			Vitreous	6	2.5–2.6		Tabular to blocky	Hardness, cleavage
Feldspar (plagioclase)	$NaAlSi_3O_8$ to $CaAl_2Si_2O_8$		None			2.6–2.8	2 CL directions at nearly 90°	Tabular to blocky	Hardness, cleavage, twinning—appears as fine parallel grooves on cleavage surface
Fluorite	CaF_2	Varied: purple, yellow, green, to colorless	White	Vitreous		3.2	4 CL directions (octahedral)	Equant	Fluorescent, phosphorescent
Galena	PbS				2 1/2			Equant	Density, luster, color
Garnet (almandine to pyrope)	$(Mg,Fe)_3Al_2Si_3O_{12}$		None		6 1/2–7 1/2	3.6–4.3		Equant, dodecahedrons	Habit, color, and luster
Gypsum	$CaSO_4 \cdot 2H_2O$	Colorless to white to pink	White	Vitreous to pearly		2.3	1 good CL direction, 2 poor	Tabular	Sectile
Halite	$NaCl$	Colorless	White	Vitreous to waxy	2–2 1/2	2.2		Equant	Salty taste
Hematite (oolitic or red ocher)	Fe_2O_3				1–6 1/2	≤5.3	F—uneven	Small (egg-shaped) ooids or fine-grained aggregate	Color and streak
Hematite (specular)	Fe_2O_3				5 1/2–6 1/2	5.3	F—uneven	Aggregate of tabular crystals, micaceous	
Hornblende	$(Ca,Na)_{2-3}{}^-(Mg,Fe,Al)_5Si_6{}^-(Si,Al)_2O_{22}(OH)_2$		None		5–6	3.0–3.4		Prismatic	Color, luster and cleavage

Continued

Table 2.5 *Continued*

Name	Chemical Formula	Color	Streak	Luster	Hardness	Density (g/cm³)	Cleavage (CL) or Fracture (F)	Crystal Habit and Form	Distinguishing Characteristic or Special Property
Kaolin	$Al_2Si_2O_5(OH)_4$				1–2 1/2	2.2–2.6	F—earthy	Very fine-grained aggregate	Smell it:
Magnetite	Fe_3O_4	Black	Black		6	5.2	F—uneven	Equant	
Muscovite mica	$KAl_3Si_3O_{10}(OH)_2$		Hard to obtain; white to colorless			2.8–2.9		Tabular	
Pyrite	FeS_2				6–6 1/2	5.0	F—uneven		"Fool's gold"
Pyroxene (augite)	$(Ca,Na)(Mg,Fe,Al)$ $(Si,Al)_2O_6$	Dark green; gray to black		Vitreous	5–6	3.2–3.4		Good crystals are rare, stubby	Cleavage and
Quartz	SiO_2		None				F—conchoidal		Hardness, fracture

15. Start filling in the information in Table 2.6 by testing the mineral properties of the unknown samples (ones with numbers only) and filling in the table. The last two columns are reserved for Lab 3. (If you feel you have identified a mineral based on your observations of Table 2.5 and Figures 2.7 and 2.8, you may lightly pencil it in, but remember that this is only a hypothesis and may change as you work Lab 3.)

Materials needed:

- Assorted unknown minerals
- Mineral Testing Kit
- Squeeze bottle of 10% hydrochloric acid solution. (Use as little as possible—one drop is enough—and clean off the sample with a tissue when you finish.)

(a) **Olivine:** an aggregate of grains showing the typical green color

(b) **Almandine garnet:** *left:* a trapezohedral single crystal, equant habit; *right:* a broken sample showing hackly fracture and adamantine luster

(c) **Augite (pyroxene):** a fragment showing two cleavages at nearly 90°. See also Figure 2.5a.

(d) **Hornblende (amphibole):** fragments illustrating vitreous luster and two cleavages at an oblique angle. See also Figure 2.5b.

(e) **Talc:** aggregate showing pearly luster

Figure 2.7   Silicate minerals

(f) **Kaolin:** an extremely fine-grained aggregate showing earthy luster and no visible cleavage

(g) **Biotite mica** (black) has resinous luster and one perfect cleavage. The white grains in this sample are quartz.

(h) **Muscovite mica:** *top:* pearly luster on the flat cleavage surface; *bottom:* cleavages seen edge on; *inset:* thin cleavage sheets illustrate the one perfect cleavage of mica.

(i) **Potassium feldspar** (**orthoclase** or **microcline**) varies in color from white to salmon pink and has two cleavages at 90°.

(j) **Plagioclase feldspar** of different chemical compositions. Ca-rich ones are usually darker. *Upper right:* dark gray Ca-rich; *left:* light gray Na-Ca; *lower right:* white Na-rich with twinning (stripes of reflected light). Twinning, if visible, is evidence for plagioclase.

(k) **Quartz:** hexagonal quartz crystals: *left:* Colorless quartz crystals (*rock crystal*) with columnar habit; *right:* smoky quartz crystals with columnar habit.

Figure 2.7 Silicate minerals *Continued*

(l) **Rock crystal:** a broken fragment of colorless quartz showing excellent conchoidal fracture

(m) **Milky quartz:** a broken fragment of quartz with white color and vitreous luster; *inset: milky quartz* crystals from a geode

(n) **Smoky quartz** or **cairngorm stone:** gray translucent quartz; may be nearly black as in Figure 2.7k; *inset:* typical smoky quartz (circled) in granite

(o) **Rose quartz:** a broken fragment of quartz with pink color and vitreous luster

(p) **Amethyst:** a broken fragment of quartz with purple color; *inset:* amethyst crystals from a geode

Figure 2.7 Silicate minerals *Continued*

(a) **Calcite:** a broken fragment showing three cleavages at an oblique angle. The bubbles result from reaction with dilute hydrochloric acid. See also Figure 2.6.

(b) **Gypsum:** common habits. *Upper left:* massive aggregate, *alabaster; right:* fibrous, *satin spar; lower left:* a single crystal cleavage fragment, *selenite*

(c) **Galena**—lead ore: a fragment with three cleavages at 90° (cubic cleavage). Metallic luster shows where freshly cleaved (left). Weathered surfaces (right) are submetallic.

(d) **Pyrite:** three samples showing brassy color and metallic luster. *Upper left:* Granular massive aggregate of pyrite; *right:* aggregate of crystals; *lower left:* pyrite cube.

(e) **Sphalerite**—zinc ore. Sphalerite has adamantine luster and is mixed yellows, browns, and black. *Left:* shows 4 of its 6 cleavages.

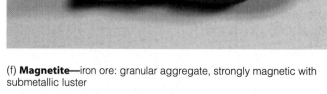

(f) **Magnetite**—iron ore: granular aggregate, strongly magnetic with submetallic luster

Figure 2.8   Nonsilicate minerals

(g) **Hematite**—iron ore. *Left: Oolitic hematite*—an aggregate of spheres of fine-grained hematite crystals. The grains and streak are reddish brown. *Right: Specular hematite* is silvery yet has a reddish brown streak.

(h) **Fluorite** has four cleavages and comes in a variety of colors. Its colors and vitreous luster are similar to quartz, but its cleavage distinguishes it from that harder mineral.

(i) **Halite:** transparent to translucent with three cleavages at 90° (cubic cleavage)

(j) **Graphite**—used in pencil leads: submetallic luster, gray streak, and low density

Figure 2.8 Nonsilicate minerals *Continued*

Table 2.6 Mineral Identification Chart of Unknown Samples

Sample Number	Color	Streak	Luster	Hardness	Cleavage/Fracture	Crystal Habit/Form (if applicable)	Distinguishing Characteristic or Special Property	Mineral Name (Lab 3)	Chemical Group (Lab 3)

Continued

Table 2.6 *Continued*

Sample Number	Color	Streak	Luster	Hardness	Cleavage/Fracture	Crystal Habit/Form (if applicable)	Distinguishing Characteristic or Special Property	Mineral Name (Lab 3)	Chemical Group (Lab 3)

Rock-Forming Minerals

OBJECTIVES

1. To develop a logical and systematic approach to mineral identification
2. To be able to recognize some of the common rock-forming minerals
3. To become familiar with the classification system of minerals, particularly the silicate minerals

Mineral Identification and Recognition

If mineral *identification* is approached in a systematic and logical manner, even a beginner can identify a surprisingly large number of minerals. The secret is to realize that minerals generally have no "fingerprint". A single property is usually not enough to distinguish one mineral from others, but you can distinguish minerals by *combinations* of physical properties. In the previous lab, you became familiar with the physical properties used in mineral identification. In this lab, you will use mineral identification tables, which list the combination of properties of some rock-forming minerals, to help you identify these minerals.

Mineral *recognition,* on the other hand, is like recognizing your friends. You learn what characteristics, such as facial features, are unique to that person, and you learn to associate these features with the person's name. For mineral recognition, you must learn to associate a set of physical properties with the name of the mineral. This takes some practice, just as it takes you a while to begin to associate names with faces when meeting a group of people for the first time.

There are more than 3000 different mineral species, but most of these are not common in rocks. After learning the technique described in this lab, if you had suitable mineral tables and plenty of time, you could probably identify a substantial proportion of the 3000. You will not be asked to do so. You are required, however, to be able to recognize some of the rock-forming minerals (listed in Tables 3.1 and 3.2), which make up the vast majority of rocks at the Earth's surface. Recognition of these minerals is necessary to identify and interpret rocks correctly. Also, you may be given a few

minerals you have never seen before and be asked to identify them using mineral identification charts to see how well you have mastered the techniques of identification.

Strategy for Mineral Identification

Successful mineral identification requires systematic application of the skills you learned in Lab 2. Use of photographs such as those in Figures 2.7 and 2.8, for example, cannot replace close examination of a mineral's physical properties. In nature (and perhaps in your tests or quizzes) there is quite a wide variety in the general appearance of different specimens of the same mineral, but many of their specific physical properties will always be the same.

The way minerals are identified is to determine their physical properties with a series of simple tests (Lab 2). Properties such as hardness, luster, streak, and cleavage are usually indicative of a mineral. Beware of relying on a property such as color, because color is highly variable in some minerals, and different minerals can come in the same color. Would you identify a particular car as a Porsche, say, just because it was painted red? In the lab don't try to guess the mineral and then look up its properties. This would be like buying a red car, calling it a Porsche, and then being surprised when it won't go 150 mph.

The correct strategy is to use the Mineral Tables (3.3 and 3.4; see pp. 49–53) systematically, as illustrated in the maze in Figure 3.1. Determine the luster of the mineral first and then go to the appropriate set of tables. For metallic minerals, the streak can be quite useful, so test the streak next, proceed to identify the mineral from Table 3.3. For nmetallic minerals test the hardness with glass (5.5) look for cleavage; then go to the appropriate section 3.4. You can probably identify the mineral out remain, these remaining choices simply by checki refining hardness, lus- tests. If you follow the maze ter, and cleavage. If ar ies as you pass a shaded area in be checked by a f sults to help you decide which path in Figure 3.1 identification is thus a process of elim- described eed in a methodical manner it is easy. In

Table 3.1 Silicates

MINERAL	CHEMICAL FORMULA	SUBCLASS
1. **Olivine**[1]	$(Mg,Fe)_2SiO_4$	Nesosilicates (isolated Si tetrahedra)
Topaz	$Al_2SiO_4(F,OH)_2$	
2. *Garnet*[2]	$(Mg,Fe,Ca,Mn)_3(Al,Fe,Cr)_2(SiO_4)_3$	
Varieties: almandine, grossular		
Tourmaline	$(Na,Ca)(Li,Mg,Al)_3(Al,Fe,Mn)_6(BO_3)_3(Si_6O_{18})(OH)_4$	Cyclosilicates (ring silicates)
Beryl	$Be_3Al_2(Si_6O_{18})$	
Pyroxenes:		Inosilicates (chain silicates)
3. **Augite**	$(Ca,Na)(Mg,Fe,Al)(Si,Al)_2O_6$	*Single chain*
Diopside	$CaMgSi_2O_6$	
Amphiboles:		*Double chain*
4. **Hornblende**	$(Ca,Na)_{2-3}(Mg,Fe,Al)_5Si_6(Si,Al)_2O_{22}(OH)_2$	
Actinolite	$Ca_2(Mg,Fe)_5Si_8O_{22}(OH)_2$	
Talc	$Mg_3Si_4O_{10}(OH)_2$	Phyllosilicates (sheet silicates)
5. **Kaolin**	$Al_2Si_2O_5(OH)_4$	
Chlorite	$(Mg,Fe)_3(Si,Al)_4O_{10}(OH)_2 \cdot (Mg,Fe)_3(OH)_6$	
Micas:		
6. **Biotite**	$K(Mg,Fe)_3(AlSi_3O_{10})(OH)_2$	
7. **Muscovite**	$KAl_2(AlSi_3O_{10})(OH)_2$	
8. *Feldspars:*		Tectosilicates (framework silicates)
Alkali feldspars	$(K,Na)AlSi_3O_8$	
Plagioclase feldspars	$(Na,Ca)Al_{1-2}Si_{3-2}O_8$	
9. **Quartz**	SiO_2	
Varieties: amethyst, rock crystal, smoky quartz, milky quartz, rose quartz		
Microcrystalline varieties: chert, flint, jasper, agate, chalcedony		

[1] Importa...
[2] Mineral gr...-forming minerals are numbered and shown in bold.
...es are shown in italics.

the process o...
the photograph...
any match you m...ing a mineral you may also want to use
You should test that ...2.7 and 2.8, but keep in mind that
and making sure they...he photos is only a hypothesis.
you think your sample m...v testing physical properties
...roperties of the mineral

1. Practice the systematic m...
on three samples provide...
3.1, and write the answers in ...l identification
...e in Figure
...3.2.

2. Use the procedure in Exercise 1 for the assigned "un-known" minerals you started in Lab 2, and fill in the last two columns in the Mineral Identification Chart, Table 2.6. If you haven't already done so, as soon as you de-termine the physical properties of the mineral, write them down in Table 2.6 in Lab 2. This will help you eliminate possible choices of mineral names that are not reasonable. Make a final identification by compar-ing your unknown sample with the properties listed in the Mineral Tables 3.3 and 3.4 and with the mineral ref-

Table 3.2 Nonsilicates

MINERAL	CHEMICAL FORMULA	CHEMICAL GROUP
10. **Dolomite**	$CaMg(CO_3)_2$	**Carbonates** Basic unit: (CO_3)
11. **Calcite**	$CaCO_3$	
Varieties: Iceland spar, onyx, calcite chalk, travertine		
Malachite	$Cu_2CO_3(OH_2)$	
Azurite	$Cu_3(CO_3)_2(OH)_2$	
12. **Gypsum**	$CaSO_4 \cdot 2H_2O$	**Sulfates** Basic unit: (SO_4)
Varieties: alabaster, satin spar, selenite		
Galena	PbS	**Sulfides** S plus a metal(s)
13. **Pyrite**	FeS_2	
Sphalerite	ZnS	
Corundum	Al_2O_3	**Oxides** O plus a metal(s)
14. **Magnetite**	Fe_3O_4	
15. **Hematite**	Fe_2O_3	
Limonite	$FeO \cdot OH \cdot nH_2O$	**Hydroxides** (OH) plus metal(s)
Apatite	$Ca_5(PO_4)_3(F,Cl,OH)$	**Phosphates** Basic unit: PO_4
Fluorite	CaF_2	**Halides** Halogen ion present
16. **Halite**	$NaCl$	
Metals:		**Native elements** Occur in elemental form
Silver	Ag	
Gold	Au	
Platinum	Pt	
Copper	Cu	
Iron	Fe	
Nonmetals:		
Diamond	C	
Graphite	C	
Sulfur	S	

erence set (labeled samples) provided in the lab. The procedure will be slow at first, but with practice, it becomes much more rapid.

Strategy for Mineral Recognition

On your exams or quizzes and when identifying rocks, you will need to recognize minerals. Mineral recognition, like recognizing your friends, requires linking distinctive features (e.g., a mineral's properties) to the name. When you get to know people, you can recognize them no matter what they are wearing or how they have styled their hair. Each mineral has some natural variation such as variation in color, size, shape, and so forth. In learning to recognize a mineral, you need to learn (1) enough properties that are al-

ways present that in combination will allow you to distinguish that mineral from others, called *diagnostic properties;* (2) the mineral's name; and (3) when a property is consistent, such as the color of olivine or azurite, and when it can vary, such as the color of quartz. In Exercise 3 you will write out a list of the diagnostic properties and practice recognizing minerals.

3. Your instructor will inform you of the minerals you are required to know. Fifteen minerals are listed in Table 3.5 (which are also in Tables 3.1 and 3.2). There is room for you to add additional minerals required by your instructor. For each of these minerals, study the example found in the mineral reference set so that you will be able to recognize it. In Table 3.5, write down some way to help you remember the mineral name. Try

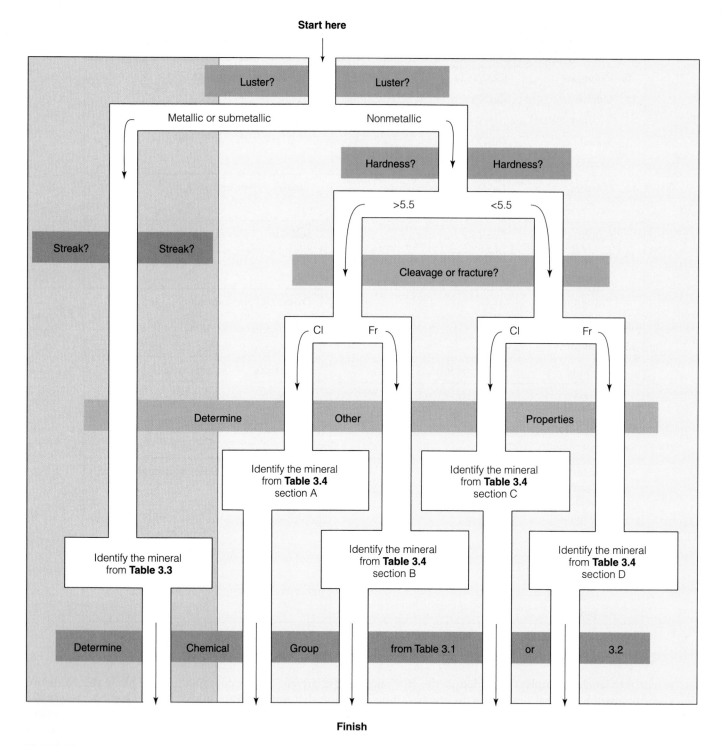

Figure 3.1 Maze for systematic mineral identification. Select a mineral sample and follow the maze with that sample in mind. At each shaded region, test the sample (or refer to the tests made in Lab 2 for that sample) and decide which path to follow based on the resulting mineral property.

to think of a word or phrase that sounds like the mineral name and possibly refers to some property of the mineral. For example, olivine is olive green, or for pyrite—pirates like gold. Also list a few diagnostic properties that will help you recognize the mineral. Practice the names that are unfamiliar.

4. How well do you know your minerals now? Your instructor will give you a set of minerals to recognize. Test the samples and write down their important or key properties and names in Table 3.6 without looking at the mineral identification tables. Watch out; there may be some tricky samples included.

Use of the Mineral Identification Tables

On the following pages are tables to assist you in identifying unknown mineral specimens. The two main subdivisions are minerals with metallic (including submetallic) lusters (Table 3.3) and minerals with nonmetallic luster (Table 3.4). Each section starts with the hardest mineral and ends with the softest. Follow the maze (Fig. 3.1) for the first few specimens so you learn the correct methodical approach. The abbreviations are CL = cleavage, P = parting, L = luster, S = streak, F = fracture, D = density (in g/cm^3).

Table 3.3 Minerals with **Metallic** and **Submetallic** Luster

(Table 3.3 is divided into two subdivisions, dark streak and medium streak, as generally, minerals with metallic lusters have streak colors that are diagnostic. Especially diagnostic properties are shown in **bold.**)

Streak	Color	Hardness	Comments	Mineral *Composition*
Dark streak: green-black	Brass yellow	6–6 1/2	"Fool's gold"; S = green-black; D = 5.0; cubic or pyritohedral crystals; F = irregular	**Pyrite** (Fig. 2.8d) *Fe sulfide*
	Black	6	**Strongly magnetic;** D = 5.2; S = black to gray-black; F = irregular; octahedral crystals; L = submetallic	**Magnetite** (Fig. 2.8f) *Fe oxide*, ore of iron
to	Iron-black	5 1/2–6	S = black to brownish red; L = metallic to submetallic; D = 4.7; may be slightly magnetic; platy crystals, common accessory in sand, sandstone, and igneous rocks	**Ilmenite** *Fe, Ti oxide*, ore of titanium
dark brown	Iron-black to brownish black	5 1/2	S = dark brown; submetallic luster; granular aggregates; D = 4.6. Color and streak are diagnostic.	**Chromite** *Fe, Cr oxide*, ore of chromium
to	Brass yellow	3 1/2–4	Called "peacock ore" because of the iridescent purple tarnish; S = green-black; D = 4.1–4.3	**Chalcopyrite** *Cu, Fe sulfide*, ore of copper
	Gun metal gray	2 1/2	CL = 3 at right angles (perfect cubic CL); S = lead gray; **D = 7.6;** very bright metallic luster	**Galena** (Fig. 2.8c) *Pb sulfide*
gray-black	Iron black/steel gray	**1–2**	Soils fingers; S = dark gray; D = 2.2; L = metallic to submetallic; greasy feel	**Graphite** (Fig. 2.8j) *Native C*
Medium streak: pale brown to red-brown to yellow-brown	Brown to black	6–6 1/2	S = pale brown; D = 4.2; prismatic crystals, sometimes in quartz (*rutilated quartz*); found in black sand	**Rutile** *Ti oxide*, ore of titanium
	Steel gray	5–6 1/2 flakes	Specular variety is bright metallic and steel gray. **S = red-brown to brown;** D = 5.3; F = irregular; aggregate of tabular crystals ("micaceous" hematite)	**Specular hematite** (Fig. 2.8g) *Fe oxide*
	Dark to brown to black	5–5 1/2	S = yellow-brown; D = 3.3–5 1/2; F = irregular. Luster may be obscured by hydration. "Limonite" is often used to name any hydrous iron oxide.	**Limonite** *Hydrous Fe oxide*, ore of iron
to	Yellowish-bronze	3 1/2–4	S = light bronze brown; octahedral parting; granular aggregates; D = 4.6	**Pentlandite** *Fe Ni sulfide*, ore of nickel
yellow	Dark brown to yellow	3 1/2–4	Variegated appearance common; L = resinous, adamantine, or submetallic; **S = yellow** to brown and lighter than sample; D = 3.9–4.1; CL = 6 good planes	**Sphalerite** (Fig. 2.8e) *Zn sulfide*, ore of zinc
to	Shades of yellow	2 1/2–3	S = gold yellow—shiny; plates, flakes, or nuggets; D = 19.3 when pure; very malleable and ductile; color yellow, paler with increasing silver content	**Gold** *Native Au*, ore of gold
coppery red	**Copper (red)**	2 1/2–3	Malleable and ductile; S = copper brown; D = 8.9; L = metallic, but surface is often tarnished and may be oxidized to blue; branching crystals.	**Copper** *Native Cu*, ore of copper

Table 3.4 Minerals with a **Nonmetallic Luster**

Table 3.4 is broken into four sections, one page each, as follows:

Section A: minerals harder than glass with cleavage or parting
Section B: minerals harder than glass with fracture
Section C: minerals softer than glass with cleavage
Section D: minerals softer than glass with apparent fracture.

Section A: Minerals that show cleavage or parting and H > glass

Streak	Color	Hardness	Comments	Mineral *Composition*
No streak	Colorless to pale yellow, color varies	**10—hardest known substance**	CL = 4 directions perfect ≠90°, octahedral cleavage; crystals usually octahedrons; D = 3.5; **L = adamantine to greasy;** high brilliance due to high index of refraction. Color commonly pale yellow to colorless but may be pale shades of red, orange, green, blue, or brown. Hardness, luster, and cleavage are diagnostic.	**Diamond** *Native C*
	Brown to pink; almost any color	9—test on a fresh surface	P = basal. **Hardness is diagnostic;** D = 3.9–4.1; L = vitreous to adamantine. Hexagonal prismatic crystals that narrow toward the ends. Gem varieties include ruby (red) and sapphire (blue).	**Corundum** *Al oxide*
	Colorless, yellow, white, pink, blue, green	8	**CL = 1** direction, basal; D = 3.4–3.6; L = vitreous. Hardness is diagnostic.	**Topaz** *Al, F, OH silicate with single tetrahedra*
	Blue-green or yellow	7 1/2–8	CL = 1 imperfect, uneven F is more apparent: D = 2.6–2.8. Some gem varieties are aquamarine, emerald (deep green), and rose beryl.	**Beryl** *Be, Al ring silicate*
	Pink reddish-brown, olive green	7 1/2	CL = 2 directions near 90°; nearly squre prisms; D = 3.2; L = vitreous; **dark cruciform cross** in cross section.	**Andalusite** *Al silicate*
	Blue to gray	**7 and 5** depending on direction	CL = 1 direction; bladed crystals; D = 3.6–3.7; L = vitreous to pearly; **H = 7** perpendicular to blades and **H = 5** parallel to blades; blue color and different hardness in different directions are diagnostic.	**Kyanite** *Al silicate*
	Colorless to brown or pale green	6–7	CL = 1 direction; fibrous to acicular; D = 3.2; L = vitreous.	**Sillimanite** *Al silicate*
	White, cream, gray, salmon to dark pink	6	**CL = 2** good planes at 90°; D = 2.5–2.6; L = vitreous. CL is diagnostic. A pink color often distinguishes K-rich feldspars. Amazonite is a rare blue-green variety of microcline.	Alkali **Feldspars** (Fig. 2.7i) *K, Na, Al Tectosilicate*
	White, gray, greenish or bluish gray	6	**CL = 2** good planes at ~90°; D = 2.6–2.8; L = vitreous. An iridescent play of colors may be seen in labradorite. Parallel, regular striations on a cleavage plane (twinning) are common in the plagioclase series. Albite (Na-rich) and anorthite (Ca-rich) are the end members.	Plagioclase **Feldspars** (Fig. 2.7j) *Na, Ca, Al tectosilicate*
	Greenish dark gray to black	5–6	CL = 2 at 90° (not perfectly planar); D = 3.2–3.3; S = pale green to gray if any; L = vitreous (slightly duller than hornblende). Stubby crystals.	**Augite** (pyroxene) (Fig. 2.7c) *Single-chain silicate*
	Black	5–6	CL = 2 at approx. 120 and 60°; D = 3.0–3.3; S = pale gray if any; **L = vitreous.** Habit = prismatic with a diamond-shaped cross-section.	**Hornblende** (amphibole)(Fig. 2.7d) *Double-chain silicate*
	Medium to dark green	5–6	CL = 2 at approx. 120° and 60° (may not be visible in fine acicular crystals); D = 3.1–3.3; S = light gray if any; L = vitreous. Color and **acicular habit** are diagnostic.	**Actinolite** (amphibole) *Double-chain silicate*

Continued

Table 3.4 *Continued*

Section B: Minerals that fracture and H > glass

Streak	Color	Hardness	Comments	Mineral *Composition*
No streak	Red-brown to brownish black	7–7 1/2	L = resinous to vitreous; D = 3.6–3.8; prismatic crystals—obtuse prisms, with common **crossing twins.**	**Staurolite** *Fe Al silicate*
	Dark brown to black common	7–7 1/2	F = conchoidal; D = 3–3.3; L = resinous to vitreous. Most commonly black prismatic with triangular cross sections. Tourmaline may be transparent green, yellow, red, pink, or blue and of semiprecious gem quality (e.g., rubellite [red or pink]).	**Tourmaline** *Chemically complex ring silicate*
	Almandine is deep red to brown	6 1/2–7 1/2	**F = hackly** to conchoidal; D = 3.5–4.3; L = adamantine to vitreous to waxy on parting. **Habit = equant** well-formed **dodecahedral crystals.** Colors vary with composition. Grossular garnet is often tan, pale yellow, pink or green.	**Garnet** (Fig. 2.7b) *Mg, Fe, Mn, Al, with single-silicate tetrahedra*
	Highly variable	7	**F = conchoidal;** D = 2.65; L = vitreous to waxy in microcrystalline varieties. The conchoidal F and H are diagnostic. Hexagonal crystals show striations perpendicular to prism (Fig. 2.7k). **Macro**crystalline quartz varieties: Rock crystal: transparent, colorless, distinct crystals (Fig. 2.7k, l) Milky quartz: translucent white (Fig. 2.7m) Smoky quartz: gray to almost black, smoky yellow to dark brown, transparent to translucent (Fig. 2.7n) Rose quartz: pink to rose colored, transparent to translucent (Fig. 2.7o) Amethyst: purple, often transparent and in crystals (Fig. 2.7p) Citrine: pale yellow, transparent/translucent Tiger eye: yellow, fibrous, and chatoyant **Micro**crystalline quartz varieties: Chalcedony; translucent, has a waxy luster; light gray to brown Agate: thin concentric layers of varying colors (Fig. 5. 27b) Onyx: parallel layers of colors Heliotrope or bloodstone: green with red patches of jasper (Fig. 5.27b) Flint: dark siliceous nodules of quartz often found in calcareous soils and rocks (Fig. 5.27b) Chert: like flint but lighter and more variable in color (Fig. 5.27a) Jasper: red, dull luster, and granular	**Quartz** (Fig.2.7k–p) *pure silica tectosilicate*
	Light to dark **olive green**	6 1/2–7	**F = conchoidal;** may appear irregular due to small grain size; D = 3.3–4.4; L = vitreous. Habit is commonly granular aggregates.	**Olivine** (Fig. 2.7a) *Mg, Fe with single-silicate tetrahedra*
White	Brown to black (rarely yellow or white)	6–7	Imperfect cleavage; D = 6.8–7.1; L = adamantine to submetallic to dull. **High density for a nonmetallic mineral.**	**Cassiterite** *Sn oxide,* ore of tin

Continued

Table 3.4 *Continued*

Section C: Minerals that show cleavage and H < glass

Streak	Color	Hardness	Comments	Mineral Composition
	Blue to gray	**7** and **5** depending on direction	CL = 1 direction; bladed crystals; D = 3.6–3.7; L = vitreous to pearly; **H = 7** perpendicular to blades and **H = 5** parallel to blades; blue color and different hardness in different directions are diagnostic.	**Kyanite** *Al silicate*
	White, colorless to gray	5–5 1/2	CL = 2 directions at 84°, splintery to hackly; prismatic to fibrous; L = vitreous to pearly; D = 2.8–2.9.	**Wollastonite** *Ca silicate*
White	Often white, yellow, violet, pale green	4	**CL = 4,** perfect, in shape of an octahedron (≠90°); D = 3.2; L = vitreous to waxy. Often transparent, fluorite is known by its CL and H. May fluoresce under black light. Color is highly variable.	**Fluorite** (Fig. 2.8h) *Ca, F halide*
Yellow	Dark brown to yellow	3 1/2–4	L = resinous to sub-metallic; CL = up to 6 good planes. Variegated appearance common.	**Sphalerite** (Fig. 2.8e) *Zn sulfide*
White	Yellowish white to pink	3 1/2–4	CL = 3 perfect rhombohedral (may be obscure due to small crystal size); D = 2.8; L = vitreous. **Effervesces in dilute HCl when powdered** (Fig. 5.24 inset). Color varies with impurities.	**Dolomite** *Ca, Mg carbonate*
Intense sky blue	Bright **azure blue**	3 1/2–4	Usually fine grained with earthy fracture; D = 3.8; L = vitreous to dull. Effervesces in HCl. Color is diagnostic and often mixed with the bright green of malachite.	**Azurite** *Cu, OH carbonate, ore of copper*
White	Highly variable, usually colorless to yellowish white to white to pink	3	**CL = 3** perfect rhombohedral (≠90°); D = 2.7; L = vitreous. **Strongly effervesces with dilute HCl.** Varieties include *Iceland spar:* transparent, shows double refraction (Fig. 2.6); *chalk:* soft, fine grained and earthy (Fig. 5.20); *onyx, marble,* and *Mexican onyx:* banded marble; *travertine:* finely layered deposits (Fig. 5.25). (Note that the fine-grained varieties may not show cleavage, but all will effervesce freely with dilute HCl.)	**Calcite** (Fig. 2.8a) *Ca Carbonate*
Light brown	Dark reddish brown to black	2 1/2–3	**CL = 1** perfectly planar; D = 2.8–3.2; **L = resinous** to vitreous. Elastic and flexible in thin sheets. Transparent. May break apart when testing streak. Marked with fingernail by creasing sheets rather than scratching.	**Biotite (mica)** (Fig. 2.7g) *K, Al sheet silicates*
White	Colorless to white to light greenish brown	2–2 1/2	**CL = 1** perfectly planar; D = 2.7–2.9; **L = resinous** to vitreous to **pearly.** Elastic and flexible in thin sheets. Transparent. May break apart when testing streak.	**Muscovite (mica)** (Fig. 2.7h) *K, Al sheet silicates*
White	Colorless to white	2 1/2	**CL = 3 perfect at 90°** (cubic); D = 2.1–2.3; L = waxy to vitreous. **Salty taste.** Transparent. **Dissolves in water and on fingers.** Tan to reddish with impurities.	**Halite** (Fig. 2.8i) *NaCl halide*
White to pale green	**Medium to dark green**	2–2 1/2	**CL = 1** perfect, but folia are small in relation to micas; D = 2.6–3.3; may disaggregate when testing streak; L = vitreous to pearly. **Thin sheets are flexible but not elastic.**	**Chlorite** *Mg, Fe, OH sheet silicate*
White	Colorless to white to pale blue	2	**CL = 3 perfect at 90°** (cubic); D = 2.0; L = waxy to vitreous. **Bitter salty taste.** Transparent. **Dissolves in water and on fingers.** Yellow or reddish with impurities. Much rarer than halite.	**Sylvite** *KCl halide*
White	Colorless to white	2	**CL = 1** perfect (in sheets), and 2 more irregular; D = 2.3; L = vitreous to pearly. Flexible in thin sheets, sectile, transparent to translucent. **Can scratch with fingernail.**	**Gypsum (selenite)** (Fig. 2.8b) *Ca sulfate*

Continued

Table 3.4 *Continued*

Section D: Minerals with apparent fracture and H < glass

Streak	Color	Hardness	Comments	Mineral *Composition*
White	Commonly green to red-brown	5	Variable color; **F = conchoidal;** poor basal cleavage. D = 3.1–3.2; L = vitreous. May occur in **hexagonal prisms** or granular aggregates.	**Apatite** *Phosphate*
White	Yellowish white to pink	3 1/2–4	CL = 3 perfect rhombohedral (may be obscure due to small crystal size); D = 2.8; L = vitreous. **Effervesces in dilute HCl when powdered** (Fig. 5.24 inset). Color varies with impurities	**Dolomite** (Fig. 5.24) *Ca, Mg carbonate*
White	Light green to dark green	3–5	Massive; L = greasy to waxy; D = 2.5–2.6; color often variegated, light green to nearly black. May be fibrous = asbestos.	**Serpentine** (Fig. 6.6f) *Mg OH sheet silicate*
Pale green	Bright green	3 1/2–4	L = adamantine to vitreous to earthy; commonly botryoidal or stalactitic; D = 3.9–4.0; recognized by its **distinctive bright green color** and **effervescence in HCl.**	**Malachite** *Cu carbonate, ore of copper*
Pale blue	Bright blue	3 1/2–4	L = vitreous to earthy; D = 3.8; recognized by its **distinctive intense azure blue color** and **effervescence in HCl.**	**Azurite** Hydrous *Cu carbonate, ore of copper*
Yellow to brown	Yellow ocher to dark brown	2–5 1/2	S = yellow-brown; D = 3.3–5.5 or less if porous; F = irregular. L = earthy. "Limonite" is often used to name any hydrous iron oxide.	**Limonite** *Hydrous Fe oxide*
Pale green	Apple green	2–3	L = greasy to waxy; D = 2.2–2.8; recognized by its **distinctive apple green color.**	**Garnierite** *Ni Mg OH sheet silicate, ore of nickel*
White	White to pink	2	Satin spar is the fibrous variety—white with **L = silky** and **F = fibrous** (cleavage is not apparent due to its fibrous habit). D = 2.3; sectile.	**Gypsum (satin spar)** (Fig. 2.8b) *Ca sulfate*
White	White to pink	2	Alabaster is **compact and massive;** L = vitreous to pearly and F = uneven (cleavage is not apparent due to its small crystals). D = 2.3; sectile.	**Gypsum (alabaster)** (Fig. 2.8b) *Ca sulfate*
Red-brown	Red brown	1–6	F = uneven; (Hardness is variable due to variations in grain size); D = 4.8–5.3; L = dull to earthy. Small (egg-shaped) ooids are usually seen; Streak and color are diagnostic.	**Oolitic hematite** (Fig. 2.8g) *Fe oxide, ore of iron*
Pale yellow	**Yellow**	1 1/2–2 1/2	F = uneven to conchoidal; D = 2.1; L = resinous to vitreous. Known by its bright yellow color (when pure) and its distinctive odor.	**Sulfur** *Native S*
Variable but commonly red-brown to yellow brown	Highly variable even in one sample white, gray, yellow-brown, red	1–5	L = dull to earthy; D = 2–2.6; recognized by its **spherical concretionary grains** (pisolites). Not really a mineral, but important as aluminum ore.	**Bauxite** *Al oxides and hydroxides, ore of aluminum*
White	White	1–2	**F = uneven;** D = 2.6; **L = dull,** earthy. **Powdery** and claylike, earthy smell. Microscopically has one perfect cleavage. **No effervescence.**	**Kaolin** (Fig. 2.7f) *Al OH sheet silicate*
White, flakes when powdered	White to gray-green	1	CL = 1 perfect (may be microscopic); D = 2.7–2.8; **L = pearly** to dull. **Greasy feel.** If compact and massive, the variety is soapstone. Talc is often associated with tremolite, an amphibole with H = 5–6, and therefore may appear harder.	**Talc** (Fig. 2.7e) *Mg OH sheet silicate*

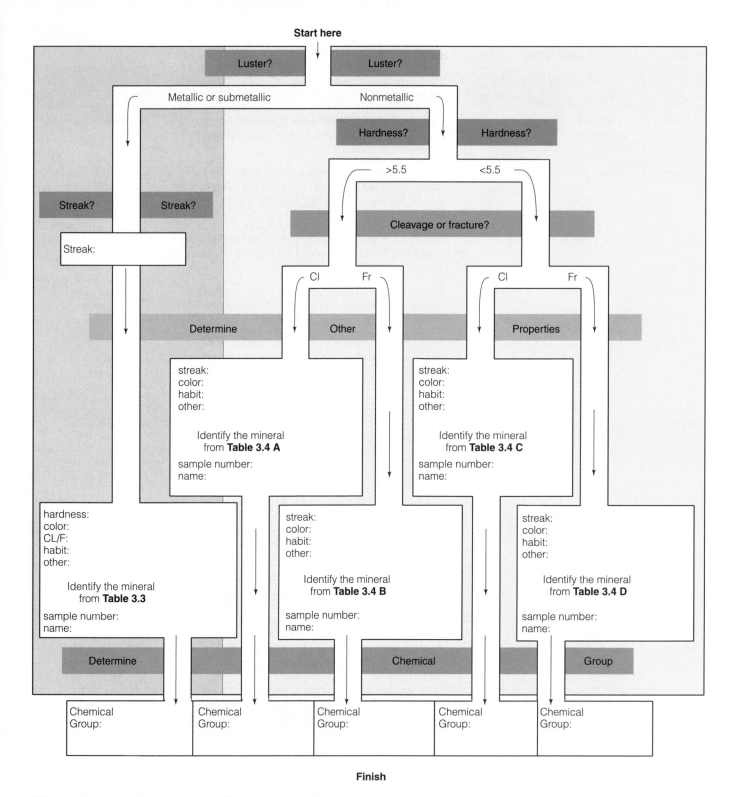

Start here

Luster?

Luster?

Metallic or submetallic

Nonmetallic

Hardness?

Hardness?

>5.5

<5.5

Streak?

Streak?

Cleavage or fracture?

Streak:

Cl

Fr

Cl

Fr

Determine

Other

Properties

streak:
color:
habit:
other:

Identify the mineral
from **Table 3.4 A**

sample number:
name:

streak:
color:
habit:
other:

Identify the mineral
from **Table 3.4 C**

sample number:
name:

hardness:
color:
CL/F:
habit:
other:

Identify the mineral
from **Table 3.3**

sample number:
name:

streak:
color:
habit:
other:

Identify the mineral
from **Table 3.4 B**

sample number:
name:

streak:
color:
habit:
other:

Identify the mineral
from **Table 3.4 D**

sample number:
name:

Determine

Chemical

Group

Chemical
Group:

Chemical
Group:

Chemical
Group:

Chemical
Group:

Chemical
Group:

Finish

Figure 3.2   Maze for mineral identification for Exercise 1

Classification of Minerals

Mineral species are classified on the basis of their chemical composition. They are grouped according to the nonmetal elements in their chemical formulas. The first part of a chemical formula usually lists the metals; the second part, the nonmetals. The exception is minerals made up of only one element; all such minerals belong to the mineral group called the *native elements.* The main groups covered here are silicates (Table 3.1), carbonates, sulfates, sulfides, oxides, hydroxides, phosphates, halides, and native elements (pure elements) (Table 3.2). Table 3.7 shows the chemical

Table 3.5 Mineral Recognition Memory Aids (Exercise 3)

Mineral	Memory Aid	Diagnostic Properties
Olivine		
Garnet		
Augite		
Hornblende		
Kaolin (or kaolinite)		
Biotite		
Muscovite		
Feldspar		
Quartz		
Calcite		
Gypsum		
Galena		
Pyrite		
Magnetite		
Halite		

Table 3.6 Mineral Practice Quiz (Exercise 4)

Sample Number	Key Property 1	Key Property 2	Key Property 3	Mineral Name

Table 3.7   Composition of Mineral Groups

Mineral Group	Chemical Makeup
Silicates	Contain silicon (Si) and oxygen (O) at least
Carbonates	CO_3 plus metal(s)
Sulfates	SO_4 plus metal(s)
Sulfides	S plus metal(s)
Oxides	O plus metal(s) without other nonmetals (no Si, C, P, S, V, or W)
Hydroxides	OH plus metal(s) without other nonmetals
Phosphates	PO_4 plus metal(s)
Halides	F, Cl, Br, or I plus metal(s) without other nonmetals
Native elements	Occur in elemental form (one element only)

Table 3.8   Group Identification (Exercise 5)

Mineral Name	Formula	Mineral Group
Sylvite	KCl	
Chalcopyrite	$CuFeS_2$	
Sulfur	S	
Tremolite	$Ca_2(Mg,Fe)_5Si_8O_{22}(OH)_2$	
Rutile	TiO_2	
Anhydrite	$CaSO_4$	

■ Use the animation in this module to gain an understanding of how atoms in magma behave and ultimately bond to make minerals.

6. What bonds form first and why?

_____

_____

7. What mineral forms first? Why?

_____

_____

■ Use the graphics on mineral structures to observe how silicate minerals assemble from separate atoms.

8. How do silica tetrahedra link to each structure?

_____

_____

■

makeup for the different mineral groups. Notice that all the mineral groups ending with -*ate* contain oxygen, and those with -*ide* endings do not contain oxygen unless *ox* is part of the name.

5. Look at the mineral formulas listed in Table 3.8, and fill in the mineral group name based on the chemical formulas. Refer to Table 3.7. Your instructor may have samples of these minerals to show you.

Introduction to the Silicates

The most important and most complex mineral group is the silicates. The basic unit for the silicates is the silicate or silica tetrahedron (SiO_4), as shown in Figure 3.3a. This basic building block is often illustrated as lines connecting the oxygen atoms, as in Figure 3.3b and as in each of the diagrams for the silicate subclasses in Table 3.1. These lines make a triangular pyramid called a **tetrahedron,** thus the name *silica tetrahedron*. Imagine an oxygen atom at each corner and a silicon atom in the center. Silicates have a subclassification based on the arrangement of the silica tetrahedra (plural) in the atomic structure of the mineral. Table 3.1 shows the silica structure subclasses diagrammatically.

 Atoms and Crystal Structure Computer Exercise

■ Start **Earth Systems Today.**
■ Go to **Topic: Earth Materials.**
■ Go to **Module: Atoms and Crystal Structure.**
■ Explore activities in this module to discover the following for yourself:

9. Construct ball and stick or ball and glue models of the silicate part of each of the silicate subclasses listed in the first column in Table 3.9. The small balls are silicon atoms; the large ones are oxygen atoms. Fill in Table 3.9 with the following information, using Table 3.1 to help you:

Column 2: How many oxygen atoms in a tetrahedron are shared in each subclass?

Column 3: What is the silicon:oxygen ratio? Count the number of oxygens per silicon atoms by counting unshared oxygens as 1 and shared oxygens as 1/2. This number is the equivalent of $4 - 1/2$ the number of shared oxygens.

Column 4: In blank spaces, name one mineral belonging to each subclass.

Column 5: What is the silicon:oxygen ratio from the chemical formula of the mineral named in Column 4?

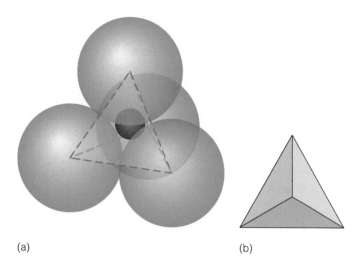

Figure 3.3 Silica tetrahedron: (a) actual configuration of atoms and (b) diagrammatic representation

(a)   (b)

How do the ratios of silicon to oxygen in the mineral formulas, Column 5, compare to the ratios in Column 3?

_____

Why? _____

_____

10. When constructing these models, which silicate subclass structures theoretically go on and on indefinitely?

_____ _____ _____

Most silicate structures have a negative charge, but complete crystals must have a neutral charge. A bunch of negative charges don't usually hang around by themselves.

11. What balances the negative charge and holds the tetrahedra together where they are not linked by sharing oxygens?

_____

(The formulas in Table 3.1 may give you a hint.)

12. The ratio of silicon to oxygen in silicates is related to the quantity of silica present in the mineral. Silicates with lots of oxygen in proportion to silicon are low in silica, and those with less oxygen compared to silicon are high in silica. What mineral from Table 3.1 has the lowest silica content? _____

What mineral has the highest? _____

13. Now let's look at feldspar in Table 3.1. It is also a framework silicate, like quartz, yet its ratio of silicon to oxygen is _____. This does not match the ratio in Table 3.9, Column 3. Augite and chlorite also have low silicon:oxygen ratios for their subclass. Some of the tetrahedra in feldspar and these other minerals have aluminum (Al) instead of silicon. Add up the Al and Si in feldspar's formula. _____

What is the ratio of Al+Si to oxygen in feldspar? _____ Now, does this match the ratio in Column 3 for tectosilicates? _____

Table 3.9   Mineral Structures (Exercise 9)

COLUMN 1	COLUMN 2	COLUMN 3	COLUMN 4	COLUMN 5
Silicate Subclass	Number of Oxygen Atoms Shared	Silicon:Oxygen Ratio in Structure	Mineral Belonging to This Subclass	Silicon:Oxygen Ratio in Chemical Formula
Nesosilicates (single tetrahedra)		1:_____		
Cyclosilicates		1:_____		
Single-chain silicates		1:_____	Diopside	
Sheet silicates		2:_____	Talc	
Tectosilicates (3D framework or network)		1:_____	Quartz	

14. Let's see how different silicate structures influence the minerals' properties. What mineral has single chains? _____ What type of cleavage do single-chain silicates have? _____ Sheet silicate mineral? _____ Cleavage? _____

Geologic Resources and Minerals

A **geologic resource** is a naturally occurring substance that can be used and comes from the Earth. For example, industrial quality diamonds are a naturally occurring mineral that can be used as an abrasive. On the other hand, concrete, made of cement and aggregate, is a product manufactured from resources such as limestone, which consists of the mineral calcite. Limestone and calcite are resources, while concrete and cement are products. So geologic resources may be an element, a mineral, or a rock (or in some cases a natural liquid substance such as petroleum), but they are not manufactured products. Resources from plants and animals are considered biological.

Minerals make up the building blocks of most solid natural substances that make up the Earth. They provide us with many resources we need to produce the common objects we take for granted in our modern society. We will explore this topic more thoroughly in Lab 17.

15. Work together in a group of five or more if possible. Discuss a simple object such as a doorknob or handle. What is the doorknob made of? Are some doorknobs made of other substances? What mineral resources are needed to produce one? The most obvious resources involved in doorknob production are ones that provide materials actually found within the doorknob. Include these in your list. But a much larger set of resources is used indirectly, in the processes of mining, manufacture, transportation, and marketing. Also include these in your list. Which of these resources are minerals?

Many minerals have several and varied uses. The concrete, steel, glass, and plaster that make up our buildings are made from such minerals as calcite, magnetite, quartz, and gypsum. You may have mentioned some of these as needed to make a doorknob. Our coins and jewelry come from other minerals such as argentite (ore of silver), native copper, native gold, diamonds, and corundum (rubies and sapphires are gem varieties of corundum).

16. Continue working in a group. List in Table 3.10 all the minerals you can think of for which you know at least one use, and list their uses.

Table 3.10   Mineral Uses (Exercise 16)

Mineral	Use

Igneous Rocks

OBJECTIVES

1. To recognize minerals in igneous rocks and learn how a rock's color is related to mineral composition
2. To recognize textural features of igneous rocks and understand their origin
3. To understand how igneous rocks form
4. To be able to identify and recognize the common igneous rocks based on texture, color, and mineral content

Igneous rocks form by solidification of molten rock, called **magma.** This may happen at the Earth's surface, in which case the magma is called **lava** and erupted fragments are *pyroclasts,* and the rock is called **volcanic** or **extrusive.** It may also occur below the surface, where the magma crystallizes into **plutonic** or **intrusive** rock. Magma and lava may have temperatures ranging generally between 800° and 1200°C. Lava may flow across the surface and cool quickly or may be scattered by explosive blasts from a volcano. Magma also intrudes at depth beneath the surface of the Earth and cools slowly. Thus, a variety of situations produce igneous rocks each of which creates a different igneous rock. By studying the rocks, we may be able to deduce what happened to create them.

Mineralogical and Chemical Composition

The minerals that make up igneous rocks are mostly silicates. They are often subdivided into two groups: light colored, or felsic minerals; and dark colored, or mafic minerals. The common *felsic* minerals in igneous rocks are

> **plagioclase feldspar,**
>
> **alkali[1] feldspar,**
>
> **quartz,** and
>
> **muscovite** (occasionally).

----

[1]Alkali feldspars are potassium and/or sodium feldspars. *Alkali* comes from the alkali metals in the first column, which includes potassium and sodium, of the periodic table of the elements.

The common *mafic* minerals are

> **olivine,**
>
> **pyroxene,**
>
> **hornblende,** and
>
> **biotite.**

The classification of igneous rocks is based on the mineral content, so it is important, if possible, to ascertain what minerals are present and their proportion when you are trying to identify an igneous rock. If only mafic minerals are present, the rock is said to be **ultramafic.** If mafic minerals predominate, the rock is said to be **mafic.** If about equal proportions of mafic and felsic minerals are present, it is **intermediate;** if felsic minerals predominate, it is **felsic.** Thus, the overall color of an igneous rock is a fairly good guide to its mineralogical composition. Dark-colored igneous rocks are therefore either ultramafic or mafic, medium colored are intermediate, and light colored are felsic (Table 4.3).

A few igneous rocks do not contain minerals (see the later discussion of glassy texture), in which case the classification is based on the chemical composition or elements in the rock instead. The minerals in igneous rocks, of course, are made of elements, too, so you could say that ultimately the classification is based on the chemical composition of the rock.

Igneous rocks and the magma they formed from contain numerous elements, but the most abundant are oxygen and silicon. When we combine these two elements together as SiO_2 **(silica),** the rocks contain somewhere between about 40% to 80% silica. The remaining 60% to 20% is made up of more oxygen, aluminum, iron, magnesium, calcium, sodium, potassium, and a myriad of other elements in small quantities. The magma contains various gases, especially water vapor, carbon dioxide, and sulfur gases that escape during solidification. The terms we used earlier—*ultramafic, mafic, intermediate,* or *felsic*—also classify the chemical composition of igneous rocks, based on the silica content. Rounding off the numbers, these correspond to about 40% silica for ultramafic, 50% for mafic, 60% for intermediate and 70% for felsic. An increase in the silica content leaves less room in the rock for other elements, resulting in a corresponding decrease in iron and magnesium. The *m* in *mafic,* in fact, stands for magnesium and the *f* for iron (chemical symbol Fe). Iron tends to impart the dark color to both the minerals and the rocks.

Silica content in magma is also closely related to viscosity of the magma. **Viscosity** is resistance to flow. Felsic lavas are much more viscous and mafic lavas more fluid. Highly viscous lavas with high gas content tend to produce dangerous explosive eruptions. So silica content influences more than the color of the rock; it also influences the behavior of the magma or lava and the nature of volcanic eruptions. Another aspect of igneous rocks that reveals more than is at first readily apparent is the rock's texture.

Textures of Igneous Rocks

The texture tells the story of the igneous rock—its journey, the places it's been. The story solves the mystery of the rock's crystallization, its gaseous encounters, its involvement in disastrous volcanic eruptions, or its quiet formation deep underground. With practice, at the end of this lab you will be able to trace the journey of the rock back to its formation and beyond using the texture of the rock. To follow this journey, we must first learn to identify and understand the significance of some basic igneous rock textures.

The term *texture* refers to the arrangement and size of mineral grains in a rock. Additional aspects of texture include the presence of glass, the proportion of glass to crystals, the presence and proportion of cavities, and the occurrence of broken rock fragments. Each of these reveals aspects of the rock's history.

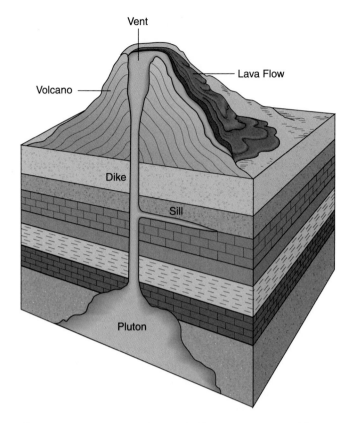

Figure 4.1 Plutons, dikes, and sills are intrusive and form coarser-grained rocks than volcanic lava flows, which tend to be fine grained.

Textures Formed by Cooling of Magmas

Rocks and their constituent minerals melt and crystallize at high temperatures. Crystals formed by the freezing of magma typically have an interlocking texture that results in a hard rock. Grain size is also an important aspect of the texture of an igneous rock because it contains information on the rock's cooling history. A plutonic rock, or one that has crystallized from magma intruded deep in the Earth's crust, will cool very slowly, allowing the crystals to become well formed and large. A volcanic rock, or one that has been extruded and cools quickly at the Earth's surface, will be so fine grained that the crystals will be invisible to the naked eye (Figs. 4.1 and 4.2). Sometimes cooling is so rapid crystals do not grow at all. Any of these rocks may have the same composition, but they are differentiated by their texture or grain size. The rocks that crystallized have interlocking crystals although you can see the crystals only in the coarser textures. The following are the principal textures found in igneous rocks that cool from magma or lava.

Coarse-Grained[2] (= Granular) Textures This type of texture has visible crystals all very roughly the same size. Such igneous rocks cooled slowly deep in the Earth where it is

warm. Coarse-grained textures are therefore typical of intrusive or plutonic rocks. The following are common plutonic rocks, with coarse-grained textures:

Granite is the most abundant plutonic igneous rock in continental crust. The dramatic cliffs of Yosemite National Park and the dome of Stone Mountain, Georgia are granite. Alkali feldspars and quartz are diagnostic for granite (Fig. 4.3). These feldspars are commonly white, light gray, and/or pink, often with two varieties visible. Quartz usually looks like smoky gray glass. Granite is a light-colored rock and is felsic (the black minerals such as hornblende and biotite may stand out, but they comprise only a small percentage of the total rock mass).

Diorite is common where oceanic crust dives under other crust, causing melting near convergent plate boundaries (see Lab 10). It is made up of dark green to black pyroxene and white to gray plagioclase feldspar in roughly equal proportions. This equal amount of felsic and mafic (light and dark) minerals often give this rock a salt-and-pepper appearance (Fig. 4.4). This makes diorite medium colored or intermediate in composition. The absence of quartz is diagnostic along with the appearance of pyroxene.

Gabbro is the most abundant plutonic rock in oceanic crust. It is a coarse, dark rock composed primarily of plagioclase feldspars, usually dark gray and translucent, and

....
[2]The technical term is **phaneritic.**

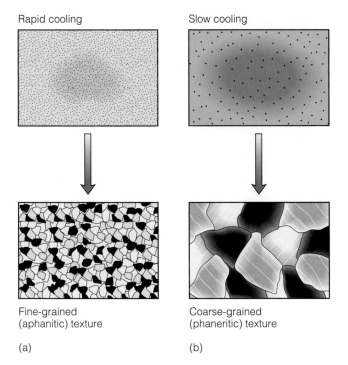

Rapid cooling   Slow cooling

Fine-grained   Coarse-grained
(aphanitic) texture (phaneritic) texture

(a)   (b)

Figure 4.2   Cooling rate and grain size in magma.
(a) Rapidly cooling magma forms fine-grained texture.
(b) Slowly cooling magma forms coarse-grained texture.

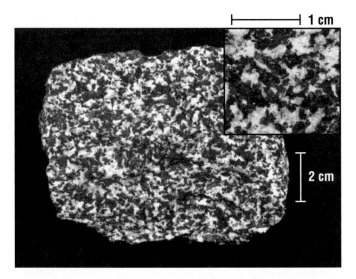

Figure 4.4   Diorite. In this sample, the black (slightly green
dark gray) grains are hornblende, and the white to pale gray
ones are plagioclase. An occasional grain of black glossy biotite appears in the inset.

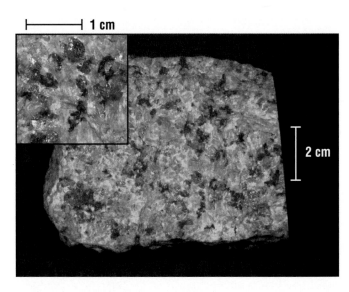

├─┤ 1 cm

2 cm

Figure 4.3   Granite. Two colors of feldspar are visible in
this sample: white K feldspar, orthoclase, and slightly greenish
pale gray Na-rich plagioclase. The other minerals are gray vitreous quartz (smoky quartz) and black biotite. The inset mag-
nifies the sample as if you were viewing it with a hand lens.

2 cm

├─────┤ 1 cm

Figure 4.5   Gabbro. This sample is made up of plagioclase
and pyroxene. Both plagioclase and pyroxene have cleavage,
so most of the rock reflects light off small, flat cleavage surfaces as one turns the sample in the light. Notice the white
rectangular grains; these are plagioclase grains reflecting light
from the cleavage. *Inset:* Euhedral tabular plagioclase grains
(rectangular in cross section) are gray and somewhat translucent. Some may have twinning, although it does not show in
this view. The pyroxene grains are more opaque, anhedral, and
greenish black.

crystals of pyroxene that are often large and irregular
(Fig. 4.5). Olivine is a minor constituent. Gabbro is mafic.

Peridotite is primarily made up of olivine and pyroxene.
Its green color and coarse-grained texture easily identi-
fies it (Fig. 4.6). Olivine has fracture, not cleavage, so this
rock can also be distinguished from gabbro by having a
lower proportion of cleavage surfaces that catch the

light. The absence of any felsic minerals means that peridotite is ultramafic. Peridotite is the rock that makes up
most of the upper mantle.

Pegmatite is a special variety of extremely coarse rocks
that may have crystals several centimeters across and is

(a)

(b)

Figure 4.6 **Peridotite.** (a) Peridotite containing olivine and pyroxene. Notice the cleavage visible on some of the pyroxene where the light is reflecting off the flat surface. Olivine does not have cleavage, so a large proportion of the rock does not reflect light in this manner. (b) A fragment of peridotite (xenolith, see Lab 7) contained in basalt. This sample has a type of peridotite with a very high proportion of olivine (dunite): olivine is green, and pyroxene is black. Basaltic magma from the mantle carried the peridotite fragment upward to be erupted with the lava.

usually found in dikes (Fig. 4.7). Rarely, pegmatite crystals can be many feet long. Granite pegmatite generally consists of orthoclase feldspar, muscovite, and quartz (Fig. 4.7 inset). This type of pegmatite is felsic.

Fine-Grained[3] Textures This type refers to rocks whose crystals are so fine grained that they cannot be seen with the naked eye. The small size of the grains means that they have cooled quickly where it is cold. Fine-grained textures are therefore typical of extrusive (volcanic) rocks that form from lava flows or lava domes. Some shallow intrusive bodies such as thin dikes and sills (see Lab 7) may have fine-grained texture because they are thin enough and near enough to the surface to cool rapidly.

Porphyritic Textures This sort of texture, shown in Figure 4.8, indicates large crystals or **phenocrysts** embedded in a more finely crystalline or glassy **groundmass.** The groundmass may be coarse grained (Fig. 4.8a), fine grained (Fig. 4.8b), or glassy. The two grain sizes must be distinctly different, not gradational. Such a texture usually indicates a *change* in the rate of cooling. At first cooling takes place slowly at depth, forming some large crystals. The magma and large crystals then rise to an environment where faster cooling takes place. This second stage of cooling results in the finer groundmass

....
[3]**Aphanitic** is the technical term.

crystals or glass without crystals. If the second stage is beneath the surface, the magma becomes completely solid without ever erupting so the rock is plutonic or intrusive, and the groundmass is likely to be coarse grained. On the other hand, if the second stage involves eruption of a lava flow at the surface with some large crystals in it, the rock is considered volcanic or extrusive, and the groundmass will be fine grained or glassy. All igneous rocks start with magma underground; it is whether the magma ever reaches the surface that determines if a rock is plutonic or volcanic. Porphyritic texture may thus occur in either plutonic or volcanic rocks.

> **Porphyry:** The noun *porphyry* is applied to a rock in which phenocrysts comprise 25% or more of its volume (see granite porphyry in Fig. 4.9).
>
> **Porphyritic:** The adjective *porphyritic* is used to describe rocks with less than 25% phenocrysts (see the later discussion on porphyritic andesite).

A crystal that displays well-formed, planar surfaces, or **faces,** is termed **euhedral;** a somewhat imperfectly formed shape is termed **subhedral;** and if the crystal lacks crystal faces, it is described as **anhedral.** Phenocrysts in porphyritic rocks are commonly euhedral (see porphyritic andesite).

Rocks with Fine-Grained or Porphyritic Textures

> **Rhyolite** is a fine-grained, light-colored rock (Fig. 4.10). **Porphyritic rhyolite** has a fine-grained groundmass. Either may be light tan, pink, beige, yellowish, or light gray, and both are felsic. Yellowstone National Park is named

H—| 1 cm

Figure 4.7 **Pegmatite** is a very coarse-grained plutonic igneous rock. Here the fluid-rich magma that formed the pegmatite intruded along cracks in the dark rocks forming numerous dikes (Lab 7). Cascades, Washington. Photo by Michael Gross.
Inset: Close-up of a rock from one of the dikes. The very large crystals in this pegmatite are K feldspar (white), quartz (pale gray and vitreous), and books of biotite (left). A few small grains of muscovite (silvery) are visible on the left.

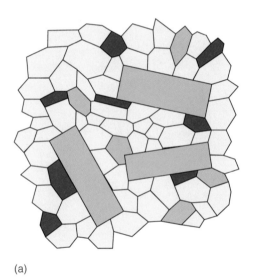

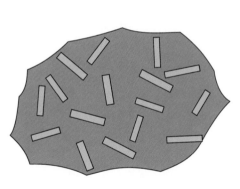

(a)   (b)

Figure 4.8 Porphyritic texture. (a) Porphyritic texture with coarse groundmass. (b) Porphyritic texture with fine groundmass.

for the yellow rhyolite in the area. Porphyritic rhyolite has at least a small percentage of quartz phenocrysts that help to differentiate it from andesite. Sometimes crystals of transparent potassium feldspar (with cleavage) are visible.

Andesite is the most common lava-flow rock in stratovolcanoes, or composite cones (Lab 7). Gray, dark tan, purple, or brown in color, andesite is intermediate in chemical composition (Fig. 4.11a). **Porphyritic andesite** has phenocrysts of plagioclase feldspar or mafic minerals such as pyroxene and hornblende (Fig. 4.11b).

Basalt makes up much of the surface of the ocean floor just beneath any sediments. It is the primary rock of shield volcanoes as well. Typically dark gray to black,

basalt is dense and mafic and has few phenocrysts (Fig. 4.12). The phenocrysts in **porphyritic basalt** are commonly olivine, which may be hard to see because of a lack of color contrast of the clear green olivine in the black rock. Basalt is likely to be slightly magnetic because of the presence of minor quantities of magnetite. **Vesicular basalt** (Fig. 4.12b) has holes (see the discussion of vesicular texture and scoria).

Glassy (or Hyaline) Textures This is the texture of glass, which has no definite minerals but does have vitreous luster. Glassy texture frequently forms from highly viscous lava (often felsic). (*Viscous* means "thick and gooey" or "resistant to flow.") Atoms within viscous lava move so slowly through the

Figure 4.9   Granite porphyry. This coarse-grained felsic igneous plutonic rock has abundant, distinctly larger phenocrysts of K feldspar. The K feldspar is pink, plagioclase is white to pale gray, quartz is medium gray, and biotite is glossy black. You can see a number of spots where light is reflecting brightly off the biotite.

Figure 4.10   Rhyolite. This sample is fine-grained pink rhyolite with a scattering of very small phenocrysts.

(a)

(b)

Figure 4.11   Andesite. A fine-grained, medium-colored groundmass is diagnostic for andesite. (a) This sample is typical of a fine-grained andesite. (b) This sample of **porphyritic andesite** has a medium gray groundmass and euhedral black hornblende phenocrysts. White plagioclase phenocrysts are also common in porphyritic andesite.

gooey stuff that they have trouble getting together to form crystals. Glassy texture may also result when magma is cooled so quickly that minerals have no opportunity to form, such as when lava comes in contact with seawater. The volcanic glass that forms in these ways lacks a crystalline structure. Therefore, volcanic glass is a natural solid that is not made of minerals (review the definition of a mineral in Lab 2). The following are common rocks with glassy texture:

Obsidian is volcanic glass with conchoidal fracture and vitreous luster, which are diagnostic, and it is often black and transparent on thin edges (Fig. 4.13). Obsidian is generally felsic, yet it is dark rather than light. It is the one rock that defies the color scheme of other felsic rocks. Since obsidian has no minerals in it and is glass, it is transparent, but the minor amount of iron in it strongly influences the color. Let's look at an analogy. Ice is transparent. If you put some ice in a white cup with, say, some seltzer water, the ice will look white or clear. But if you put ice in a glass of cola, the ice will take on the dark color of the drink and look brown. So obsidian, being basically transparent, takes on the dark color of a minor constituent (iron), thus its dark color.

Pumice also has a glassy texture (see the inset photo for pumice on p. 67), although its main texture is vesicular (discussed later).

(a)

(b)

Figure 4.12 **Basalt** is a fine-grained mafic igneous volcanic rock. (a) This sample is fine grained with very few green olivine phenocrysts that are quite difficult to spot. (b) This sample of **vesicular basalt** has holes and vesicles and is also porphyritic, but again the olivine phenocrysts are difficult to see.

Figure 4.13 **Obsidian** is a glassy felsic igneous volcanic rock with conchoidal fracture and is commonly black, as in this sample. Flow banding and red streaks may be present in some samples.

1. Using a hand lens,[4] observe each of the rock samples provided by your instructor: gabbro, porphyry, basalt, and obsidian. Indicate their texture and cooling rate in Table 4.1. Use the texture terms described earlier.

Growth of Crystals from a Melt

To gain a better understanding of the igneous processes that form the textures just discussed, let's perform some melting and crystallization experiments.

••••••••••••
[4]When using a hand lens, hold the lens close to your eye, then bring the sample up close until it is in focus. By using this method, you get a larger and clearer field of view.

It is difficult to melt and crystallize real rocks in a laboratory because of the high temperatures involved; instead, we will melt and crystallize artificial "igneous rocks" using an organic substance, thymol, that has a low melting (or freezing) point. Thymol is slightly noxious; in fact, breathing the vapors is not good for you, so experiments with liquid thymol should be done in a fume hood. If no hood is available in your lab, your instructor may provide Petri dishes for Parts 1, 2, and 3 that have already been made for you. In either case, you will need to read how the experiment is done so you can interpret the results.

Materials needed:

- Fume hood (all liquid thymol should be kept in the fume hood)
- Hot plate
- Thymol
- Three Petri dishes
- Labels
- Spatula
- Ice bath in the fume hood

Before starting the experiment, find out from your lab instructor what setting to use for the hot plate, set it and turn it on.

Part 1: Slow Cooling Use a Petri dish already containing crystals of thymol. If one is not available, with a clean spatula, place enough solid thymol in a clean dry dish to cover the bottom completely when melted. You may add more thymol if needed once melting has begun. Label the dish "slow cooling" or "part 1." *In the fume hood,* put the dish on a hot plate and watch it until the thymol has *almost* completely melted. Do **not** allow it to vaporize. If vaporization occurs easily or quickly, the hot plate is set at too high a temperature. Set the dish with liquid thymol aside in the fume hood so that it may cool slowly, undisturbed.

Table 4.1 Cooling Textures of 4 Rocks

Sample Number	Rock Name	Texture	Cooling Rate(s) (very fast, fast, slow, or slow followed by fast)
	Gabbro		
	Porphyry		
	Basalt		
	Obsidian		

Part 2: Fast Cooling Duplicate the procedure used in Part 1 with another Petri dish, but do not set it aside. Instead, remove it to the ice bath and observe the crystallization. Be careful not to get water into the dish. Look back at Part 1 occasionally to see how it is progressing.

Part 3: Slow Followed by Fast Cooling Repeat the procedure for Part 1 and observe the formation of the first crystals. After a few large crystals have formed but some liquid is still left, transfer the dish to the ice bath, in the fume hood.

2. Determine the relative crystal sizes (small, medium, or large) and briefly describe the crystals from Parts 1, 2, and 3. Use the experimental products you just made or ones provided for you. Are the crystals euhedral, subhedral, or anhedral? What texture do you observe? Use the texture terms you have learned.

Part 1 (slow cooling): Size: _____

Description and texture:

_____

_____

Part 2 (fast cooling): Size: _____

Description and texture:

_____

_____

Part 3 (slow followed by fast cooling):

Sizes: _____ and _____

Description and texture:

_____

_____

3. How does the rate of cooling affect crystal size and shape?

_____

_____

_____

4. Are your observations of cooling rates and crystal sizes for Part 1 and Part 2 consistent with your observation of Part 3? Explain.

_____

_____

5. Why did we do this experiment? The experiment is intended to be an analogy for natural processes in the Earth. What processes are these? What would happen naturally in the Earth to make similar events, results, and textures? For Part 1?

_____

_____

For Part 2?

_____

_____

For Part 3?

_____

_____

6. Using relative crystal size as a determining factor, determine which rock sample from Table 4.1 the thymol in Part 1 most closely resembles. In Part 2? In Part 3?

Part 1 (slow cooling) resembles sample number: _____

Name: _____

How and where might this rock have formed? Is it volcanic or plutonic?

_____

_____

Part 2 (fast cooling) resembles sample number: _____

Name: _____

How and where might this rock have formed? Is it volcanic or plutonic?

_____

_____

Part 3 (slow followed by fast cooling) resembles sample number: _____. Name: _____

How and where might this rock have formed? Is it volcanic or plutonic?

_____

_____

Clean up from your experiments as instructed.

Textures Formed by Gases in Magmas

In addition to the textures already described, other textures may be found in igneous rocks. Gases in the magma may cause holes in the resulting rock or even explosions within the volcano when gases are trapped. These explosions fragment or spray the magma as it comes out of the volcanic vent.

Vesicular Textures   This type of texture refers to the presence of small cavities called **vesicles,** which were originally gas bubbles in the magma. The gas was initially dissolved in the magma but came out of solution because of the pressure decrease as the magma rose before eruption. This process is similar to the formation of bubbles and foam when you open a warm bottle of a carbonated soft drink. Vesicular textures are typically found in extrusive igneous rocks. Generally the pressure is too high at depth for the gas to come out of solution where intrusions form. If vesicular texture is a minor feature of a volcanic rock, maybe about like the holes in Swiss cheese, the word *vesicular* is used to modify the rock name, such as *vesicular basalt* (Fig. 4.12b). Rocks where the vesicular texture is a more prominent feature have special names:

Scoria: The vesicles making up a high proportion of the volume of this rock give it a spongelike appearance. Scoria ranges from reddish brown to black (Fig. 4.14) and is usually mafic.

Pumice: A rock composed of more vesicles than matrix, pumice is essentially solidified lava "foam" (Fig. 4.15). Because the rock consists of more void than rock, specimens will often float on water. Generally, pumice is light gray to tan. The rock part of pumice is glassy; use a hand lens to confirm this (see Fig. 4.15 inset and the earlier discussion on glassy texture). Pumice is usually felsic, having the same chemical composition as obsidian, granite, and rhyolite.

Sometimes the vesicles may be filled with secondary minerals deposited long after the solidification of the original rock. The filled area is an **amygdule** and the rock is said to have an **amygdaloidal** texture.

Pyroclastic Textures   Volcanic ash and larger rock fragments are violently ejected from a volcano during some eruptions. This debris and the rocks that form from it have a **pyroclastic texture.** Such eruptions are dangerous, so recognition of pyroclastic texture helps volcanologists to assess areas of future volcanic hazards. Rocks with pyroclastic texture do not have the compact, interlocking-grain texture of rocks that crystallized directly from a magma, but instead may have a more "powdery" texture or appear to be made up of broken pieces stuck together. Pyroclastic texture may resemble the clastic texture in a sedimentary rock (see Lab 5). Pyroclastic rocks may consolidate by cementation or lithification (in the manner of a sedimentary rock), but contain volcanic ash or other fragmented volcanic pieces. In some cases, instead of cementation (see welded tuff in the following list), the fusing

Figure 4.14   **Scoria** is a very vesicular, mafic to intermediate igneous volcanic rock. This sample is mafic. Scoria may be reddish brown but is more commonly black, as shown here. The vesicles constitute a major proportion of the volume of the rock so the sample has quite low density, although not as low as pumice.

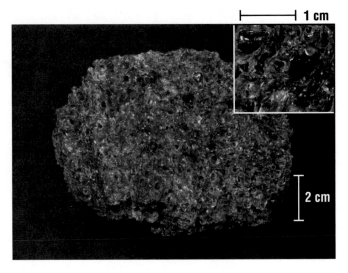

Figure 4.15   **Pumice** is a highly vesicular and *glassy*, felsic to intermediate igneous volcanic rock. This sample has unusually large vesicles and shows the glassy texture very well, but in many samples, the vesicles are much smaller, and the glassy texture can only be seen with a hand lens.

of loose pyroclastics into rock happens while the components are still hot. Although pyroclastic material may be mafic to felsic, it is more commonly intermediate to felsic, because the magmas are more explosive due to high viscosity and gas content. Specific types of pyroclastic substances include the following:

Volcanic ash: loose material consisting of silt- to sand-sized fragments explosively erupted from a volcano (Figs. 4.16 and 4.17)

Tuff: a rock consisting of silt- to pebble-sized ash and/or pumice fragments (Fig. 4.18)

Figure 4.16 Eruption of Mount Pinatubo sent volcanic ash circling the globe. Eventually this material settled out as fine dust and was incorporated in sediments.

(a)

(b)

Figure 4.17 **Volcanic ash** is loose pyroclastic volcanic particles of silt to sand size. (a) The central eruption of Mount St. Helens on May 18, 1980, produced billowing clouds of ash that later fell over a large swath of Washington, Idaho, and Montana. (b) This layer of volcanic ash and cinders erupted from Volcán Colima, Mexico. Photo by Grenville Draper. The close-up of volcanic ash is from Mount St. Helens.

Figure 4.18 **Tuff** is a pyroclastic volcanic rock made of silt-to pebble-size pyroclastic material. This sample contains numerous light gray pumice fragments embedded in a matrix of beige ash.

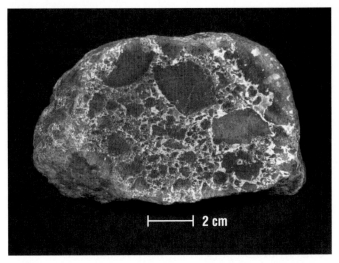

Figure 4.20 **Volcanic breccia** is a coarse clastic volcanic rock with angular fragments either cemented together or held together by lava. This sample shows angular basalt clasts cemented together by calcite cement. Cooling crusts broken off the top and front of lava flows as they move may produce some volcanic breccias.

Figure 4.19 **Pyroclastic flow** churning at more than 100 km (60 mi) per hour down the slope of Mount St. Helens. Welded tuff results from such flows.

Welded tuff: a special kind of tuff that was deposited when the ash particles were still molten. As a result, it often has visible areas with a glassy and streaked appearance. The streaks are commonly pumice fragments that have flattened and lost their gas after the ash was deposited. Welded tuff results from high speed, pyroclastic flows (Fig. 4.19).

Volcanic breccia: a rock consisting of large broken volcanic rock fragments (Fig. 4.20)

7. Examine the samples listed in Table 4.2, and indicate their texture and origin. By "origin" we mean the natural processes or method by which the texture formed. The rock texture descriptions in the text include information about the origin of the textures.

 Origin of Igneous Rocks Computer Exercise

- Start **Earth Systems Today.**
- Locate **Topic: Earth Materials.**
- Go to **Module: Rocks and Rock Cycle.**
- Explore activities in this module to discover the following for yourself:

Table 4.2 Textures of Some Samples That Result from Gas in the Magma

Sample Number	Sample Name	Texture	Origin
	Pumice		
	Volcanic ash		
	Tuff		

8. Where and how does magma originate?

_____

_____

■ Using the "rock laboratory" portion of this module, ob-serve the formation of both phaneritic (coarse-grained) and porphyritic igneous rocks through the process of crystallization.

9. Based on what you see in the "rock laboratory," write a description of how a phaneritic rock crystallizes.

_____

_____

10. Describe how porphyritic rock crystallizes.

_____

_____

11. Use what you learned from the computer to address this point: Rocks in Figures 4.9 and 4.11b are both por-phyritic, yet they formed in different environments.
a. What is different about their textures?

_____

_____

b. What was different about the environment where they solidified?

_____

_____

12. What is the origin of "interlocking texture" among crys-tals in an igneous rock? How does forming from a liquid contribute to development of this texture?

_____

_____

_____

13. How do plutonic igneous rocks, which solidify below-ground, end up at the surface where we can view and col-lect them today?

_____

_____

Classification and Identification of Igneous Rocks

Igneous rocks are classified according to a combination of their _composition_ and _texture_ as indicated in Table 4.3. To identify an igneous rock correctly, use the following strategy in conjunction with Table 4.3:

■ Determine the color (dark, medium, or light) of the spec-imen, and identify its minerals to locate the appropriate _column_ in Table 4.3.

■ Examine the texture of the rock—the grain size or other texture. Is it coarse or fine grained, or does it have a mix-ture of grain sizes (porphyritic)? Is it glassy, vesicular, or pyroclastic? Locate the appropriate _row_ in Table 4.3.

■ The intersection of the color column and texture row in-dicates a possible identification of the rock. Confirm the identification, however, by determining the specific min-erals in the rock (e.g., gabbro and peridotite are both dark, coarse grained rocks, but the presence of plagio-clase feldspar will confirm that the rock is gabbro, not peridotite).

14. Without looking at the rock names, examine a few of the known rocks in the Igneous Rocks Display. Use Table 4.3 and the method just described to identify each rock. Check your answers. When you understand how to use the table, continue to the next exercise.

15. Examine the unknown rocks. In each case, fill in Table 4.4, "Igneous Rock Identification Form," as indicated here:

Texture: Use appropriate terminology from earlier in this lab; see Table 4.3.

Color: Indicate whether the rock is light, medium, or dark, and also note down the actual color. For example, rhyolite is _light_ colored and might be _pink._

Composition: If the rock is coarse grained, note which minerals are present and the color of each mineral. If it is porphyritic, identify the phenocryst minerals (the vis-ible minerals) and the color of each. For all samples, in-dicate the composition classification from Table 4.3 (ul-tramafic to felsic); remember that obsidian is felsic even though it is black in color.

Rock name: Use Table 4.3 and the rock's composition and texture to determine the rock name. You will find the name of the rock where the composition column intersects the texture row in Table 4.3. Also, indicate whether the rock is volcanic (extrusive) or plutonic

(_text continues on p. 74_)

Table 4.3 Composition and Texture of Common Igneous Rocks

Knowledge of the relationship of color to composition and texture (grain size) in igneous rocks contributes to the correct identification of rock samples. Note the relationships of the items shown in the table before proceeding with identification of an unknown sample. Rocks and textures in the shaded area are plutonic, whereas those below, in the unshaded area, are volcanic. Boldface type indicates rocks you are more likely to see.

Color	Tint	**DARK** (dark greenish)	**DARK** (dark gray or black)	**MEDIUM**	**LIGHT**
Composition	Silica Content	<45%	45–53%	53–65%	>65%
	Composition Classification	Ultramafic	Mafic	Intermediate	Felsic
	Mineral Constituents	Olivine, Pyroxene, Ca Plag. Feldspar	Ca Plag. Feldspar, Pyroxene, Olivine	CaNa Plag. Feldspar, Hornblende, (pyroxene), Quartz, Biotite	K-feldspar, Na Plag. Feldspar, Quartz, Hornblende, Biotite, Muscovite

Mineral Constituents graph (100% – 50% – 0%): Olivine; Ca / Pyroxene; Plagioclase Feldspar; K-feldspar ± Muscovite; Quartz; Na; hornblende; Biotite

Texture and Rock Name			**DARK** (dark greenish)	**DARK** (dark gray or black)	**MEDIUM**	**LIGHT**
coarse ⋯⋯	Coarse-grained		Peridotite	**Gabbro**	**Diorite**	Pegmatite (very coarse)
						Granite
	Porphyritic (two grain sizes)	with coarse groundmass		Porphyritic Gabbro	Porphyritic Diorite	Granite Porphyry or Porphyritic Granite
fine ⋯⋯		with fine groundmass		Basalt Porphyry or Porphyritic Basalt	Andesite Porphyry or **Porphyritic Andesite**	Rhyolite Porphyry or Porphyritic Rhyolite
	Fine-grained			**Basalt**	Andesite	**Rhyolite**
	Special Textures	Glassy		Most volcanic glass is black and felsic. Color is not diagnostic for chemical composition.	**Obsidian** (black in color)	
		Vesicular		**Scoria**	**Pumice**	
		Pyroclastic		**Tuff** and Breccia / Volcanic ash (loose material)		

(Ultramafic volcanic rocks are rare)

PLUTONIC

VOLCANIC

Table 4.4 Igneous Rock Identification Form

| Sample Number | Texture | Color (light/medium/dark and actual color[s]) | COMPOSITION | | Rock Name | ORIGIN | |
			Visible Minerals and Their Colors	Ultramafic/Mafic/Intermediate/Felsic		How Did the Rock Form?	Volcanic/Plutonic

Continued

Table 4.4 *Continued*

Sample Number	Texture	Color (light/medium/dark and actual color[s])	COMPOSITION		Rock Name	ORIGIN	
			Visible Minerals and Their Colors	Ultramafic/Mafic/Intermediate/Felsic		How Did the Rock Form?	Volcanic/Plutonic

(intrusive). Note in Table 4.3 that rocks with coarse-grained texture and porphyritic texture with a coarse grained groundmass are plutonic, and all other textures are volcanic.

Origin: Briefly indicate how the rock formed and whether the rock is volcanic or plutonic in origin. Remember that texture tells you a lot about origin or how the rock formed. For example, for a vesicular rock, you should mention the presence of gas in the magma. You may want to review your answers in Tables 4.1 and 4.2 to help with this part.

16. Practice identifying the rocks so you can readily recognize them. Once you feel you know the rocks, use the second set of unknown rocks to quiz yourself without using Table 4.3. Fill in Table 4.5 with the rock names and their key characteristics.

Table 4.5   Practice Quiz

Sample Number	Key Characteristic 1	Key Characteristic 2	Key Characteristic 3	Rock Name

Sedimentary Rocks

Formation of Sedimentary Rocks

Sedimentary rocks form in two stages. First, a sediment is formed; second, the sediment is **lithified** (literally, turned to stone) to become a rock.

Sediment includes loose material from rock and mineral particles, from chemical precipitation at the Earth's surface, such as salt deposits in a dry lakebed, and from organisms and their remains. These three types of sediment are as follows:

1. **Clastic** or **detrital**[1] **sediment** forms by mechanical weathering and erosion of other rocks into sediment made up of broken rock or mineral pieces, followed by transport and deposition of the sediment. This type of sediment can be recognized as consisting of loose grains separated from each other by space. When they become rocks, the grains are held together and the space may be partly filled with natural **cement** or matrix. See Figure 5.1 and conglomerate or breccia on page 87. Clastic sediments are further described by determining the mineral makeup and size of these particles.

2. **Chemical sediment** forms by chemical precipitation of compounds out of a water solution (often seawater). These substances in solution result from chemical weath-

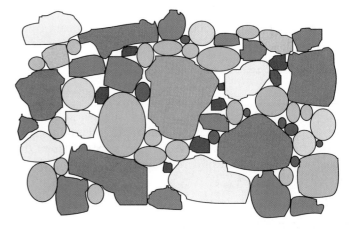

Figure 5.1 Example of clastic sediment. Notice the space between grains.

ering of rocks. In many chemical sediments, crystals with irregular boundaries fit together so there is virtually no space between the grains. Travertine limestone and rock salt on pages 92 and 93 are rocks consisting of chemical sediment. Some chemical sediments are made of sand-size spheres that were chemically precipitated—for example, ooliths in oolitic limestone (p. 93).

3. **Biochemical sediment** results from the accumulation of parts of organisms. While the organisms live, they extract material from their surroundings and produce shells, skeletons, or wood and living tissue. In many cases, the material extracted is available in solution from chemical weathering of rocks. When the organisms die, the harder parts are transported and deposited as sediment. In some cases transport does not occur, and the sedimentary rock forms right where the shells formed. These sediments may vary widely in appearance depending on what organism is preserved in the sediment, but generally they can be recognized by the abundance of organism remains and an absence of clastic grains.

The three types of sediment may remain separate or they may be mixed in various combinations. Once sediment has accumulated, it is buried and lithified into a sedimentary rock by compaction and/or cementation.

----

[1]Wherever we use the term *clastic sediment* or *clastic sedimentary rock* in this lab, you may substitute *detrital sediment* or *detrital sedimentary rock* if you prefer.

Experimental Formation of Chemical Sediment

Evaporite   Place a watch glass on a hot plate and half fill it with seawater. Allow the seawater to evaporate without boiling. While you are waiting for it to evaporate, start the next experiment.

1. After the seawater has evaporated, examine the residue on the watch glass under a microscope or with a hand lens and describe it.

    a. Describe the texture and general appearance.

    _____

    _____

    b. Where did the solid material in this sediment come from?

    _____

    _____

Evaporite minerals may include halite, gypsum, anhydrite, and sometimes calcite. An interlocking **crystalline** texture is common for evaporites. This is very similar to the coarse-grained texture of igneous rocks, but the minerals are totally different.

2. What natural sedimentary processes and depositional environment do you think this experiment is supposed to simulate? What might be a natural heat source for this process? You may want to discuss this answer with your lab partners.

    _____

    _____

    _____

    _____

Lithographic Limestone   Stick a straw into a half-full beaker of saturated $Ca(OH)_2$ solution, and cover the beaker around the straw with plastic wrap or paper to keep the solution from splattering. Blow gently through the straw into the solution. ***Don't suck.*** Calcium hydroxide is harmful if swallowed. Don't hyperventilate; get some of your lab partners to help blow. Care should also be taken to ensure none splatters into anyone's eyes.

3. What are you adding to the solution by blowing through the straw? (*Hint:* What gas do you breathe out?) _____

    _____

4. What happens to the solution as you blow into it? _____

    _____

    _____

5. This is a chemical sediment in suspension in the water. What would happen if you let this cloudy beaker sit for a

    while? _____

6. **a.** What do you think is the chemical composition of the compound you have produced?

    Hint: Review the chemistry of the solution and your answer to Exercise 3. _____

    _____

    b. If this compound had been produced naturally from seawater, what would its mineral name be? (If you need to, look back at the chemical formulas of minerals in either Lab 2 or 3.) _____

    c. What test would confirm this? _____
    Do it. What happened?

    _____

    _____

If the sediment from the prior question were turned into a sedimentary rock, it would become fine-grained limestone such as lithographic limestone. **Fine grained,** as for igneous rocks, means the grains are too small to see with the unaided eye.

7. Where in nature might you expect a similar chemical

    reaction to occur? _____

Don't forget to check how the evaporation is proceeding in Experiment 1, and complete your answer.

Lithification

Sediment becomes a sedimentary rock through burial and lithification. **Lithification** of a sediment involves bringing the grains closer together—**compaction**—and gluing them together—**cementation.** Burial of the sediment by more sediment provides the pressure for the compaction to take place. Compaction can make a rock out of material, such as clay, that has a tendency to stick to itself when squeezed together, but compaction is insufficient to cause either sand to hold together to form sandstone, or pebbles to cohere to form the rock conglomerate. In these situations, cementation is needed. Cementation occurs when water containing dissolved substances moves through the rock and precipitates some of the dissolved material between the grains. This helps hold the grains together. These "gluing" minerals are known as **cement** and are commonly quartz, calcite, or iron oxide.

8. Use a hand lens or microscope to examine the sandstone provided. Can you see any cement holding the sand

    grains together? _____ If so, what type of

    mineral is the cement? _____ You may need

to do some tests to decide the type of mineral. What tests did you perform, and what was the result of those tests?

_____

_____

9. Continue viewing the sample with a hand lens or microscope, and sketch the sand grains and cement, and label each in your sketch.

![CD icon] **Origin of Sedimentary Rocks Computer Exercise**

- Start **Earth Systems Today.**
- Go to **Topic: Earth Materials.**
- Go to **Module: Rock and Rock Cycle.**
- Explore activities in this module to discover the following for yourself.
- Using the "rock laboratory" portion of this module, observe the formation of a clastic sedimentary rock (sandstone) through the process of lithification.

10. Summarize the steps in the formation of sandstone.

_____

_____

_____

_____

11. What role does water play in the lithification process?

_____

_____

12. Based on what you see in the "rock laboratory," describe how a sandstone is lithified.

_____

_____

_____

_____

13. Use the animation to observe how compaction and cementation work together to form sedimentary rock. Summarize how this works.

_____

_____

_____

_____

_____

14. How are igneous and sedimentary rocks different in their texture and mode of formation?

_____

_____

_____

_____

_____

■

Sedimentary Environments

Sedimentary rocks contain a wealth of information about Earth's surface environment where the sediments accumulate (Fig. 5.2). Since some sedimentary rocks date back millions and even billions of years, they can become a kind of time machine, giving us a glimpse of the past before humans ever walked the Earth. Where you now sit may once have been a mountain range or the bottom of the sea, complete with strange creatures. If you know how to decipher them, various features and structures of sedimentary rocks can sometimes divulge secrets of these past environments.

Structures of Sedimentary Rocks

The most distinctive feature of sedimentary rocks seen from a distance is **bedding,** which forms by the deposition of sedimentary particles one layer after another. Igneous rocks may have flow banding or layers of lava or ash; metamorphic rocks (Lab 6) may have layers of minerals, or *foliation;* but it is usually fairly easy to distinguish sedimentary bedding from igneous or metamorphic layering. Other structures are also common in sedimentary rocks and may give us additional hints about the environment of deposition.

Bedding is the arrangement of a sediment or sedimentary rock in flat parallel layers formed at the time of deposition (Fig. 5.3). Bedding may be visible as a result of obvious changes in grain size, constituents, or color (Fig. 5.3a). Conversely, it may result from subtle changes or simply details of the arrangement or orientation of grains. Bedding is generally deposited in horizontal layers because of

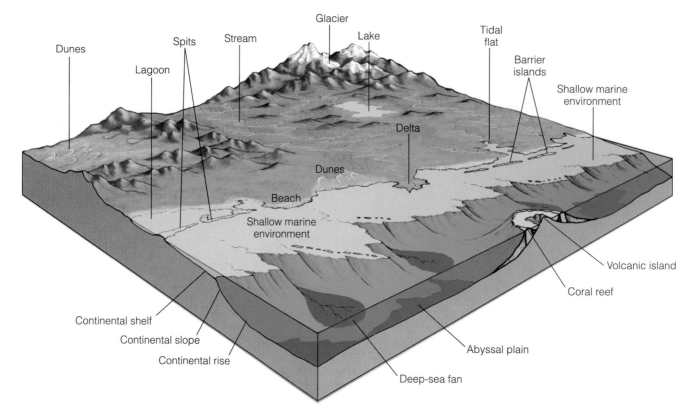

Figure 5.2 Some of the environments in which sediments may be deposited. Each environment has its distinctive characteristics in terms of sediment composition, size, sorting, structures, and organism.

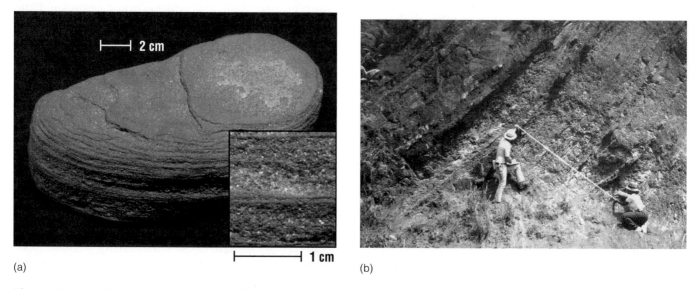

(a)

(b)

Figure 5.3 Bedding. (a) A hand sample of bedded sandstone. *Inset:* Close-up of sandstone hand sample showing details of the layers. Notice the layers are distinctly marked by variations in grain size, degree of cementation, and color. Other more subtle factors may add to the layering. (b) Tilted bedding in sandstones. Students are measuring the thickness of a particular set of layers. Photo by Grenville Draper.

gravity (principle of original horizontality; see Lab 7), so nonhorizontal bedding indicates either cross bedding (discussed next) or tilting of the rocks (Fig. 5.3b). Sedimentary rocks without bedding are said to be **massive** (Fig. 5.4).

Cross bedding is a kind of inclined bedding in which certain layers formed at low angles to the major bedding

(Fig. 5.5). The action of currents, either wind or water, produces cross bedding in medium to coarse sediments. Migrating sediment in underwater bars or wind-blown dunes form cross bedding that slopes downward in the direction of the current and accumulates on the side of a dune or bar that is protected from the current. So cross bedding is as indicator of ancient current directions

(Fig. 5.5). The layers of cross bedding taper downward toward the bottom and are truncated at the top where the wind or current has swept away the top part of the bed. As a result, cross bedding is also an **up indicator** (Fig. 5.6); that is, since the top is different from the bottom, cross bedding can tell you which way was up when the rock originally formed. Why one might want to know this will become more apparent in Lab 7 when we look at deformation of rocks.

Graded bedding is a *single* layer or bed of sediment or sedimentary rock in which the largest grains are concentrated at the bottom, gradually decreasing in size upward to the smallest at the top of the bed (Fig. 5.7). An ebbing (diminishing) current deposits the whole bed in a single episode of deposition. Graded bedding might occur where a river enters a lake or as a result of a submarine landslide. Graded bedding is an up indicator (Fig. 5.6).

Other sedimentary structures include ripple marks, mud cracks, and raindrop prints. These can also indicate something about the environment in which they formed and can be up indicators (Fig. 5.6).

15. Examine the samples of sedimentary structures provided by your instructor. Fill out Table 5.1 with the name and a sketch of the structure. Also include in the table what other information you can gain from structure: What sedimentary environment formed the structure? Can the structure be used to tell which direction was up when the sediment was deposited? (If so, indicate which way was up on your sketch.)

Make Your Own Graded Bedding

Materials needed:

- Graduated cylinder
- Water
- Sediment of various sizes

Fill the graduated cylinder roughly halfway with sediment and most of the way with water, or use a cylinder already filled with sediment and water. Cover the graduated cylinder with your hand, hold it over a sink or basin, and shake it until the sediment is well mixed with the water. Then set the cylinder down. Observe what happens.

Figure 5.4 Massive sandstone. Massive means the absence of bedding. This sample is a quartz-rich sandstone with iron oxide-stained cement. White specs are places where the cement is not stained.

Figure 5.5 Cross bedding is most common in sandstone but may also occur, as shown here in Nassau, Bahamas, in oolitic limestone. The lower cross-bedded layer is 1.5-m thick and shows that the current that deposited the oolite sand was flowing from left to right. Photo by Florentin Maurrasse.

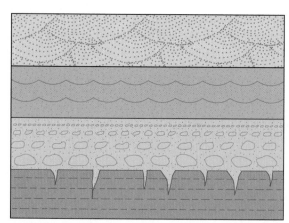

Undisturbed layers

Cross-beds: Often concave upward and cut off at top

Wave-formed ripple marks: Sharp crests and smooth troughs

Graded bedding: Largest particles on bottom

Mud cracks: Open toward top of layer

Figure 5.6 Sketch of the side view of some sedimentary structures that are also up-indicators. The cross bedding, ripple marks, graded bedding, and mud cracks shown each have some aspect that distinguishes the top from the bottom.

Figure 5.7 Graded bedding of a sandstone layer in vertical beds tells which way was formerly up. The bottom was toward the left where the layer contains small pebbles. These grade toward the right (formerly upward) to coarse and then fine sand.

16. a. Sketch the layer you have deposited in the graduated cylinder, paying particular attention to the grain-size variation within the layer (sketch should include a scale bar). This is called *graded bedding*.

b. What does this experiment tell you about the speed of large and small grains as they fall through water?

_____

_____

c. What natural environment could produce the mixing of sediment you achieved by shaking the cylinder?

_____

d. What natural environment would allow mixed material to settle? _____

Fossils

Fossils are naturally preserved remains or traces of life preserved in a sedimentary rock at the time the rock formed. Fossils may include shells, impressions of shells called *molds*, petrified wood, footprints, animal burrows, and so forth. A rock containing fossils is said to be **fossiliferous.** Figure 5.8 shows a number of examples of fossils. Many fossils are large enough to see with the unaided eye (Fig. 5.8a–j), but some are much too small to see without a microscope and are called *microfossils* (Fig. 5.8k). A rock made up exclusively of fossils has a **skeletal texture** and would be a biochemical sedimentary rock.

A fossil can tell you a lot about the rock in which you find it. You know the time of sedimentation must have been a time when the organism was inhabiting this environment.[2] For example, the trilobite in Figure 5.8a lived during the Cambrian Period, which was 545 to 505 million years ago. Critters of this kind went extinct 505 million years ago, so for a rock such as the one in the picture to have a trilobite preserved in it means the rock is at least that old! You are looking back in time hundreds of millions of years. (The closest living relative of trilobites is the horseshoe crab.) Another example is the ammonoid in Figure 5.8b. This particular species lived during the Mississippian Period, about 350 million years ago. This is not as amazingly old as the trilobite but

----

[2]An exception to this is that the fossil could have been **reworked.** A reworked fossil is one that was originally preserved in a previous rock, was eroded from that rock, and became part of a new rock. For this discussion of fossils we assume the fossil is not reworked.

Table 5.1   Sedimentary Structures

Sample Number	Structure or Feature	Sketch	Sedimentary Environment and Other Information

(a) Trilobite—*Olenellus robsonensis.*

(b) Ammonoid, relative of the squid and octopus, which lived during the early Mississippian (approximately 350 million years ago, Fig. 8.21); *Imitoceras rotatorium* from lower Mississippian Rockford Limestone, Indiana. This species shows the coiling and fairly simple suture pattern of one of the forebears of ammonites (Fig. 8.3).

(c) Leaf fossil, example of *carbonization,* Oligocene alder leaf, John Day Beds, Oregon. The leaf's length is about 6 in.

(d) Mammoth bones, from La Brea Tar Pits, California, Late Pleistocene, are *body fossils.*

(e) Fossiliferous limestone with *recrystallized* clam shells, seen mostly in cross section. The clam shells can be recognized as arc shapes.

Figure 5.8 Fossils

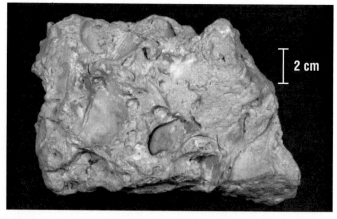

(f) *Molds* of clam shells in limestone. The actual shell has dissolved away, leaving its impression as a clam shell-shaped hole in the limestone.

(g) Fossil casts of tree stumps of *Stigmaria* from the Carboniferous, Victoria Park, Glasgow, Scotland. The Carboniferous (known as the Mississippian and Pennsylvanian in the United States) was between 360 and 286 million years ago (Fig. 8.21). The casts formed when trees partly buried in sediment died and rotted away and the holes filled with sand.

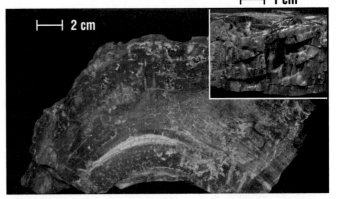

(h) *Petrification or replacement*. A slice of petrified wood across the trunk of a tree showing tree rings and well-preserved bark. A close-up of the outside edge shows details of the bark structure.

(j) These *trace fossils* preserved on the under side of a sedimentary layer are fossil burrows. Photo by Grenville Draper.

(i) Hominid footprints in a 3.4-to-3.8-million-year-old volcanic ash deposit, Tanzania. Two adults and possibly a child walked upright across the ash. Footprints are a type of trace fossil.

Figure 5.8 Fossils *Continued*

(k) *Microfossils*. Electron photomicrographs of (top) coccoliths from Gulf of Mexico, (middle) diatoms from Java, and (bottom) dinoflagellates from Alabama (left) and Gulf of Mexico (right). The symbol μ stands for a micron, which is 1/1000 of a millimeter. Coccoliths form calcium carbonate shells and are abundant in some chalks. Diatoms form silica shells and are abundant in some cherts. Dinoflagellates form organic cysts that can be preserved as fossils. See also Figures 5.20b and c and 5.27a, inset.

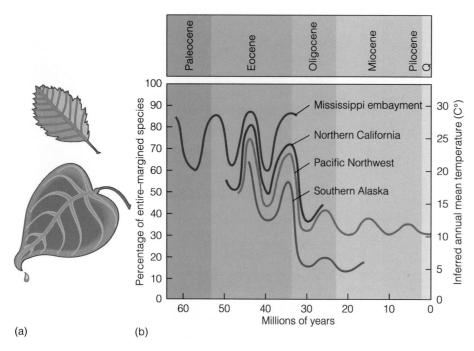

Figure 5.9 Fossils can be climate indicators. (a) *Top:* Plants from cool climates tend to have small leaves with serrated edges or incised margins. *Bottom:* Leaves on tropical plants tend to have smooth margins with a drip tip and are considerably larger— referred to as "entire-margined" leaves. (b) Climatic changes in four different regions of North America are reflected in the percentage of plants with entire-margined leaves (tropical type).

it is a venerable old age. If you already know the age of the rock, fossils can tell you about the types of organisms that lived at that time in the past. If you don't already know the age of the rock, a fossil might tell you the age (see Lab 8).

A fossil can also tell you the environment of deposition of the sediment with which it accumulated. Let's take the trilobite again. It crawled or swam around on the shallow bottom of a warm ancient sea, but it can now be found in sedimentary rocks in mountainous areas. Finding a trilobite in this sedimentary rock tells you that the area was a warm shallow sea. Since then, the sediment was buried, turned into rock, then shoved upward by the forces of the Earth into mountains. A trilobite is an ancient creature you will never see alive today, but some fossils are more familiar. In fact, we can use knowledge about organisms of the present to help us to interpret older organisms. Another example is a simple leaf fossil as in Figure 5.8c. "Nothing special here," you say? Notice the small size and incised margins of the leaf. Study of plants of the present from tropical versus cooler climates indicates that plants with small serrated leaves grow in cooler climates and plants with large smooth-edged leaves ending in a drip point are from warmer humid climates (Fig. 5.9a). The leaf in Figure 5.8c tells us at least two things: it is from a land plant, so the sediment is likely to be nonmarine; and the climate was cool, not tropical. Paleobotanists, people who study plant fossils, have used such leaves to document major climate cooling during the early Oligocene Epoch, about 30 million years ago (Fig. 5.9b).

Can you deduce anything about the sedimentary environment from a fossil? In this next exercise, we will use fossils that are fairly easy to recognize, so you have a good chance to infer something about the environments in which they lived.

17. Examine the rocks containing a fossil or fossils provided by your instructor. Filling in Table 5.2, describe and/or sketch each fossil. What can you tell about the environment in which this (these) organism(s) lived and therefore the environment of deposition of the rock? Some possible envi-

ronments to consider might be a marine environment (in the sea) in deep, shallow water or coral reef, a nonmarine or terrestrial environment, river, lake, desert, or beach. Was the environment warm or cold, wet, or dry? You may not have information about all of these things but write down what you do know from the fossil in each rock in Table 5.2. What organism is it, or is it similar to an organism living now? Do you know the name of the organism?

Since fossils are defined in part as evidence of life, that evidence can be preserved in many different ways. A fossil can be the actual remains of the organism or some part of the organism, or it can be an impression of some sort left behind or some sign of that organism's activity. Here are some types of preservation:

Body fossils: These are actual remains of the organism, or parts of the organism, such as the mammoth bones in Figure 5.8d.[3]

Recrystallization: Original materials such as aragonite ($CaCO_3$) or opal recrystallize into more stable minerals such as calcite or quartz. The clam shells in the fossiliferous limestone in Figure 5.8e have been recrystallized.

Molds: Molds form after burial when the organism's remains dissolve away, leaving a cavity such as the mold of a clam shell in Figure 5.8f.

Casts: If the cavity of a mold fills with material, the resulting fossil is a cast (Fig. 5.8g).

Petrification: The original material is replaced by other substances (Fig. 5.8h). Petrified wood may be so well preserved by the replacement process that minute details of its structure may be still visible.

....

[3]Bone fossils of land animals are actually quite rare, expecially complete skeletons, because there is little deposition of sediment where most of these organisms live.

Table 5.2 Fossils

Sample Number	Fossil Name	Description/Sketch	Sedimentary Environment and Other Information

Carbonization: The original organic matter has been reduced to just its carbon (Fig. 5.8c). These fossils will be brown to black and are commonly leaves.

Trace fossils: These are marks or traces of an organism, not any part of the organism itself. Many trace fossils are possible. **Footprints** and **trails,** like the hominid tracks in Figure 5.8i, are types of trace fossils. Some other examples include burrows (Fig. 5.8j), teeth marks, and broken shells of other critters that were eaten by the organism.

All fossils, but especially trace fossils, are only preserved a small percentage of the time because of their fragile nature, the long time between burial and observation, and the myriad of forces acting on the sediments and rocks during that time.

18. Examine the fossils provided for this exercise, and decide what type of preservation you see.

Sample: _____ Preservation: _____

Sample: _____ Preservation: _____

Sample: _____ Preservation: _____

Sample: _____ Preservation: _____

Clastic Sedimentary Rocks

If a rock is clastic (or detrital), its texture must be described by noting the particle size. Table 5.3 gives the most commonly used rock names for clastic sedimentary rocks based on fragment or clast size. Most students already have a good idea of the size of each of the types of clasts named in the table. For those who do not know words such as *cobble* or *granule,* the grain diameters are included in the table. Although students may not be required to memorize these precise sizes, the values may help in understanding the clast terms.

Detrital Sedimentary Rocks Computer Exercise

■ Start **In-TERRA Active.**
■ Click **Materials.** Then click **Sedimentary Rocks.** Then click **Explore.**

19. Do the exercise, then fill in Table 5.4 with the correct answers.

■

Some common clastic sedimentary rocks are further explained in the following sections. You can examine many of these rocks in the Sedimentary Rocks Display.

Coarse Clastic Rocks

Conglomerate has particles that are greater than 2 mm in diameter and are rounded (Fig. 5.10). If rounded particles in a rock are larger than sand, call the rock a con-

Table 5.3 Classification of Clastic Sedimentary Rocks

Size of Clast	Grain Size	Grain Diameters	ROCK NAME	
			Rounded Clasts	Angular Clasts
Boulder	Very coarse	>256 mm	Conglomerate	Breccia
Cobble	Coarse	64–256 mm		
Pebble	Moderately coarse	4–64 mm		
Granule	Medium coarse	2–4 mm		
Sand	Medium	1/16–2 mm	Sandstone	
Silt	Fine	1/256–1/16 mm	Siltstone	
Clay	Very fine	<1/256 mm	Shale (or mudstone)	

Table 5.4 Detrital Sedimentary Rock Computer Exercise

Energy	High	Fairly High	Medium	Low
Sediment				
Process				
Rock				

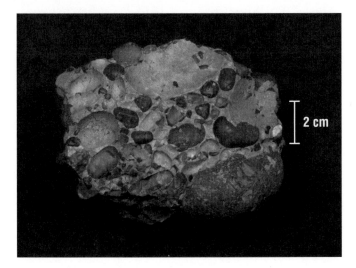

Figure 5.10 **Conglomerate** is a coarse-grained clastic sedimentary rock with rounded clasts. Shown here is a pebble conglomerate in which pebbles are the largest clasts, surrounded by a matrix of cemented orange-yellow sand. The color comes from iron oxide (limonite), a sign of oxidation usually indicative of a nonmarine environment. Most of the pebbles are chert in this sample.

Figure 5.11 **Breccia** is a coarse-grained clastic sedimentary rock with angular clasts. This sample is a pebble to cobble breccia with very angular gray chert clasts in a matrix of cherty material. Iron staining colors parts of the sample orange.

glomerate. The larger clast size in conglomerate indicates the clasts probably were transported only a short distance, yet the rounded shape suggests farther travel than that for breccia (discussed next). Streams can tumble rocks to make large rounded clasts. Tumbling in water can also have a polishing effect on the clasts. Some conglomerates were once streambed gravels; others formed from wave-washed boulders at the base of a sea cliff.

Breccia also has clasts larger than 2 mm, but these particles are angular instead of rounded (Fig. 5.11). Most breccias are massive (unbedded). Breccia is likely to form in an environment where sediment has not been transported or has traveled only a very short distance. Angular clasts are not as likely to have been carried or moved by water as rounded clasts. Some breccias are landslide deposits; others, called *tillite,* were once deposited by glaciers.

Sandstones (Medium-Grained Clastics)

Quartz sandstone is a rock made of quartz sand grains cemented together (Fig. 5.12). The quartz particles appear translucent and sometimes vitreous when viewed with a hand lens (Fig. 5.12, inset). They may be stained red or orange by iron oxide. The rock breaks between grains and in some cases may easily crumble. Sandstones feel like sandpaper. Because quartz is resistant to chemical and physical weathering, it lasts a long time in sedimentary environments. Sandstone with a high percentage of quartz is called *mature.* A **mature** sediment has

traveled far and is deposited a long distance away from its source, the rock from which this sediment weathered.

Arkose is similar to quartz sandstone but contains feldspar in addition to quartz (Fig. 5.13). Feldspar grains are usually weathered and appear opaque, chalky, and white, light pinkish, yellowish, or grayish in color. Since feldspar weathers to clay in moist environments, the occurrence of arkose indicates a dry climate and/or a relatively short transport distance. Weathering of granite is a likely sediment source for the quartz and feldspar grains.

Graywacke is a "dirty" sandstone characterized by angular grains of quartz, feldspar, and small fragments of rock set in a matrix of clay-sized particles (Fig. 5.14). Graywacke indicates a short transport distance because it contains grains that would weather away if transported a long distance. Graywacke is considered an **immature** sedimentary rock.

Fine Clastics

Siltstone is a fine-grained clastic sedimentary rock containing silt-sized particles. It has a slightly gritty feel to the touch. In comparison, shale lacks this gritty feel, and sandstone feels very gritty.

Shale and mudstone are soft, fine-grained clastic rocks made up of clay compacted together. These rocks are usually a gray to black, buff, greenish, or reddish color. Shale is **fissile,** which means it splits into platy slabs

Figure 5.12 **Quartz sandstone** is a medium-grained clastic sedimentary rock in which the clasts and cement are quartz. *Inset:* When viewed with a hand lens, the quartz sand grains look like minute grains of clear glass.

Figure 5.13 **Arkose** is a sandstone with quartz and feldspar sand grains predominant. The feldspar grains in this arkose are pinkish. Muscovite grains are also present. The dark red color of this particular arkose results from the presence of hematite (iron oxide), common for desert sediment. This arkose formed from the weathering of granite in a dry environment.

Figure 5.14 **Graywacke** is a sandstone with abundant rock (lithic) fragments and high matrix content. In this sample, the lithic fragments are a bit too large for a sandstone, making this technically a graywacke granule breccia, but it does show the lithic fragments well. Graywackes typically have angular grains.

Figure 5.15 **Shale** is a well-bedded, fine-grained clastic sedimentary rock made up primarily of clay. The samples here are light greenish gray and dark red brown, but shale may be a number of different colors, including gray, tan, or black. The bedding is best observed when looking at a sample on edge, as for the sample at the left. The fine-grained texture is more discernible on the flat bedding plane of the sample on the right.

more or less parallel to bedding (Fig. 5.15). Mudstone is a clay or clay and silt rock that does not split into layers but breaks into chunky pieces (Fig. 5.16). Although your lab samples have probably been selected so that they do not fall apart, many shales or mudstones crumble or fall apart when water is added. Shale or mudstone may contain well-preserved fossils.

Biochemical and Chemical Sedimentary Rocks

Classification

Biochemical and chemical sedimentary rocks are classified on the basis of their chemical or mineralogical composition.

Figure 5.16 **Mudstone** is a massive fine-grained clastic sedimentary rock made up primarily of clay and/or silt. This is a hand sample of silty greenish mudstone.

For example, *limestone* is a general name given to any sedimentary rock made up almost entirely of calcite regardless of whether it is made of shells, tiny calcite spheres (ooliths), limestone boulders, or limey ooze. The material that makes up the non-clastic rocks may be calcite, dolomite, halite, gypsum, silica, or carbonaceous substances. Table 5.5 lists some examples of rocks classified by composition.

Table 5.5 Classification of Biochemical and Chemical Sedimentary Rocks

Composition	Rock Name	Origin
Calcite (or aragonite)	Limestone	Biochemical, chemical, or (clastic)
Dolomite	Dolomite	Chemical
Halite	Rock salt	Chemical
Gypsum	Gypsum	Chemical
Quartz	Chert	Chemical or Biochemical
Carbonaceous material	Coal	Biochemical

Biochemical and Chemical Sedimentary Rocks Computer Exercise

- Start **In-TERRA Active.**
- Click **Materials;** then click **Sedimentary Rocks;** then click **Explore;** then click **Chemical.**

20. Do the exercise, then fill in Table 5.6 with the correct answers.

■

Origin

Although nonclastic sedimentary rocks are classified by composition, they generally form by biochemical and/or chemical processes. The names of the following rocks may reflect both the composition of the rock and the origin. You can examine many of these rocks in the Sedimentary Rocks Display.

Rocks with Predominantly Biochemical Origin

Fossiliferous limestones:

Coralline limestone is limestone composed almost entirely of the continuous calcium carbonate skeletons secreted by coral polyps. **Coral polyps** are small colonial animals that build coral reefs. Platelike divisions of the individual coral's body, called *septa,* are often visible in coralline limestone (Fig. 5.17). Coralline limestone indicates a warm, shallow-sea environment.

Bryozoan limestone is a rock made up primarily of the calcareous remains of a marine organism, which, like coral, is a colonial animal. Bryozoans appear as porous calcitic crusts on sea grasses, mangrove roots, and other marine objects. They are discernible from corals by the lack of septa, smaller size, and a less regular, lumpier-appearing colony (Fig. 5.18). Bryozoans live in a warm, shallow, quiet marine environment such as a lagoon.

Coquina is limestone composed almost entirely of fossil limestone material, usually in the form of coarse shells and shell fragments (Fig. 5.19). The shell fragments commonly accumulate in a beach environment where wave action breaks the shells. In contrast to lithification of most sedimentary rocks, lithification of the shells into coquina may occur right on the beach without burial.

Table 5.6 Chemical Sedimentary Rock Computer Exercise

Environment	Salt Flat or Desert Lake	Swamp	Coral Reef	Deep Sea
Sediment				
Process				
Rock				

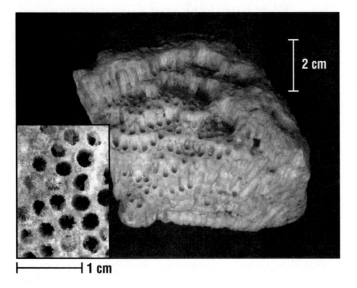

Figure 5.17   **Coralline limestone** is made up of calcium carbonate coral skeletons. *Inset:* Septa have been preserved in parts of this sample.

Figure 5.19   Coquina is a biochemical limestone made up of shells or shell fragments cemented together.

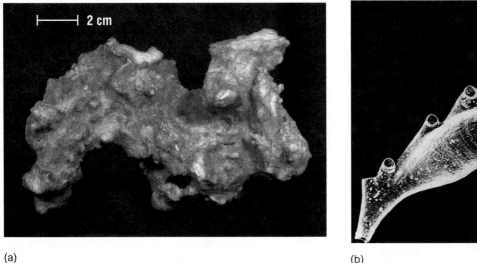

(a)

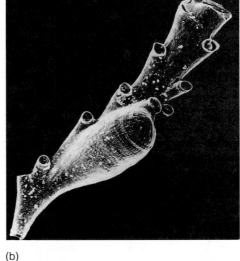

(b)

Figure 5.18   Bryozoan limestone is made up of calcium carbonate skeletons of bryozoan colonies. (a) Bryozoan limestone hand sample with typical lumpy branching bryozoan colonies. (b) Electron photomicrograph (electron microscope image) of a segment of bryozoan *Crisia*.

Chalk is limestone made up in part of biochemically derived calcite in the form of shells or fragments of microscopic oceanic plants and animals. These biochemical remains are found mixed with very fine-grained calcite deposits of either biochemical or inorganic chemical origin. Chalk is white or very lightly colored, and it is soft and porous. Notice in Figure 5.20 that chalk is fine grained, but when viewed under a microscope, its microfossils are visible.

Coal is rock made up of the altered remains of plants. It is a fossil fuel often used in power plants to generate electricity.

■ **Peat** is dark brown, fibrous, decomposed, and compacted residue of swamp plants. Because it is generally porous and crumbly, you will need to be careful with the samples so future students can view them. Peat forms as a step in the formation of coal but is gener-

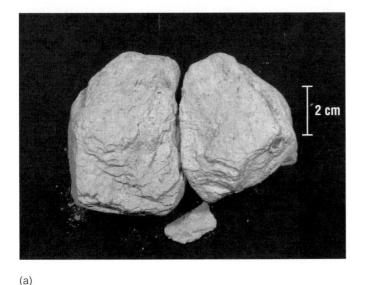

(a)

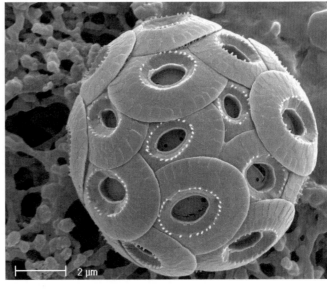

2 µm

(b)

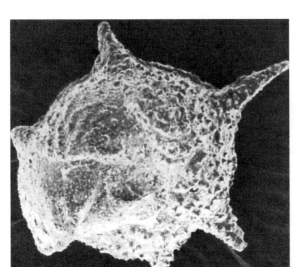

(c)

Figure 5.20 **Chalk** is a biochemical limestone made up of microscopic calcium carbonate shells loosely cemented to produce a powdery rock. (a) Hand sample of chalk. (b) Scanning electron microscope image of a coccosphere of interlocking coccoliths (circles) that form a hard shell for a single cell organism, *Umbilicosphaera hulburtiana*. The scale bar is 0.002 mm. Coccoliths are a major constituent of many chalks, including the White Cliffs of Dover, England. (c) Scanning electron microscope image of a foram (foraminifera) from Cretaceous Pecan Gap Chalk, Texas

ally not considered coal itself. It is also considered a fossil fuel and can be burned to provide heat.

■ **Lignite** is brown to black coal, a smooth-surfaced, further alteration of peat. Softer than bituminous coal, lignite commonly has dusty, dirty surfaces that soil fingers. It is also called "dirty coal," because it puts out a lot of pollution when burned.

■ **Bituminous coal,** or "soft coal," yields bitumins (hydrocarbons) upon heating. It is dark brown to black (Fig. 5.21) and is glossy but not as lustrous as anthracite. It is harder than lignite but fractures and falls apart more readily than anthracite. It results from deeper burial of lignite, with the corresponding higher pressure and temperature.

■ **Anthracite** is "hard coal" with a high percentage of carbon. It is black and has an adamantine to submetallic luster. It is often grouped with metamorphic rocks, because in deposits of anthracite, slate (a low-grade metamorphic rock; see Lab 6) is a common neighboring rock. Anthracite forms by low-temperature metamorphism of other coal.

Rocks with Primarily Chemical Origin

Evaporites:

Rock salt is made up of halite; formed by precipitation of salt from water in which sodium chloride has been concentrated by evaporation. This may occur where seawater is trapped in a confined basin or in desert lakes.

Figure 5.21 **Bituminous coal** is a biochemical sedimentary rock made up mostly of carbon from the remains of swamp plants. This variety of coal is lustrous, commonly banded, and fairly fragile. Banding is not visible in this photo because of the orientation of the sample.

Figure 5.22 **Rock salt** is a chemical sedimentary rock made up of halite. It is commonly translucent to transparent, as in these samples. Impurities can impart various colors; here iron oxide imparts a reddish color.

Figure 5.23 **Gypsum** is a chemical sedimentary rock made up of the mineral gypsum.

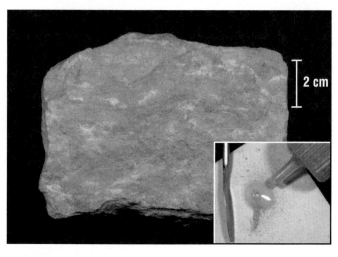

Figure 5.24 **Dolostone** is a chemical sedimentary rock made up of calcium-magnesium carbonate (the mineral dolomite). *Inset:* powdered dolomite effervesces in dilute hydrochloric acid.

Rock salt can vary in color but generally is translucent and light colored (see Fig. 5.22). Use the properties of the mineral halite to help you to identify rock salt.

Gypsum is a common mineral of evaporites. Where gypsum makes up a high percentage of a rock, the rock is called **rock gypsum** or just gypsum (Fig. 5.23). Use your mineral identification skills to recognize gypsum; remember it is softer than a fingernail. Gypsum is a hydrous calcium sulfate with three varieties: alabaster, selenite, and satin spar. Anhydrite, the anhydrous (lacking water) calcium sulfate, is commonly massive in evaporite beds.

Chemically derived carbonates:

Dolostone (Fig. 5.24) is very similar to limestone except that it reacts much more slowly with dilute hydrochloric acid. Powdered dolomite, such as the powder left on a streak plate, reacts more vigorously (Fig. 5.24, inset). Dolostone forms by sedimentary chemical alteration of limestone or calcareous sediment. The limestone texture may be obscured or obliterated: fossils appear as faint shadows or as molds; color is commonly tan due to small amounts of iron in the dolomite.

Figure 5.25 **Travertine** is a chemical limestone made up of calcium carbonate precipitated from fresh water in caves. Holes are commonly included parallel to the layers as the travertine forms.

Figure 5.26 **Oolitic limestone,** also called **oolite,** probably has a combined chemical and biochemical origin. It is made up of calcium carbonate spheres precipitated in moving seawater, probably with the aid of algae. The inset shows spherical sand-size ooliths cemented together with calcite.

Travertine is limestone that is deposited from solution (commonly forming stalactites[4] and stalagmites[5] in caves). It generally has a crystalline texture with crystallized layers visible (Fig. 5.25). If the travertine was part of a stalactite or stalagmite, the layers may be concentric where the stalactite or stalagmite was broken off. Depending on how the travertine formed—for example, hanging vertically—the layers may not have formed horizontally (an exception to the principle of original horizontality, Lab 7).

Rocks with Mixed or Uncertain Origins

Lithographic limestone (or **micrite**)is a limestone that is exceptionally fine grained and homogeneous and exhibits subconchoidal fracture. It was formerly in demand for use in lithography (a method of printing), hence its name. Such fine-grained limestones may form as chemical precipitates or as lithified biochemical ooze of microscopic calcium carbonate shells (Fig. 5.8k, top).

Oolite or **oolitic limestone** contains small spherical bodies, or **ooliths,** usually 1/2 to 1 mm in diameter, cemented together (Fig. 5.26). Ooliths have radiating crystals; they may be chemically precipitated, or algae may aid the precipitation biochemically. Although it looks somewhat like sandstone (Fig. 5.26), oolite is easily distinguished from sandstone by testing with dilute HCl.

Chert comprises a hard, silica-rich, fine-grained rock that comes in a variety of colors and has conchoidal frac-

....

[4]Stalactites are "icicles" hanging from the roof of a cave.

[5]Stalagmites are upright calcite "posts" on the floor of a cave.

ture (Fig. 5.27a). It is composed of cryptocrystalline[6] varieties of quartz and/or noncrystalline silica, opal. Chert may form chemically as precipitates of silica or silica replacement of limestone. It may form biochemically as deposits of microscopic silica shells of diatoms or radiolaria (Fig. 5.27a, inset and 5.8k, middle). Or it may combine these origins. Some colored cherts have special names: **flint** is dark; **jasper** is red; **carnelian** is yellow to orange and translucent (Fig. 5.27b).

Identification and Description of Sedimentary Rocks

The maze in Figure 5.28 will help you to determine the name of a sedimentary rock. Begin where it says **Start Here.** At each fork in the maze, decide which way to go by observing a property of the rock or making the test requested. You may want to try using the maze with a known sample or two from the Sedimentary Rocks Display first.

21. Examine each unknown rock specimen with and without a hand lens. Fill in Table 5.7, "Sedimentary Rock Identification Form," as indicated here:

Texture: Record the *grain size* or size of fossils making up the rock. Also choose one of the following textures: (1) **clastic,** (2) **skeletal** (the whole rock is fossils without clastic material between the fossils), (3) **crystalline**

....

[6]*Cryptocrystalline*, with the same root as *cryptic*, means the grains are so small they cannot be seen even with a standard light microscope.

(a)

(b)

Figure 5.27 **Chert** and **flint** are names for fine-grained biochemical or chemical sedimentary rock made of micro- to cryptocrystalline quartz. Many of these rocks may have started out as silica ooze, sediment made of microscopic silica shells. (a) Hand sample of chert showing typical conchoidal fracture. *Chert* is a general name and is usually used for tan, beige, or gray varieties of microcrystalline silica rocks. *Inset:* Electron photomicrograph of a radiolarian shell such as is found in many cherts. Chert may also consist of diatom shells (Fig. 5.8k, middle). (b) Slices of chert of different colors. *Jasper* refers to a red variety of chert, *agate* is translucent and has bands, *flint* refers to black chert, and *carnelian* is a name given to yellow to orange varieties. *From left to right:* agate, carnelian, flint, carnelian, and bloodstone.

(interlocking crystals), or just (4) **fine grained** (the grains are too small to see with the naked eye; this includes cryptocrystalline).

Structure: Note any sedimentary structures or other features. These may include bedding (parallel, cross, graded), fossils, mud cracks, ripple marks, raindrop prints, and so forth. Describe any *fossils* present, identifying what animal or plant is preserved, if you can.

Color: Briefly describe the color, such as light pinkish tan. The color of a sedimentary rock may simply help to distinguish the rock from another similar rock, or it may have a greater significance. For example, reddish to yellowish colors indicate the presence of oxidized iron and may suggest a nonmarine origin for the rock. Black often results from abundant organic matter, and greenish color hints at a lack of oxygen in some lake or marine environments.

Composition: Indicate minerals or rock clasts in the rock or its chemical composition. For example, limestone is predominantly calcite, a chert conglomerate contains chert clasts, and coal is made of carbonaceous material. Composition is especially important for the classification of nonclastic sedimentary rocks. Determine composition by testing for the distinguishing properties of the following minerals commonly found in sedimentary rocks: quartz, calcite, dolomite, gypsum, halite, feldspar, and muscovite. For coarse clastic rocks, you should attempt to identify the clasts; one common clast is chert fragments,

because chert is so hard and chemically resistant that it tends to survive well in sedimentary settings.

Rock name: Identify the rock using Figure 5.28, and check your identification against the samples provided in the Sedimentary Rocks Display.

Origin: Briefly indicate how the rock formed and whether it has a clastic, biochemical, or chemical origin.

22. In each row in Table 5.8, circle one texture or structure, choose a rock from the previous exercise that has the texture or structure, and explain how the rock formed. If you found an outcrop of this rock on a mountaintop, what would the rock's presence suggest about the history of the Earth at this location? Try to embellish your answer with as many details as you can. What can you say about the source rock for the sediments in this sample? Maybe the minerals or clasts give you a clue. Or where might dissolved substances that formed the sediment originate? Then work your way through the steps that formed this rock. Don't forget to include what you can deduce from the texture or structure about the sedimentary environment.

23. Study the sedimentary rocks so you will be able to recognize them on a quiz or exam. Then take a practice quiz with another set of unknown samples. This time just jot down the name and the key features of the rock that help you to recognize it. Fill in Table 5.9 with this information.

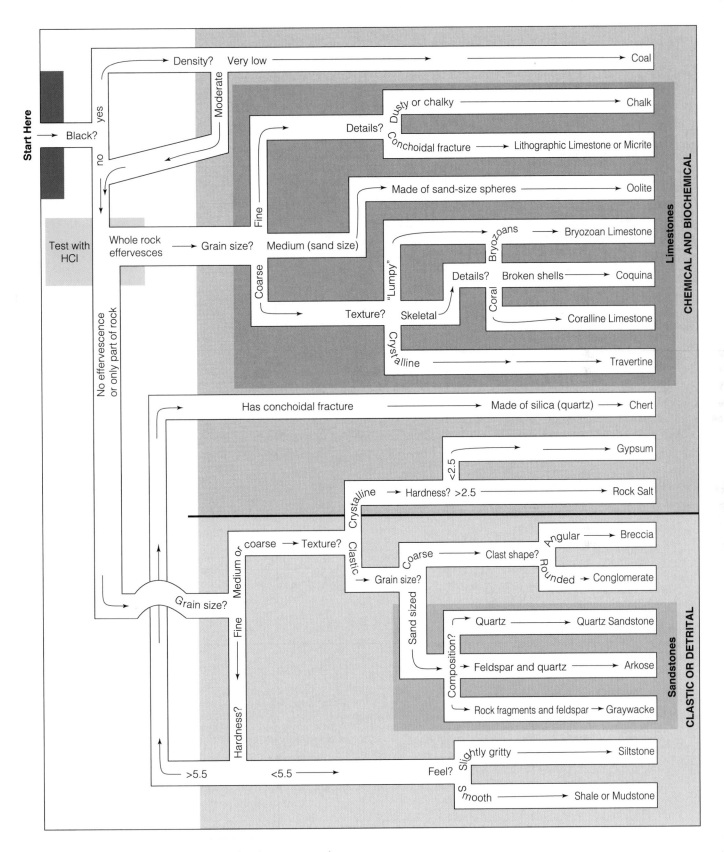

Figure 5.28   Maze for identification of sedimentary rocks

Table 5.7 Sedimentary Rock Identification Form

Sample Number	Texture and Grain Size	Structure (bedding, fossils, etc.)	Color	Composition (minerals or chemical makeup)	Rock Name	ORIGIN	
						How Did the Rock Form?	Clastic/Biochemical/ Chemical

Continued

Table 5.7 *Continued*

Sample Number	Texture and Grain Size	Structure (bedding, fossils, etc.)	Color	Composition (minerals or chemical makeup)	Rock Name	ORIGIN	
						How Did the Rock Form?	Clastic/Biochemical/ Chemical

Table 5.8   The Origin of Sedimentary Textures and Structures

Texture or Structure	Rock Number	Rock Name	Formation and History of the Rock
Clastic or crystalline			
Skeletal or fossiliferous			
Bedding, cross bedding or graded bedding			

Table 5.9   Practice Quiz

Sample Number	Key Characteristic 1	Key Characteristic 2	Key Characteristic 3	Rock Name

Metamorphic Rocks

OBJECTIVES

1. To understand how metamorphic rocks form
2. To recognize and understand the origin of metamorphic textures
3. To understand what the minerals in a metamorphic rock indicate about the origin of the rock
4. To understand the concepts of metamorphic grade, zones, and facies
5. To recognize and be able to identify the major metamorphic rock types

Rocks in the crust, either igneous or sedimentary, may exist unchanged for vast amounts of time. Eventually however, the continuing motion of the lithospheric plates will cause many of them to be dragged into a subduction zone and/or involve them in a continental collision. Such tectonic activity subjects the rocks to substantial increases in temperature and pressure. At the new temperature and pressure, the original minerals in the rocks are usually unstable, break down, and recrystallize, without melting, into a new set of minerals. This process is known as **metamorphism** and creates metamorphic rocks. Both the minerals found in a rock and the rock's texture change as a result of metamorphism. As the new minerals in a metamorphic rock are frequently only stable at the new temperatures and pressures, they can be used as indicators of the temperature and pressure (depth) of metamorphism. These minerals and the new textures may reveal where and how the rock was metamorphosed.

Because metamorphism takes place at high pressure underground, we never see it actually happening, which makes metamorphic rocks the most mysterious of the three classes of rocks. The energy from Earth's interior that drives plate tectonics (see Lab 9) also causes metamorphism. As a result, the stories told by metamorphic rocks are especially useful to Earth scientists who study plate tectonics and rock deformation (a.k.a. *structural geologists*).

The temperature of metamorphism can range from about 200°C to about 900°C. The upper limit of metamorphism is melting, which produces igneous rocks. The melting temperature depends on pressure and chemical composition, so some rocks can remain solid to 900°C and even higher. Temperatures in the Earth increase about 25°C for every kilo-

meter of depth, and it is this increase in temperature with depth that is one of the major heat sources for metamorphism. The pressure experienced by a rock at depth, called the **confining pressure,** is caused by the weight of the overlying mass of rock. Confining pressure is directly related to depth of metamorphism and is similar to the pressure experienced by divers in deep water. The pressure is measured in kilobars (kb), where a kilobar is about 1000 times atmospheric pressure. Three kilobars corresponds to a depth of 10 km. The range of pressures is usually 1–12 kb (or 3.3–40 km deep). Metamorphism does take place at greater depths than 40 km, but the resulting metamorphic rocks rarely make it to the Earth's surface.

Confining pressure from the weight of rock above results in a type of pressure that squeezes the rocks from all directions (Fig. 6.1). If confining pressure is applied without much increase in temperature, the volume of the rock decreases, resulting in an increase in density with the growth of more compact, higher-density minerals. Movement of lithospheric plates (Lab 9) can produce another kind of pressure—**directed pressure** (also called **differential stress;** see Fig. 6.1), in which the forces on the rock are not equal in every direction. As directed pressure increases, the rock flattens, resulting in the formation of a texture known as **foliation.** Thus, the texture of a metamorphic rock depends mainly on the

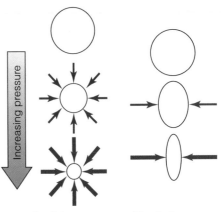

Confining pressure   Directed pressure

Figure 6.1 The difference between confining pressure and directed pressure

presence or absence of directed pressure, and mineral content of the rock depends on the original rock composition, temperature, and confining pressure.

Textures of Metamorphic Rocks

As we have seen for igneous and sedimentary rocks the word *texture* refers to shapes, sizes, orientation, and relationships between grains within the rock and is closely related to the rock's history. For all rocks, including metamorphic rocks, classification depends in part on the rock's texture. In fact, texture is the chief property used to identify rocks that have undergone directed pressure. Confining pressure without directed pressure results in different metamorphic textures. The following sections describe some textures found in metamorphic rocks.

Foliation Textures

Directed pressure during metamorphism causes the arrangement of platy minerals such as micas to orient parallel to planes in the rock (Fig. 6.2). This mineral alignment or preferred orientation is known as **foliation** (Fig. 6.3a). Figure 6.4 shows rocks with foliation. Another type of preferred orientation, very similar to foliation, is **lineation.** A **lineated** rock has elongate minerals such as hornblende arranged parallel to lines within the rock (see amphibolite).[1]

Foliated rocks:

Slate is fine grained. It breaks along fairly smooth planes of foliation called **slaty cleavage** (Fig. 6.4a).

Phyllite is made of fine to very small medium grains. The slaty cleavage of phyllite has a sheen and may have a slightly rippled surface (Fig. 6.4b).

Schist is medium to coarse grained. The grains are large enough to see. Mica grains are oriented parallel to each other, forming a foliation. The foliation surface looks sparkly like glitter and is known as **schistosity** (Fig. 6.4c–e; see also amphibolite).

Gneiss is medium to coarse grained, *banded* rock. Gneisses have alternating segregations or bands of granular minerals and flaky or elongate minerals, often making light and dark streaks, where many of the segregations are typically wider than about 2 or 3 mm. The segregations or layers that make up the foliation are known as **gneissosity** (Fig. 6.4f and g) and are the defining characteristic of gneiss.

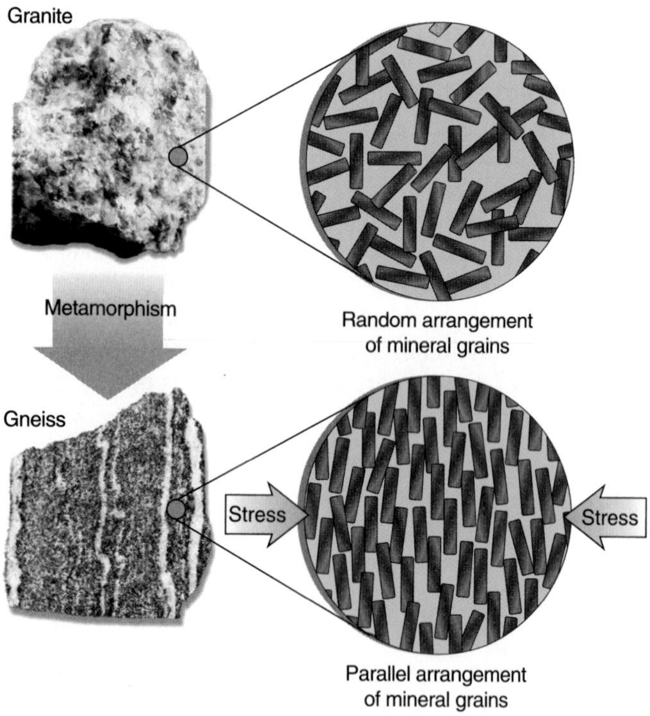

Experimental Formation of a Foliation Take a half-fist-size blob of Play-doh® and push a number of pennies, buttons, or washers (which will substitute for platy materials) into it at a random orientation. (See Figure 6.5 on p. 103). Next, squeeze the Play-doh against the tabletop with your hand. This is a simulation of deformation during metamorphism of a rock.

1. What type of pressure did you apply to the Play-doh?

   (Refer to Fig. 6.1.) _____

2. Sketch and describe the orientation of the pennies/buttons/washers in the Play-doh. What is the name of the texture you created?

Figure 6.2 Igneous rocks (top) develop randomly oriented mineral grains. This also applies to metamorphic rocks subjected to confining pressure but not directed pressure. Metamorphic rocks subjected to directed pressure (bottom) develop foliation, represented by platy and elongate minerals arranged parallel to each other, at right angles to the direction of pressure (stress).

....

[1]A foliation is similar in geometry to a bunch a playing cards scattered on top of each other on a table or a bunch of pencils similarly scattered on a table every which way. A lineation would be more similar to a bunch of pencils tied up in a bundle. Both foliation and lineation would be present if the pencils in the bundle were untied and let to roll out onto the table.

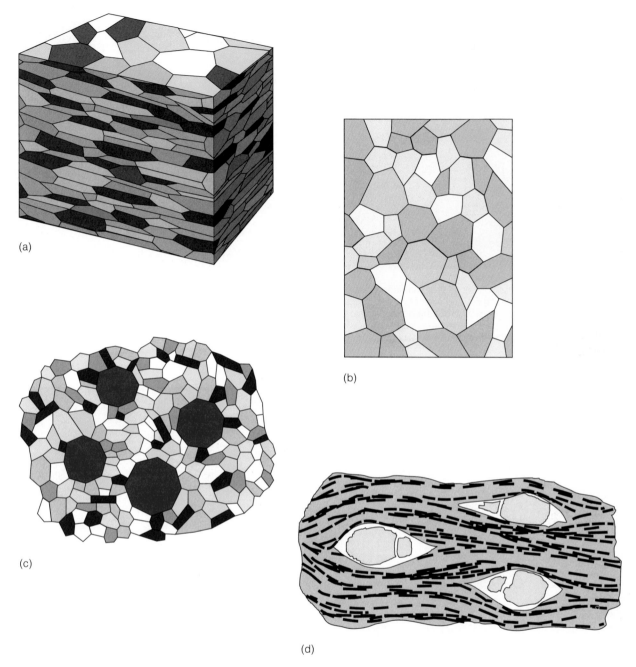

(a)

(b)

(c)

(d)

Figure 6.3 Metamorphic textures. (a) Example of *foliation*—interlocking mineral grains showing parallelism to planes in the rock. (b) Example of *granoblastic* texture—interlocking grains arranged at random in the rock. (c) Example of *porphyroblastic* texture—large geometric crystals surrounded by smaller crystals. (d) Another example of *porphyroblastic* texture—in a foliated rock.

3. Compare your pennies/buttons/washers in Play-doh with marble and schist in the Metamorphic Rocks Display. Which of these two rocks exhibits some similarity to your simulated Play-doh rock? _____

Why? _____

_____

_____

Granoblastic Texture

Not all metamorphic rocks have minerals with preferred orientation. If the mineral grains are visible and are randomly oriented in the rock, the texture is called **granoblastic** (or *nonfoliated;* see Fig. 6.3b). Granoblastic texture is visible in the marble and quartzite in Figures 6.6a, b, and c. Granoblastic texture occurs in rocks that have formed under conditions of confining pressure not directed pressure. It may also form if directed pressure is weak and if the rock contains no minerals that are easily arranged into a foliation. Blocky minerals such

(a) **Slate** is a low-grade metamorphic rock. Slate's fine grain size prevents identifying minerals in hand sample and even makes identification difficult under a microscope. Foliation allows slate to break into flat slabs along the foliation planes. Slate has a large number of possible colors. Some of the common colors—gray, red-brown, or gray-green—are shown here.

(b) **Phyllite** is a low-grade metamorphic rock that has a sheen along foliation planes from muscovite grains that are almost large enough to see with the unaided eye. Chlorite imparts a slightly green cast to this particular phyllite.

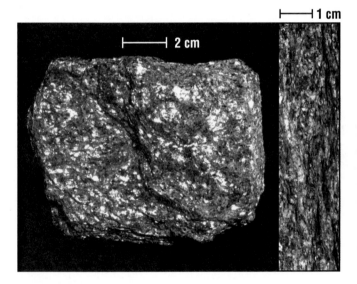

(c) **Mica schist** is a foliated metamorphic rock with two types of mica: biotite (black) and muscovite (silvery/pearly). In this view, looking at the foliation surface, the bright white spots are where light is reflecting from either muscovite or biotite. *Inset:* an edge view looking along the foliation where, in addition to the micas, quartz and feldspar are visible. The number of reflections of light from the surface are much reduced because the view is along the edge of the mica cleavages.

(d) **Garnet mica schist** is a foliated metamorphic rock with garnet and two types of mica. Commonly, as in this sample, garnets are red-black equant porphyroblasts. Muscovite is much more abundant than biotite in this sample. The inset is a view parallel to the foliation and shows the garnets more clearly.

Figure 6.4 Foliated metamorphic rocks

as calcite, quartz, and feldspar are not easily foliated. Metamorphic rocks that are commonly granoblastic include *quartzite, marble,* and *eclogite.*

Porphyroblastic Texture

This type of texture (Fig. 6.3c and d) occurs in metamorphic rocks that contain large crystals (**porphyroblasts**) surrounded by smaller ones, and it is similar in appearance to porphyritic texture in igneous rocks. Porphyroblastic texture can occur in either a foliated or a granoblastic rock. Porphyroblastic texture generally forms in rocks when the temperature and pressure are correct for the growth of a particular mineral that is not plentiful and tends to grow rapidly. Especially good porphyroblast formers are minerals that have trouble starting new crystals (nucleating). If conditions are right for this mineral to grow rapidly, rather than having many smaller crystals, the few crystals that form will grow large and become porphyroblasts.

(e) **Blueschist** is a foliated metamorphic rock with blue amphiboles. In this sample, blue amphibole and muscovite are clearly visible, both of which form the foliation. Blueschist signifies high-pressure metamorphism (blueschist facies; Fig. 6.11).

(f) **Gneiss** is a foliated metamorphic rock with bands or streaks of commonly light and dark minerals. In this sample, the black minerals are biotite and hornblende, and the white and light gray minerals are plagioclase feldspar and quartz.

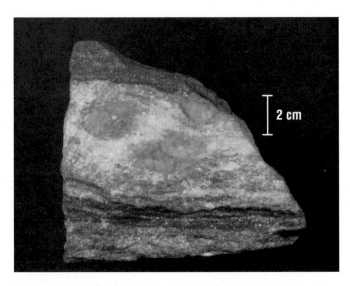

(g) **Augen gneiss** is a foliated metamorphic rock with bands or streaks of light and dark minerals and large lens-shaped alkali feldspar porphyroblasts. The buff-colored feldspar augen, in this sample, are embedded in a matrix of quartz, plagioclase, and more alkali feldspar. The dark layers here are biotite and hornblende.

Figure 6.4 Foliated metamorphic rocks *Continued*

(h) **Stretched-pebble conglomerate** is a foliated metamorphic rock in which pebbles contained within a former conglomerate have been stretched out of shape during metamorphic deformation. In the field photograph, notice how elongate the pebbles are. In hand sample (*inset*) the pebbles are visibly flattened due to deformation, and the foliation is wrapping partly around them. The pebbles may once have been chert but are now quartzite. Field photo by Grenville Draper.

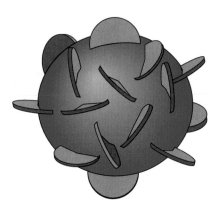

Figure 6.5 Play-doh® and pennies, buttons, or washers

(a) **Marble** is a metamorphic rock composed of calcite (or sometimes dolomite). It may be unfoliated (a) or foliated (b). *Inset:* a close-up of the crystalline, interlocking *granoblastic* texture of the calcite in sample. (a).

(b) Foliated *marble*. Green amphibole and black biotite produce the foliation in this sample. Calcite here is pink, white, and gray.

(c) **Quartzite** is a metamorphic rock composed primarily of quartz. Notice the grains have vitreous luster and conchoidal fracture. Quartzite may have a variety of different colors, as seen in the inset.

Figure 6.6   Metamorphic rocks defined by their mineral content

(d) **Amphibolite** is a metamorphic rock made up of plagioclase and hornblende. This sample is both foliated and lineated. The larger photograph is a view of the foliation surface where the lineation is visible as hornblende crystals are arranged approximately parallel to each other, up and down. (The hornblende is black, with grains reflecting the light looking bright white.) *Inset:* the edge of the sample where the layering of the foliation is visible.

Examples of porphyroblastic rocks:

Augen gneiss is a type of gneiss that has large feldspar crystals. In some cases, the feldspar porphyroblasts are tapered at the edges, which makes the rock look like it has eyes (*Augen* in German; see Fig. 6.4g).

Garnet mica schist is schist (see schist on p. 100) in which the garnets are commonly porphyroblasts, which cause the foliation to be uneven where the micas bend around them (Fig. 6.4d).

Eclogite may have prominent garnet porphyroblasts (see the discussion of eclogite in composition of metamorphic rocks later and Fig. 6.6e).

Spotted hornfels is a hornfels (see the later description of hornfels in contact metamorphism) with porphyroblasts. You will see larger grains in a fine-grained and commonly black matrix (see Fig. 6.7b).

(e) **Eclogite** is a metamorphic rock with green pyroxene and red garnet. Here the garnets are porphyroblasts, and a few grains of muscovite are visible. Eclogite represents high-pressure metamorphism in eclogite facies (Fig. 6.11).

(f) **Serpentinite** is a fine- to medium-grained metamorphic rock composed of serpentine minerals. These minerals are soft (H = 3–5) and green to black. This sample is serpentinite with *slickensides,* front surface, showing a region of highly foliated serpentine that formed along a fault zone or a zone of high shear stress. The scaly or mottled coloration common to serpentinite is visible along the upper-left edge of the sample.

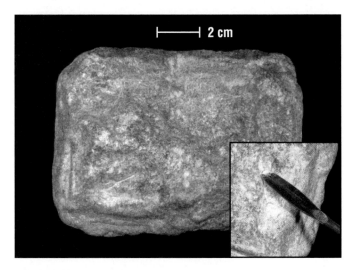

(g) **Soapstone** is a metamorphosed ultramafic rock composed of talc, for which the metamorphism was slightly hotter than for serpentinite. Recall that talc has a hardness of 1; thus, the sample can easily be carved with the spatula (*inset*) and even scratched with a fingernail.

Figure 6.6 Metamorphic rocks defined by their mineral content *Continued*

Composition of Metamorphic Rocks

The **parent rock** or **protolith** (rock from which the metamorphic rock forms) significantly influences the composition and mineral makeup of the final metamorphic rock. Sometimes the metamorphic rock is named for its parent rock by putting the prefix *meta-* in front of the rock name. You would use this naming system to emphasize the parent rock rather than other aspects of the metamorphic rock.

Rocks named for their protolith:

> **Metabasalt** is metamorphosed basalt (e.g., Figs. 6.4e, 6.6d and e).

> **Stretched-pebble conglomerate** (or **metaconglomerate**) is metamorphosed conglomerate in which the pebbles or clasts are still visible but stretched out into a lineation or foliation (Fig. 6.4h).

> **Metapelite** (or **metashale**) is metamorphosed shale (e.g., Figs. 6.4a–d and 6.7a and b).

Certain minerals in the original parent rock and in the resulting metamorphic rock reveal the chemical composition of the rock. Table 6.1 lists some compositions and minerals that reflect those compositions. Table 6.2 compares minerals developed in metamorphosed basalts with those developed at the same temperature in metamorphosed shales. Aluminum-rich minerals develop in metashales (metapelites) and belong to the different temperature Barrovian zones labeled in the table. Mafic minerals, such as epidote, actinolite, and hornblende, form in metabasalts at the same series of temperatures and pressures (bottom row in Table 6.2). If a quartz-rich rock such as quartz sandstone or chert is metamorphosed, it becomes quartzite, a rock containing 95%–100% quartz. Marble (a rock containing 95–100% calcite) forms by metamorphism of limestone over a large range of temperatures. A number of metamorphic rocks are defined based on the minerals that comprise them.

Rocks named for their mineral content:

> **Marble** is a medium- to coarse-grained rock made of calcite (or dolomite, in which case the metamorphic rock

(a)

(b)

Figure 6.7 **Hornfels** is a fine-grained or porphyroblastic unfoliated metamorphic rock formed by contact metamorphism. (a) Biotite hornfels, as in this sample, may be finegrained and black and thus is difficult to distinguish from basalt. (b) Porphyroblastic hornfels is commonly called *spotted hornfels*. This sample has cordierite porphyroblasts that are paler blotches on the otherwise black rock. The porphyroblasts are not distinct because they are full of inclusions of other minerals due to their rapid growth caused by the heat of the intrusion.

name is *dolomite* marble). The grains are large enough to see. Marble may have a foliation but generally does not (Fig. 6.6a and b). Marble forms by metamorphism of limestone or dolostone, but it is distinguished from these by more prominent crystalline grains and a lack of sedimentary textures.

Quartzite is a fine- to coarse-grained rock made of crystalline quartz (Fig. 6.6c). Most quartzites are medium grained. Some quartzites may be foliated. The parent rock of quartzite is a rock rich in quartz, such as quartz sandstone or chert.

Table 6.1 Parent Rocks and Composition

Composition	Minerals	Parent Rocks	Metamorphic Rocks
Carbonate	Calcite and dolomite		
High aluminum	Clay, chlorite, muscovite, biotite, garnet, staurolite, kyanite, and sillimanite		
Silica-rich	Quartz		
Felsic	Feldspar and quartz		
Mafic	Hornblende, pyroxene garnet, chlorite, epidite, and actinolite		
Ultramafic	Olivine, serpentine, and talc		

Table 6.2 Mineral Assemblages in the Barrovian Zones and Comparable Mafic Assemblages

Shown here are the temperature and mineral assemblages in metashales for each Barrovian zone. The first appearance of a mineral with increasing temperature is marked in bold. For the equivalent temperature range, the assemblages for mafic rocks are also listed.

Metamorphic Zone	Chlorite Zone	Biotite Zone	Garnet Zone	Staurolite Zone	Kyanite Zone	Sillimanite Zone
Temperature Range	300°–425°C	425°–500°C	500°–530°C	530°–550°C	550°–700°C	≥700°C
Mineral assemblage in metashale (metapelite)	**Chlorite** Muscovite Albite* Quartz	Chlorite Muscovite **Biotite** Albite* Quartz	Chlorite Muscovite Biotite **Garnet** Albite* Quartz	Muscovite Biotite Garnet **Staurolite** Plagioclase Quartz	Muscovite Biotite Garnet Staurolite **Kyanite** Plagioclase Quartz	Biotite Garnet **Sillimanite** Plagioclase Quartz
Mineral assemblage in metabasalt (mafic) at equivalent temperatures	**Chlorite** Actinolite† Albite*	Chlorite **Epidote** Actinolite† Albite*	Chlorite Epidote Actinolite† Albite*	Chlorite Epidote **Garnet Hornblende Plagioclase**	Epidote Garnet Hornblende Pyroxene Plagioclase	Garnet **Pyroxene** Hornblende Plagioclase

*Na-plagioclase feldspar.
†A green amphibole.

Amphibolite is a medium- to coarse-grained rock made of hornblende and plagioclase feldspar. Amphibolite may have a foliation or, more commonly, a lineation, so is sometimes called *hornblende schist.* However the name amphibolite is more prevalent even for foliated varieties (Fig. 6.6d).

Eclogite is medium- to coarse-grained rock containing a green pyroxene called *omphacite* and red garnets (Fig. 6.6e). Muscovite may also be visible. Eclogite is mafic rock, such as gabbro or basalt, metamorphosed at high pressure.

Serpentinite is a rock made up of serpentine minerals. Serpentinite is fine grained and green and has a greasy luster and a hardness of about 4. It can be unfoliated and very dark green but commonly has smooth slick surfaces that are a form of foliation called *slickensides* with a lighter or mottled green color (Fig. 6.6f). Serpentinite forms from metamorphism of ultramafic igneous plutonic rocks.

Soapstone is a very soft fine-grained rock made of talc. The color of soapstone can vary quite a bit within one sample (Fig. 6.6g). Soapstone, with its high magnesium content from talc, results from the metamorphism of peridotite, at a slightly higher temperature than serpentinite.

4. In Table 6.1, list igneous or sedimentary parent rocks that have the compositions indicated in each row. Then fill in the metamorphic rocks that have similar compositions and therefore could form from these parent rocks. Place one rock name in each blank box.

Types of Metamorphism

Regional Metamorphism

Mountain building and plate tectonic processes acting over a large region deep within the Earth's crust produce **regional metamorphism.** Regional metamorphism commonly results at convergent plate boundaries (Lab 9). Directed pressure combined with a wide range of temperatures and confining pressure at moderate to great depths produces this type of metamorphism. Recall that the deeper in the Earth, the hotter the temperature and the higher the confining pressure. Rocks that have undergone regional metamorphism are commonly foliated and are slates, phyllites, schists, or gneisses. However, if a quartz sandstone or a limestone undergoes regional metamorphism, they may not develop an obvious foliation because of the blocky nature of their minerals.

Contact Metamorphism

Heat from an intrusion produces **contact metamorphism,** also known as *thermal metamorphism* (Fig. 6.8). The metamorphic effect, or **aureole,** of the intrusion may be a few centimeters to several kilometers thick, depending on the size, composition, and depth of the intrusion. The greater the size of the intrusion and the higher the temperature of the intrusion and/or its surroundings, the thicker the contact aureole or metamorphosed area will be. Contact metamorphic rocks are often fine grained and are not usually foliated. Most contact metamorphic rocks are hornfelses.

Hornfels is fine grained or porphyroblastic and a very tough rock because its minerals tend to be intricately

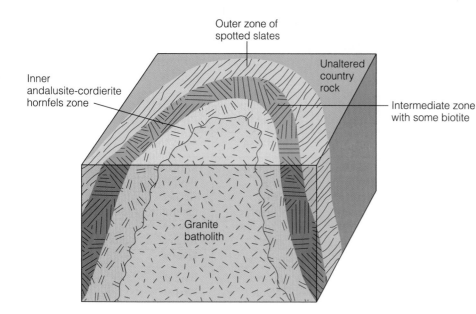

Outer zone of spotted slates

Unaltered country rock

Inner andalusite-cordierite hornfels zone

Intermediate zone with some biotite

Granite batholith

Figure 6.8 Contact metamorphism. The heat from an intrusion metamorphoses the surrounding rocks, known as a *contact aureole*. Zones of hornfelses showing increasing temperature toward the intrusion contain distinct mineral assemblages. The assemblages shown here are for shale country rock.

interlocking. Because the rock is fine grained, this feature is only visible under a microscope. Contact (or thermal) metamorphism forms hornfels, so the presence of an intrusive igneous rock nearby is one way to confirm its identity. Hornfelses are a bit difficult to recognize in a lab because they are rather nondescript—fine grained and unfoliated (Fig. 6.7a). Their color may vary but they are commonly black and may be mistaken for basalt. Basalt is generally denser and has more magnetite (which makes basalt slightly magnetic) than hornfels. In some cases hornfels has porphyroblasts, in which case the term *spotted hornfels* applies (Fig. 6.7b).

5. With your lab partners, examine samples provided by your instructor. Without knowing anything more about the metamorphism of these rocks than what you can see, what type of metamorphism may have produced each rock? Explain your reasoning.

_____

_____

_____

_____

_____

_____

 Protoliths and Types of Metamorphism Computer Exercise

■ Start **In-TERRA-Active.**
■ Click **Materials;** then click **Metamorphic Rocks;** then click **Apply.**

6. Click **Next** until you see a "known situation" in the upper right corner that matches a box in the "known" row in Table 6.3. Follow the instructions until you have the correct answer. Then fill the appropriate column in Table 6.3 with the correct combination of process, temperature, pressure, and finishing rock. Continue to click **Next** until you get to each of the known situations, so you can fill in all the remaining columns in the table.

■

Temperature and Pressure

As the temperature and pressure of metamorphism increase, the **metamorphic grade** of the rock also increases. You can think of metamorphic grade as an approximate measure of the amount or degree of metamorphism. The three metamorphic grades are closely tied to the temperature of metamorphism: **low grade** is mainly low temperature and low pressure, **medium grade** is moderate temperatures and pressures, and **high grade** is for rocks that have undergone metamorphism at high temperatures and pressures. Pressure is of less importance than temperature, so a rock metamorphosed at high temperature and moderate pressure, for example, would still be high grade. One of the most important concepts to understand about metamorphic rocks is that different minerals form at different temperatures and pressures. As temperature and pressure change during metamorphism, new minerals grow within the metamorphic rock, replacing preexisting minerals. This means that the minerals can be clues about the metamorphic history of the rock.

Minerals Reveal Metamorphic Temperature

A group of minerals that grow or coexist together at the same temperature and pressure is a **mineral assemblage.**

Table 6.3 Parent Rocks and Types of Metamorphism in Computer Exercise

PARENT ROCK (STARTING ROCK)	SANDSTONE	GRANITE	LIMESTONE	SHALE	SHALE
Known:	This sandstone is undergoing folding at shallow depth.	This granite is undergoing folding at great depth.	This limestone is next to a body of magma.	This shale is next to a body of magma.	This shale is undergoing folding at moderate to great depth.
Process					
Temperature					
Pressure					
Metamorphic rock (finishing rock)					

We can therefore use mineral assemblages as our thermometer to measure the temperature of metamorphism. In fact, when metamorphic petrologists (people who study metamorphic rocks) use minerals to determine temperatures of metamorphism, they call them *geothermometers.* Mineral assemblages operate like the oral thermometer used to measure your temperature when you have the flu in that it is designed to record the maximum temperature that it experienced. Mineral assemblages also record the maximum temperature experienced.

Certain mineral assemblages are characteristic of specific ranges in temperature (see Fig. 6.9a and Table 6.2). A distinctive assemblage represents a **metamorphic zone.** Simply, a metamorphic zone refers to the group of minerals that would form at certain temperature range in a rock of a certain composition. The zones are named for an especially useful **index mineral,** which is stable at that particular temperature. Clay-rich shales are particularly sensitive in showing progressive mineral changes with increasing temperature during regional metamorphism, and they may display what are known as **Barrovian zones** (Fig. 7.23). Each Barrovian zone is delineated by the first appearance of a distinctive mineral such as chlorite, biotite, or garnet, but you should note that this index mineral generally persists into higher temperature zones. The boundaries between the different zones are called **isograds.** If the region is uplifted and eroded, these assemblages can be viewed and mapped at the surface of the Earth (see Fig. 6.9b and c). The rocks are now cold, but the minerals remain to show us the maximum metamorphic temperatures experienced.

Some minerals or mineral assemblages are stable over an extensive variety of temperatures and pressures. Quartzite may contain the same mineral (quartz) if metamorphosed at 200°C or 800°C. Calcite in marble is stable over almost as wide a range of temperatures. Pressure can also vary widely for the formation of these rocks.

7. Examine the metashales in the Barrovian Zones Display. For each rock, determine what minerals are present and match the rock with one of the Barrovian zones in Table 6.2 (see also Fig. 6.9). Enter the rock number and the range of temperature of metamorphism of the appropriate zone in Table 6.4.

Hint: In some cases the rock name tells you what important mineral is in the rock.

Note: Check with your instructor to find out if you need to learn to recognize any new minerals such as staurolite.

8. Do you think calcite or quartz would make good index minerals? _____ Would they make good indicators of pressure (geobarometers)? _____ Why or why not? _____

_____

_____

_____

_____

Grain Size and Metamorphic Temperature (Grade)

9. Among the metashales, is there a relationship between temperature and grain size? _____ If so describe it. _____

_____

_____

Cooling rate would not be relevant to a metamorphic situation as it was for grain size of igneous rocks. Instead, what other factor would be more pertinent to metamorphic grain size than cooling rate?

Hint: It also has to do with time.

_____

_____

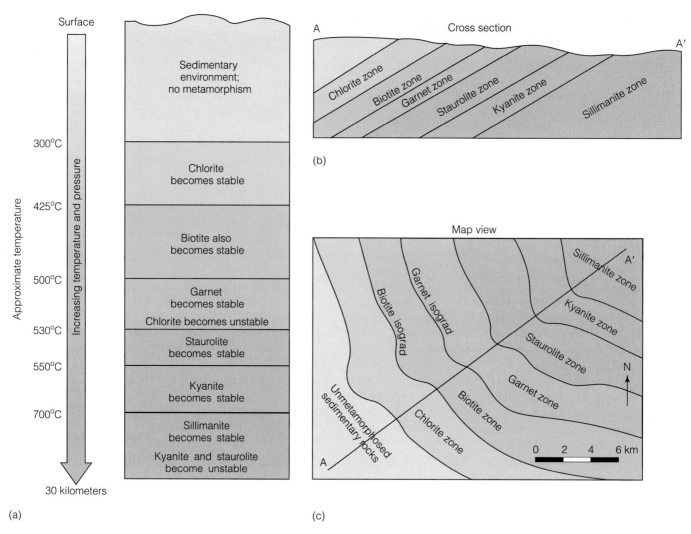

Figure 6.9 Diagrams illustrating the variation in metamorphic zones with depth of burial. (a) Geologic column of a deeply buried succession of shale, with temperature and pressure increasing systematically with depth. Metamorphism is taking place in these rocks, producing different mineral assemblages represented by the metamorphic zones. (b) Cross section of an area with more rapid uplift on the right than the left side; erosion on the right exposed deep crustal levels. Minerals that characterize various zones occur at the surface in a progressive sequence from low to high temperature, from left to right. (c) Map view of the area shown in b, with low-grade, low-temperature metamorphism at the southwest and high-grade, high-temperature metamorphism at the northeast. Lines termed *isograds* connect the first appearance of metamorphic minerals. Moving along an isograd leads you along equal metamorphic temperatures, whereas crossing isograds moves you through changes in metamorphic temperatures.

Table 6.4 Temperature of Some Barrovian Metashales

Rock Name	Rock Number	Metamorphic Zone	Temperature of Metamorphism
Staurolite mica schist or staurolite quartzite			
Chlorite schist			
Kyanite mica schist or kyanite quartzite			
Mica schist			
Sillimanite gneiss			
Garnet mica schist			

As temperature increases, grain size of minerals in the rocks also tends to increase. This is because atoms move faster at higher temperatures and thus can travel farther to attach themselves to a mineral grain. The farther atoms can travel to arrive at one mineral grain, the more atoms can reach the grain and the larger the grain can grow. At low metamorphic temperatures (*low grade*), atoms move slowly and the rocks are generally fine grained (e.g., slate). As grade increases, the grain size increases to produce phyllite, then schist or marble with medium-size grains (*medium grade*). Finally, at the highest temperatures of metamorphism (*high grade*), atoms within the solid rock become sufficiently mobile to allow for segregation of different minerals into separate bands, such as those seen in gneiss, or to allow for growth of large crystals in a coarse-grained marble. At these temperatures, the rocks usually have the coarsest grain size. Other variables, such as the rock's fluid content, different rates of growth of different minerals, and how much time the metamorphic temperatures and pressures persist, also influence grain size, so there may be considerable variation in the size of grains in rocks of the same temperature.

10. Based on the grain size of the porphyroblasts in the metashale samples in Table 6.4, which minerals seem to grow at the fastest rate?

_____

_____

Pressure and Density

Certain metamorphic rocks are formed under special high-pressure conditions deep (>20 km) in the Earth's crust and have different minerals than the more commonly seen kind of regionally metamorphosed rocks. Remember the concept of mineral assemblages as thermometers; we find that they can also be barometers to measure pressure of metamorphism. Blue amphibole, garnet, and omphacite (a green pyroxene) are high-pressure minerals.

Blueschist is a high-pressure/low-temperature metamorphic rock characterized by the presence of the blue amphibole glaucophane. The blue amphibole gives the rock a bluish-black appearance (Fig. 6.4e). Other unusual minerals such as **jadeite** (the main constituent of the gem jade) are sometimes found in these rocks. These rocks are formed near and within subduction zones[2] at a depth of 20–40 km and between the oceanic trench and the volcanic arc. In the subduction zone, the natural temperatures in the Earth are cooled by subduction of cold oceanic crust; thus, the subduction zone setting creates the especially high-pressure/low-temperature con-

....
[2]A *subduction zone* is an area at some convergent plate boundaries (Lab 9) where oceanic crust moves downward into the mantle. Oceanic trenches occur where the crust first starts to descend, and volcanic arcs occur where the subducting plate generates magma that rises to form the volcanoes at the surface.

ditions that form blueschists (see the graph showing blueschist facies on p. 112).

Eclogite is a rock of basaltic composition that has recrystallized to garnet and a green pyroxene called **omphacite,** which unlike other pyroxenes is rich in sodium and aluminum. With two high-pressure minerals, this red and green rock (Fig. 6.6e) is formed by metamorphism at high pressure. The temperature of metamorphism of eclogite may vary considerably, as shown in the graph on page 112.

11. Remember that density is mass divided by volume. How would you expect the density of a rock to change during

high-pressure metamorphism? _____

Now you will test this hypothesis by determining the density of basalt and eclogite.

12. Measure the mass of basalt and eclogite, and write it down. Be sure the units are grams.

Basalt: sample number _____   mass = _____ g

Eclogite: sample number _____   mass = _____ g

Since density is mass/volume, you now need to measure the volume of the rock. Do this by displacing water in a graduated beaker or measuring cup.

13. Measure the volume of basalt and eclogite as follows: Carefully place the rock being measured in a 1000-ml beaker or metric measuring cup. Fill the beaker/cup with water to a line above the rock, being as accurate as possible. Read the water level at the bottom of the meniscus (Fig. 6.10). Record this level as first level in the space provided. Now remove the rock without spilling any of the water and allow the rock to drip into the beaker for a few seconds. Read the new level as accurately as possible by interpolating between the lines or using a graduated cylinder to measure the change in level. Record the new level as second level. Next, find the volume of the rock by subtracting the second level from the first, and record the volume. Note that milliliters are

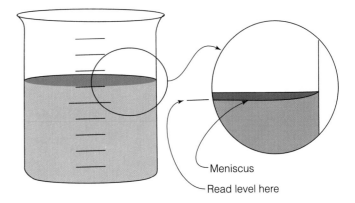

Figure 6.10   Beaker of water showing the meniscus. Read the water level at the bottom of the meniscus.

equivalent to cubic centimeters. Determine the density by dividing the mass from Exercise 12 by the volume.

Basalt: 1st level _____ ml

minus 2nd level _____ ml

volume = _____ ml,

density = _____ g/cm^3

Eclogite: 1st level _____ ml

minus 2nd level _____ ml

volume = _____ ml,

density = _____ g/cm^3

14. Does the result in Exercise 13 match your hypothesis in Exercise 11? _____

Metamorphic Facies

Metamorphic zones are based on mineral assemblages. Since different original rocks may produce different mineral assemblages, it is difficult to compare metamorphic zones of, for instance, metashales and metabasalts. Unlike metamorphic zones, metamorphic facies are not restricted to the chemical composition of one rock type. A **metamorphic facies** is the set of all mineral assemblages that may be found together in a region where the rocks have different chemical composition but were all metamorphosed at the same conditions of temperature and pressure. Whether a rock is a metashale or a metabasalt (or for that matter a metalimestone), if it formed over a particular range of pressure and

temperature, it belongs to the same facies as any other rock formed at the same conditions. The estimated temperature and pressure of formation of different facies are shown in Figure 6.11 along with the conditions for the Barrovian zones of metashales.

15. By reading the graph in Figure 6.11, determine the metamorphic facies of each rock listed here:

Blueschist _____

Chlorite muscovite schist _____

Staurolite mica schist _____

Greenschist _____

Kyanite mica schist _____

Sillimanite gneiss _____

Amphibolite _____

Mica schist (biotite and muscovite) _____

Eclogite _____

Garnet mica schist _____

Classification of Metamorphic Rocks

The most commonly used metamorphic rock names are based on two different classification systems. One system is based on the texture or foliation of the rock and is not used for unfoliated rocks. The other system is based on the miner-

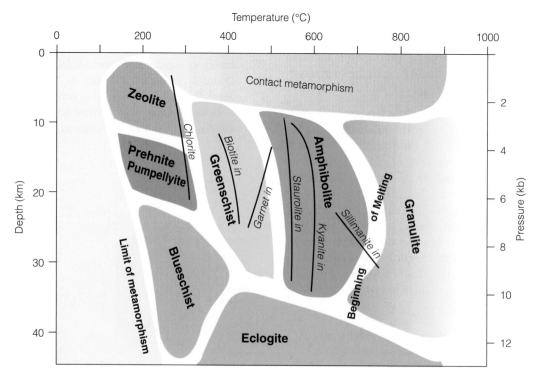

Figure 6.11 Approximate pressure-temperature fields of metamorphic facies. Note also the fields of the Barrovian zones, marked by isograds such as *garnet in.*

Table 6.5 Foliated Metamorphic Rocks (including lineated rocks)

Grain size	Color or Luster	Texture	Other Identifying Characteristics	Rock Name and Grade	Common Parent Rock
Very fine grained	Dull luster, gray, black, red, tan, or green	Good slaty cleavage	Generally flat	**Slate;** low	Shale
Fine grained	Silky sheen on flat surfaces	Cleavage may not be as well developed as in slate	Chlorite and muscovite barely discernible with a hand lens	**Phyllite;** low	Shale
Medium (to coarse) grained	Glossy if fresh, especially on flatter surfaces	Micas oriented parallel to foliation; tends to split parallel to foliation (may be wavy or folded)	Minerals visible: biotite, muscovite, *may* see quartz, garnet, or hornblende	**Schist** Mica schist Garnet mica schist; medium grade Hornblende schist; medium grade	Shale Shale Basalt
Medium grained	Blue or blue-gray	Foliated or lineated	Blue amphibole	**Blueschist;** high pressure	Basalt
Coarse (to medium) grained	Bands or segregations of black and white or pink	Layered, banded or streaked; does not break parallel to foliation	Light colored bands of feldspar and quartz, black bands of biotite or hornblende	**Gneiss** high grade	Granite or shale or basalt

als in the rock; since this system does not consider texture, these rocks can be foliated or not, although many of them are not. The tricky part for students is to decide when to use which system. Here are some general guidelines:

- If the rock is well foliated and made up of a variety of minerals, use the foliated-rock system of classification (see the section "Foliated Rocks" and Table 6.5).
- If the rock is well foliated but either you cannot see the minerals or only one or two types of minerals are present, check both systems to find the best match. For example, slate is foliated and fine grained and belongs to the foliated system (Table 6.5), but soapstone is commonly fine grained and foliated but is defined in the mineral-based system (Table 6.6) based on the presence of the soft mineral talc.
- If the rock is not foliated, use the mineral-based system (see "Composition of Metamorphic Rocks" and Table 6.6).
- If the rock is predominately made of one mineral, use the mineral-based system.
- If you are having trouble getting a rock to match a rock name in one system, check the other system.

Actually, this process is not as hard as it sounds if you use the maze in Figure 6.12 and Tables 6.5 and 6.6 to help you to classify metamorphic rocks. Some of the rocks in Table 6.6 may have a foliation or a lineation; in such cases, a reasonable rock name can be used from either table. A typical example is an amphibolite that is lineated or foliated. If it has black and white bands greater than about 3 mm wide, it could be called a hornblende gneiss or an amphibolite—either name is acceptable. If not banded, hornblende schist or amphibolite are permissible names.

16. Examine the samples in the Metamorphic Rocks Display, and note the features of metamorphic rocks as already

discussed. In particular, note the textural features of the rocks. Pick three samples and try classifying them without looking at their names, using the maze in Figure 6.12. Did you get the correct names? _____ If not, try a couple more.

17. Examine the unknown rocks and fill in the information in Table 6.7, Metamorphic Rock Identification Forms. Use Figure 6.12 and Tables 6.5 and 6.6 to determine the rock name. Use the same tables, Figures 6.9 and 6.11, and Table 6.2 to determine the metamorphic grade, temperatures, and/or pressures, if possible. For some rocks, it is not possible to determine the metamorphic grade using minerals. For example, a quartzite with no minerals besides quartz could have formed at almost any temperature and pressure. Its mineral assemblage (quartz) is not diagnostic of temperature or pressure. However, even for a quartzite, the grain size can be used to hypothesize the metamorphic grade as low, medium, or high.

18. In each row in Table 6.8, circle one texture term, choose a rock from the previous exercise that has that texture, and explain how the rock formed. If you found an outcrop of this rock on a mountaintop, what would the rock's presence suggest about the history of the Earth at this location? Don't forget that the beginning of the story is the history of formation of the parent rock.

19. Study the metamorphic rocks so you will be able to recognize them on a quiz. Then take a practice quiz with another set of unknown samples. This time just jot down the name and the key features of the rock that help you to recognize it. Fill in Table 6.9 with this information.

Table 6.6 Metamorphic Rocks Classified by Mineral Content

Hardness	Grain Size	Color	Minerals	Other Characteristics	Rock Name and Grade	Common Parent Rock
H > 5 1/2	Medium to coarse grained	Bright to dark green with red-brown spots	Omphacite (green pyroxene) and garnet (red)	High density for a rock (D = 3.3)	**Eclogite;** high pressure	Basalt
	Medium grained (fine to coarse)	Light shades, various colors possible—tan, pink, green	Quartz (>90%, commonly 100%)	Generally has one homogeneous color throughout	**Quartzite;** low to high grade	Quartz sandstone or chert
	Medium to coarse grained	Black and white	Hornblende and feldspar	Commonly lineated	**Amphibolite;** medium grade	Basalt
	Fine grained	Dark to medium gray to tan—variable	May have spots	Very tough	**Hornfels;** medium to high grade	Any
2 1/2 < H < 5 1/2	Medium to coarse grained	Light shades; white, white with gray streaks, or pink are common	Calcite (>90%)	Rarely fossils may be preserved (reacts with HCl)	**Marble;** low to medium high grade	Limestone
	Fine grained	Light to dark green to black, commonly streaked or mottled	Serpentine	Waxy, greasy luster; some varieties fibrous (asbestos)	**Serpentinite;** low grade	Peridotite
H < 2 1/2	Fine grained	Blue-green, gray	Talc	Waxy	**Soapstone;** medium grade	Peridotite

Note: Some of these rocks may have foliation. Remember: Hardness of glass = 5 1/2, fingernail = 2 1/2.

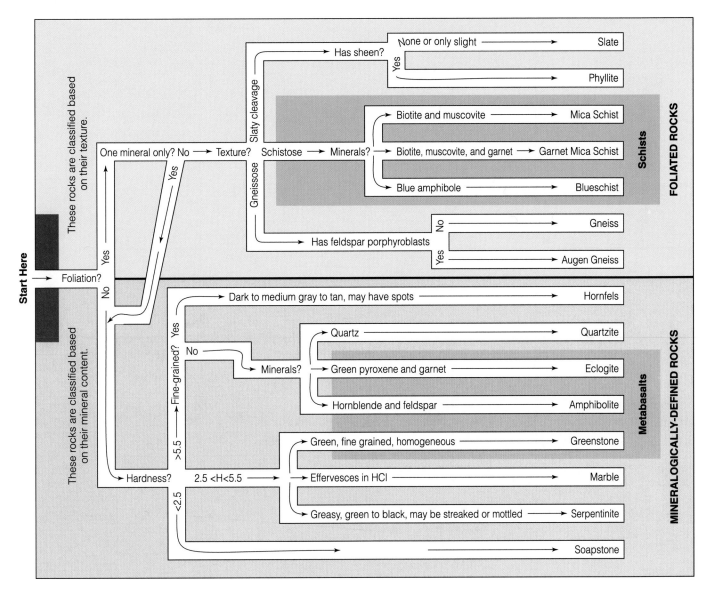

Figure 6.12 Metamorphic rock maze for use in identifying metamorphic rocks

Table 6.7 Metamorphic Rock Identification Form

Sample Number	Texture (including foliation and grain size)	Color(s)	Composition (minerals and their colors; porphyroblasts)	Rock Name	Parent Rock	Conditions of Metamorphism (metamorphic grade or best guess for pressure and temperature)

Continued

Table 6.7 *Continued*

Sample Number	Texture (including foliation and grain size)	Color(s)	Composition (minerals and their colors; porphyroblasts)	Rock Name	Parent Rock	Conditions of Metamorphism (metamorphic grade or best guess for pressure and temperature)

Table 6.8 The Origin of Metamorphic Textures

Texture	Rock Number	Rock Name	Formation and History of the Rock
Slaty cleavage or schistosity or gneissosity			
Nonfoliated or granoblastic			

Table 6.9 Practice Quiz

Sample Number	Key Characteristic 1	Key Characteristic 2	Key Characteristic 3	Rock Name

Rock Masses

OBJECTIVES

1. To learn some basics of geologic maps and cross sections
2. To learn the different types of bodies of igneous, sedimentary, and metamorphic rocks and their relationships to each other
3. To understand the laws of stratigraphy and cross-cutting relationships
4. To learn types of rock behavior and forces of deformation
5. To learn types of folds and faults
6. To understand how an unconformity forms

The three types of rocks you have learned in the previous labs combine to make up the crust of the Earth in various shapes and masses. Some are layers, some are massive bodies that cover large regions; others have irregular shapes. Sometimes various processes in the Earth, especially plate tectonics, earth movements, and metamorphism, change the shape or form of rocks after they have formed. The general term for these changes is **deformation.** In this lab we will look at the different types of rock masses, their relationships to each other, and their deformation.

Introduction to Geologic Maps and Cross Sections

A geologic map endeavors to explain the underlying geology of an area by showing the placement of rock *formations* and their *structures, contacts,* forms and shapes.

- A **formation** is a continuous or once continuous layer of rock—igneous, sedimentary, or metamorphic—that can be easily recognized in geologic fieldwork and is the basic unit shown on geologic maps.
- **Contacts** are the boundaries between rock units or formations, where one formation gives way to the next, depicted as a thin black line on most maps.

- **Structures** are the physical arrangements of rock masses and include intrusive bodies, unconformities, orientation of rock layers, and deformational features such as faults and folds. These structures will be discussed later in this lab and more in Labs 8 and 16.

On each geologic map, you will find an explanation of the symbols used for the structures, colors, and patterns on that particular map. The key or legend of the map is an explanation of rock formations organized by rock type and is in chronological order with the youngest rocks at the top and the oldest rocks at the bottom (Fig. 7.1a). Colors are usually used to differentiate different rock formations. On maps that are not printed in color, the cartographer uses different patterns to indicate the different rock units. Some common examples of patterns for different rock types are shown in Figure 7.2.

Cross sections, also known as *structure sections,* are included with many geologic maps to show the arrangement and history of the rocks, as in Figure 7.1b. These sections are like a slice of layer cake—they cut through the layers, exhibiting the arrangement and compositions of the interior from a side view much as the Colorado River has cut through the rock layers forming the Grand Canyon (Fig. 7.3). The line along which a cross section has been drawn is generally labeled on the accompanying map with letters such as A-A′ or B-B′ (Fig. 7.1a).

1. Examine the Geologic Map of Bright Angel Quadrangle, Grand Canyon, Arizona, and the Explanation in Figure 7.1a.
   a. What type of rock does the symbol Mr represent?

   _____

   b. What do you think the M in Mr stands for?

   _____

   c. How about the r in Mr ? _____
   d. In what way does color aid a person reading this map—that is, what do the colors represent?

   _____

   _____

Figure 7.1 (*see next page*) (a) Part of the Geologic Map of the Bright Angel Quadrangle, Grand Canyon, Arizona. (b) Cross sections for the lines A-A′, B-B′, and C-C′ on the map.

Geology by John H. Maxson

Topography by Francois E. Matthes, 1902-1903
United States Geological Survey
Culture as of 1962

From: Grand Canyon Natural History Association

KAIBAB
PLATEAU

112°5'W

36°10'

36°5'N

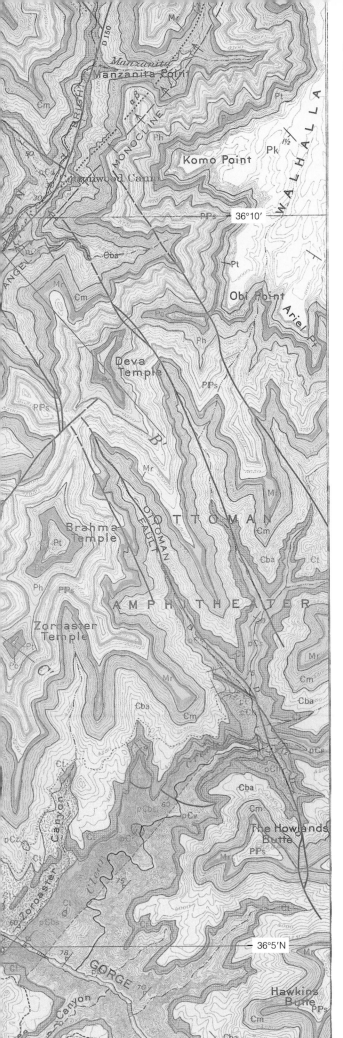

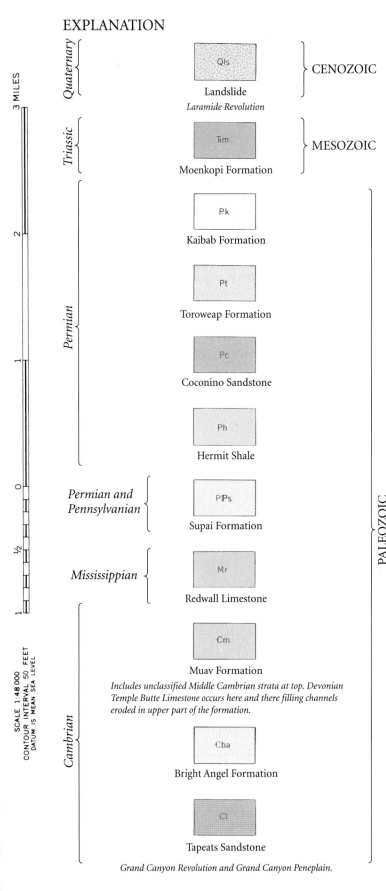

GEOLOGIC MAP OF THE BRIGHT ANGEL QUADRANGLE, ARIZ.

EXPLANATION

Quaternary — Qls — Landslide — CENOZOIC

Laramide Revolution

Triassic — Ťm — Moenkopi Formation — MESOZOIC

Permian
- Pk — Kaibab Formation
- Pt — Toroweap Formation
- Pc — Coconino Sandstone
- Ph — Hermit Shale

Permian and Pennsylvanian — PⅡPs — Supai Formation

Mississippian — Mr — Redwall Limestone

Cambrian
- Єm — Muav Formation

Includes unclassified Middle Cambrian strata at top. Devonian Temple Butte Limestone occurs here and there filling channels eroded in upper part of the formation.

- Єba — Bright Angel Formation
- Єt — Tapeats Sandstone

Grand Canyon Revolution and Grand Canyon Peneplain.

PALEOZOIC

SCALE 1:48 000
CONTOUR INTERVAL 50 FEET
DATUM IS MEAN SEA LEVEL

(a)

STRUCTURE SECTIONS

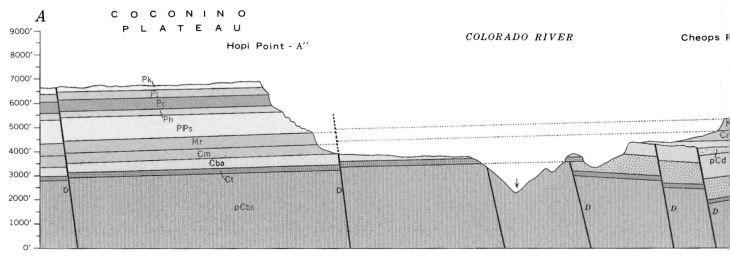

A

C O C O N I N O
P L A T E A U

Hopi Point - A''

COLORADO RIVER

Cheops F

Pk, Pt, Pc, Ph, PPs, Mr, Ꞓm, Ꞓba, Ꞓt, pꞒbs

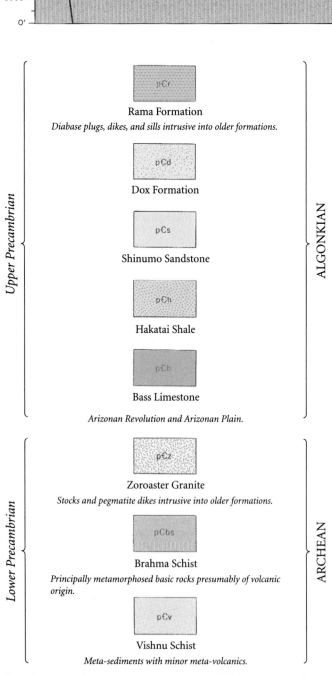

Upper Precambrian

pꞒr

Rama Formation
Diabase plugs, dikes, and sills intrusive into older formations.

pꞒd

Dox Formation

pꞒs

Shinumo Sandstone

pꞒh

Hakatai Shale

pꞒb

Bass Limestone

Arizonan Revolution and Arizonan Plain.

ALGONKIAN

Lower Precambrian

pꞒz

Zoroaster Granite
Stocks and pegmatite dikes intrusive into older formations.

pꞒbs

Brahma Schist
Principally metamorphosed basic rocks presumably of volcanic origin.

pꞒv

Vishnu Schist
Meta-sediments with minor meta-volcanics.

ARCHEAN

Note: The terms Archean and Algonkian are used as subdivisions of Precambrian time. Their long established use in describing the classic geology of Grand Canyon makes this desirable. The United States Geological Survey and the United States National Park Service do not recognize any formal subdivisions of Precambrian time.

MAP SYMBOLS

Contact

Gradational contact

───── 160 ── ── ── ·· ··
Fault
Dashed where approximately located; dotted where inferred beneath Paleozoic strata. Direction and amount of dip indicated.

─── D 100 ───
Ep-Algonkian Fault
D, downthrown side and amount of displacement.

─── D 30 ───
Laramide Fault
D, downthrown side and amount of displacement.

α, first Ep-Algonkian Bright Angel thrust fault.

β, second Ep-Algonkian Bright Angel thrust fault.

γ, Laramide Bright Angel normal (oblique slip) fault.

Fault with strike slip component

Bedding plane overthrust fault

Monocline
Arrows show direction of inclination. Dashed where inferred beneath Paleozoic strata.

Strike and dip of beds

Strike and dip of overturned beds

Strike and dip of foliation

Strike of vertical foliation

TRUE NORTH MAGNETIC NORTH 15°

APPROXIMATE MEAN
DECLINATION, 1961

122   Lab 7

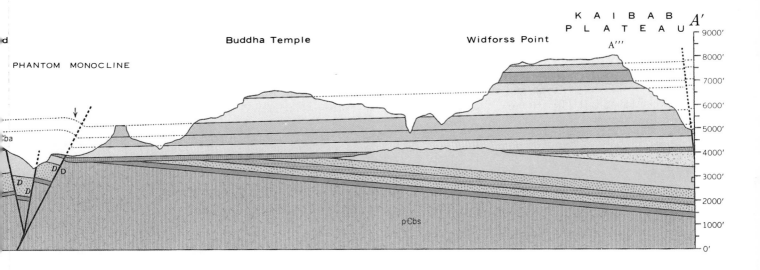

PHANTOM MONOCLINE

Buddha Temple

Widforss Point

K A I B A B
P L A T E A U   A'

A'''

p€bs

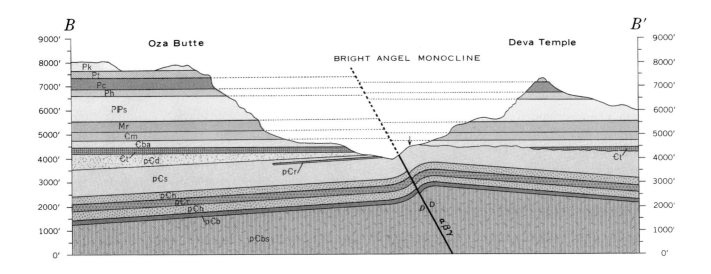

B

Oza Butte

BRIGHT ANGEL MONOCLINE

Deva Temple

B'

9000'
8000'
7000'
6000'
5000'
4000'
3000'
2000'
1000'
0'

Pk
Pt
Pc
Ph
PIPs
Mr
€m
€ba
€t   p€d
p€s
p€h
p€r
p€h
p€b
p€bs

p€r

€t

D
D
α,β,γ

9000'
8000'
7000'
6000'
5000'
4000'
3000'
2000'
1000'
0'

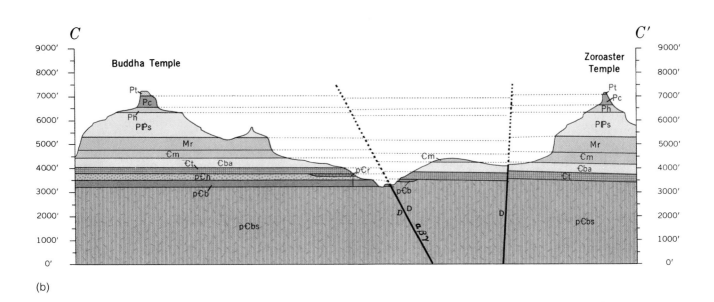

C

Buddha Temple

Zoroaster
Temple

C'

9000'
8000'
7000'
6000'
5000'
4000'
3000'
2000'
1000'
0'

Pt
Pc
Ph
PIPs
Mr
€m
€t   €ba
p€h
p€b
p€bs

p€r

€m

p€b
D
D
α,β,γ

D

Pt
Pc
Ph
PIPs
Mr
€m
€ba
€t
p€bs

9000'
8000'
7000'
6000'
5000'
4000'
3000'
2000'
1000'
0'

(b)

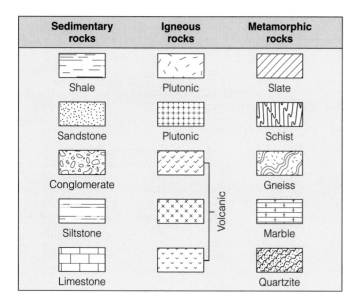

Figure 7.2 Some map patterns used for different types of rocks

e. Locate a contact and name the formations on either side of it. _____

_____

f. List two structures labeled on the map, using the key to help you. _____

_____

Plotting Contacts on Maps

Most geologic maps show the locations of rocks on the basis of outcrops (Fig. 7.4). **Outcrops** are areas of rocks exposed at the surface that are not obscured or covered by foliage, soil, sediment, or manmade structures (Fig. 7.4a). Geologists infer the locations of most rocks or formations from the outcrops that are available. The map is then drawn as if the area were made entirely of outcrops (Fig. 7.4b).

2. The map in Figure 7.5 shows the locations of outcrops of three formations. Imagine you are the geologist who made this map. Turn this into a geologic map by inferring the location of the contacts and coloring or labeling the remaining areas with the formations inferred to be at those locations. Where the position of a contact is precisely known, draw it in as a solid black line. Where it is approximately located or uncertain, draw it in as a dashed line. Include the symbols you use for the contacts on the map key.

Sedimentary Rock Masses

Layering

In Lab 5, we saw that some sedimentary rocks have bedding, a kind of planar feature in the rock. Not only do these rocks

Figure 7.3 The Colorado River has cut a cross section through horizontally layered sedimentary rocks at the Grand Canyon, Arizona. Foliated metamorphic rocks are visible in the deepest parts of the canyon.

have bedding at the scale of a hand sample such as Figure 5.3a, but they also have larger-scale layering (Fig. 5.3b and 7.3). A layer of sedimentary rock that is visually separable from other layers is called a bed or **stratum.** The plural form of the word, **strata,** is more commonly used. Because of the way they form, beds or strata generally follow or obey certain laws called the *laws of stratigraphy.*

Stratigraphic Laws

Stratigraphy is the study of strata, or sedimentary layers. The principle of **original horizontality** states that sedimentary layers are deposited horizontally or nearly so. Even after much time has passed, some of the rocks in the Grand Canyon still have their original horizontal orientation (Figs. 7.1b and 7.3). When inclined beds are found (excluding cross beds), this principle implies that tilting of the layers occurred after deposition. Deformation, caused ultimately by movements of plates, can cause bedding to become tilted from horizontal. Sometimes deformation is so great that tilting is greater than 90°, in which case we say the rocks are **overturned.**

Sedimentary rock layers are deposited in sequence one on top of the other, so that the oldest rocks are at the bottom

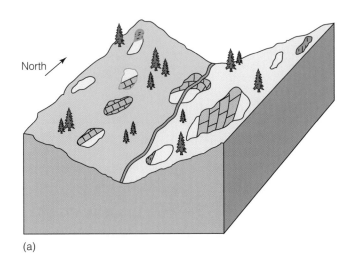

North

North

33

35

North

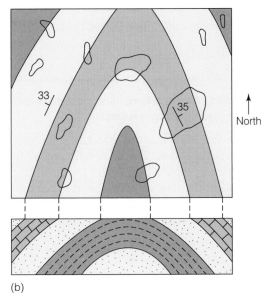

(a)

(b)

Figure 7.4 Making a map and cross section from outcrops. (a) A region with scattered outcrops of limestone sandstone and shale. (b) Map and cross section of the same region inferred from the outcrops.

Key

☐ Limestone

☐ Conglomerate

☐ Shale

Figure 7.5 Outcrop map of a region of sedimentary rocks

of a sequence and the youngest rocks are on top (Figs. 7.1b and 7.3). This is the principle of **stratigraphic superposition**. It is true only if the sequence has not been overturned, which is a rare occurrence (Fig. 7.6). Certain sedimentary features, such as graded bedding, cross bedding, and ripple marks (Fig. 5.6), may indicate whether a sequence has been overturned after deposition and lithification.

3. Without looking at the map explanation, list the rocks between Pk and Mr in the cross section A-A′ in Figure 7.1b of the Grand Canyon in order from oldest to youngest. _____

_____

Check your answers with the Explanation.

Figure 7.6 The rocks in the center of this photo are overturned, or tilted more than 90°.

4. Examine the cross sections in Figure 7.1b. What does the principle of original horizontality tell you has happened to the layers from the Bass Limestone to the Dox Formation? _____

_____

5. Review graded bedding and cross bedding from Lab 5 if necessary. Locate the graded bedding and the cross bedding in the outcrop of rock illustrated in Figure 7.7.

    a. What do these layers tell you about this outcrop?

    _____

    b. How could this have happened?

    _____

6. On Figure 7.7 number the rocks in order from oldest to youngest.

7. List the rocks in Figure 7.33 in order from oldest to youngest. _____

Igneous Rock Masses

The bodies or masses of igneous rocks have different characteristics and shapes if they are extruded than if they are intruded.

Figure 7.7 Strata containing graded and cross bedding layers

Extrusives

Molten lava solidifies into lava flows, lava domes, or pyroclastic material depending on how it erupts.

■ **Lava flows** are extruded magma that has solidified in tongue shapes and as sheets. Fluid lava will tend to flow farther forming long tongues (Fig. 7.8a) or occasionally to spread out in wide flat lava sheets of great aerial extent called **flood basalts**. More *viscous* lava will make thick short tongues covering smaller areas (Fig. 7.8b).

(a)

(b)

Figure 7.8 Lava flows. (a) Fluid lava makes thin flows. (b) Viscous lava makes thick flows.

Figure 7.9 Lava domes at Mono Craters, California, in foreground and Sierra Nevada in the background

Figure 7.10 The shield volcano in the distance is Fernandina in the Galápagos Islands.

Figure 7.11 Eldfell cinder cone, near Vestmannaeyjar Iceland, formed in 1973, growing 100 m in 2 days. A second cinder cone, Helgafell, is visible to the right.

- Sometimes lava is so viscous that it does not spread out but domes up instead. When this happens, a **lava dome** is created (Fig. 7.9).
- When volcanic eruptions are explosive, a spray of magma and particles of rock spew out of the volcano, producing **volcanic ash** and other pyroclastic deposits (Fig. 4.17) that may lithify into tuff (Fig. 4.18). Volcanic ash is a layered deposit (Fig. 4.17b) that may extend over a wide region and become buried within volcanic or sedimentary sequences. Volcanic ash or tuff deposits may be interlayered with lava flows depending on the sequence and type of eruption.

Lava flows and volcanic ash obey the law of stratigraphic superposition.

Where a volcanic vent (the opening where volcanic eruptions occur) erupts frequently or repeatedly, a hill or mountain known as a **volcano** can build up. A volcano may be built entirely of lava flows, entirely of pyroclastic deposits, or a mixture of both.

- A **shield volcano** is one made up of basaltic lava flows and very little ash (Fig. 7.10). Shield volcanoes can be very massive and are gently sloping with an angle of about 10°. The tallest (not highest) and most extensive mountain on Earth, Mauna Loa in Hawaii, is a shield volcano with its base at the bottom of the sea.
- A **cinder cone** is a small volcano made up of pyroclastic material, ash, and cinders (Fig. 7.11). It does not contain lava flows.
- A **stratovolcano** or **composite volcano** is made of interlayered pyroclastic deposits and lava flows (Figs. 7.12 and 4.17a). Stratovolcanoes make impressive snow-capped volcanic peaks with steep, about 30°, slopes (Fig. 7.13) but are generally smaller in volume than shield volcanoes.

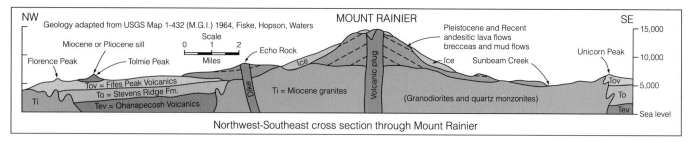

Figure 7.12 Northwest-southeast cross section through Mount Rainier stratovolcano

Figure 7.13 Stratovolcano—Mount St. Helens, Washington, before it erupted in 1980. This view is from the shores of Spirit Lake.

8. Examine the Geologic Map of Mauna Loa in Figure 7.14.
   a. What types of rock masses make up the area?

   _____

   b. Measure the length of the youngest rock masses.

   _____

   c. Do you think this lava was fluid or viscous?

   _____

   d. Why? _____

   _____

   _____

e. Which lava flows would you expect to find on top of all the others? _____

9. Compare the scales and slopes of Mauna Loa, Mount St. Helens, and Capulin Mountain in Figure 7.15.
   a. What type of volcano is each?

   Mauna Loa is a _____,

   Mount St. Helens is a _____,

   and Capulin Mountain is a _____.

   b. How do you know? _____

   _____

   _____

   c. On Figure 7.15 roughly resketch Mount St. Helens and Capulin Mountain profiles next to and at the same scale as Mauna Loa. Only the part of Mauna Loa above sea level is shown in Figure 7.15!

Intrusives

Intrusive igneous rock masses have a variety of shapes and sizes and are classified based on these characteristics (Fig. 7.16). Some intrusives such as dikes and sills are planar in form, sheet shaped or tabular.

- **Dikes** are planar bodies that cut across layers or through unlayered rocks (Fig. 7.17).
- **Sills** are planar bodies that intrude parallel to layers (Fig. 7.18).
- **Laccoliths** also intrude parallel to layers but they bulge upward to make a three-dimensional body, often doming the layers above them (Fig. 7.16).

  Some intrusive bodies are more similar in all three dimensions:

- **Stocks** are roughly equidimensional intrusions of small size (Fig. 7.16), with an outcrop area less than 100 km^2 (or 40 mi^2).
- **Batholiths** are rough equidimensional intrusions of large size, with an outcrop area greater than 100 km^2 (or 40 mi^2).

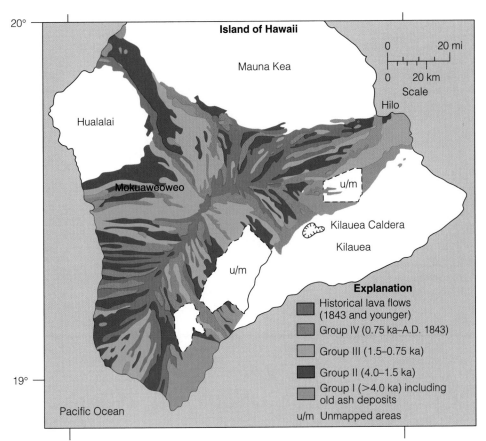

Figure 7.14 Geologic map of Mauna Loa. The age unit *ka* means thousands of years. 0.75 ka = 750 years; 1.5 ka = 1500 years.

Modified from J. P. Lockwood and P. W. Lipman, "Holocene Eruption History of Mauna Volcano," Chapter 18 in R. W. Decker et. al., eds., *Volcanism in Hawaii*, USGS Professional Paper 1350, 1987.

Explanation

- Historical lava flows (1843 and younger)
- Group IV (0.75 ka–A.D. 1843)
- Group III (1.5–0.75 ka)
- Group II (4.0–1.5 ka)
- Group I (>4.0 ka) including old ash deposits
- u/m Unmapped areas

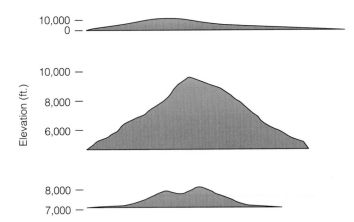

Figure 7.15 Side views or unexaggerated topographic profiles of three volcanoes: Mauna Loa (top), Mount St. Helens before its eruption in 1980 (middle), and Capulin Mountain (bottom). The three volcanoes are drawn at three different scales, as indicated by the elevation information on the left.

Batholiths in western North America (Fig. 7.19) are quite extensive and include the rocks of Sierra Nevada, California, such as those seen in Yosemite National Park (Fig. 7.20).

Where intrusions come in contact with the **country rock** (the rock they intrude), the edges are cooled more quickly than the centers, producing generally finer-grained igneous rock along the margin of the intrusion called **chilled margins.** The country rocks are metamorphosed to hornfels where they come in contact with an intrusion (Fig. 6.8). The intrusion sometimes envelopes pieces of the country rock or brings up rock pieces from below. These bits of foreign rock embedded in igneous rock are known as **xenoliths** (Fig. 4.6b).

Cross-Cutting Relationships

A geologic feature may cut across another feature, indicating that this feature is younger than the rocks it cuts. This is the principle of *cross-cutting relationships.* A structure X that cuts across a structure Y indicates that X is younger than Y. Dikes, stocks, and batholiths, discussed earlier, are cross-cutting igneous intrusions. Sills and laccoliths are intrusions that parallel the bedding and are more difficult to recognize, but they, too, are younger than the rocks they intrude. Later we will see additional cross-cutting features such as faults and unconformities.

10. List the rocks in the cross section in Figure 7.21 in order from oldest to youngest. _____

11. Label directly on the photo in Figure 7.17 the order of formation from oldest (1) to youngest (3) of the country rock and the two white dikes.

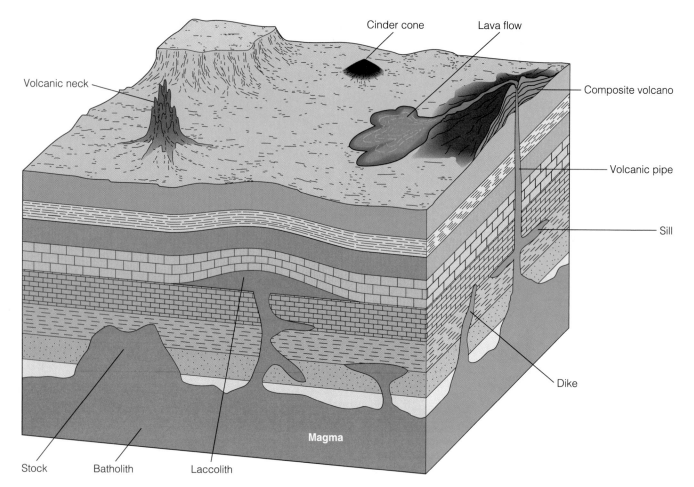

Labels on Figure 7.16:
Volcanic neck — Cinder cone — Lava flow — Composite volcano — Volcanic pipe — Sill — Dike — Stock — Batholith — Laccolith — Magma

Figure 7.16 Block diagram of igneous bodies. Volcanic neck, cinder cone, lava flow, and composite volcano (or stratovolcano) are volcanic features. Stock, batholith, laccolith, dike, sill, and volcanic pipe are plutonic. Some plutons are discordant and cut across layers, while others are concordant, intruding parallel to layers.

Figure 7.17 Different episodes of intrusion of light-colored rhyolite dikes cutting through a granodiorite pluton of the central mountains of the Hispaniola (Dominican Republic). Photo by Grenville Draper.

Figure 7.18 Basaltic sill intruded between sedimentary beds in Big Bend, Texas

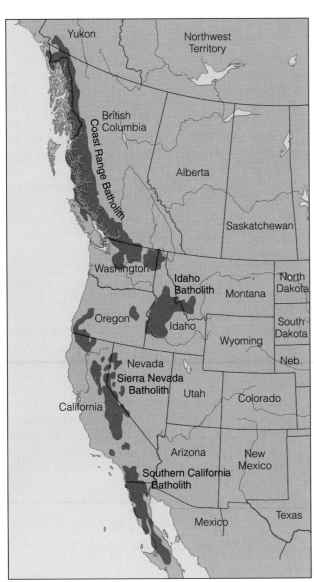

Figure 7.19 Map showing the extent of batholiths in Western North America

12. Directly on the photo in Figure 7.18 label the rocks in order from oldest (1) to youngest.

13. Examine pЄr in cross sections B-B′ and C-C′ in Figure 7.1b.
    a. Based on its geometry in the cross sections and its age compared with neighboring rocks, what type of rock mass is it? _____

    b. What rocks has it intruded? _____

    c. How can you tell it is not a lava flow? _____

    _____

    d. What would you look for in the field to confirm that it is not a lava flow? _____

    _____

14. Examine the geologic map of part of Cross Mountain quadrangle, California in Figure 7.22.
    a. What are the pink rocks, Tr, on this map? _____

    Are they intrusive or extrusive? _____

    Do you think the magma was fluid or viscous?

    _____

    b. Read the explanation for Tri. What evidence is there for chilled margins for these intrusive rhyolites?

    _____

    c. How many different kinds of dikes are shown, and how does one distinguish between them on this map?

    _____

    d. Does Tri intrude gd or vice versa? _____

    How can you tell? _____

    _____

Figure 7.20 Batholith at Yosemite National Park, California. Virtually all the rocks visible from this viewpoint are part of the Sierra Nevada Batholith.

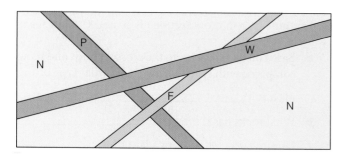

Figure 7.21 Cross section showing a number of dikes that intruded at different times

e. What igneous rocks are found in Tsv?

_____

f. What rock makes up Tsb? _____

Is it intrusive or extrusive? _____

g. Which volcanic material spread out most during its

eruption, Tr, Tsv, or Tsb? _____ Why is

this the case? _____

_____

h. Which spread out the least? _____ Why?

_____

Metamorphic Rock Masses

Recall that some metamorphic rocks have **foliation,** or a layered structure caused by the arrangement of minerals in the rock. This foliation tends to develop perpendicular to the direction of compression. Some metamorphic rocks retain original structures from their protoliths or parent rocks. Bedding in a sedimentary rock may be preserved even after metamorphism and is known as **relict bedding.** For example, the color bands in the marble in Figure 6.6b most likely resulted from chemical differences inherited from sedimentary bedding.

In areas near igneous intrusions, **contact aureoles** develop where contact metamorphism occurs. This area is subjected to the intense heat and fluids associated with magma. The metamorphism gradually increases in intensity as one approaches the intrusive contact (Fig. 6.8).

In areas of regional metamorphism (Fig. 6.9) and in contact aureoles (Fig. 6.8), it may be possible to map gradations of metamorphism by the mineralogical changes in the rocks. Recall that a **metamorphic zone** is a region marked by a particular set of minerals usually distinguished by one or two **index minerals.** The boundaries between metamorphic zones are called **isograds.** Isograds are drawn as lines on a geologic map, similar to contacts.

15. Examine the Geologic Map of Washington in Figure 1.3.
  a. Mount Rainier National Park is on this map. See whether you can find it using what you know about the geology of Mount Rainier from Figure 7.12. Label the approximate location of the summit of Mount Rainier on the map.
  b. Are there any batholiths in the map area?

  _____ If so, what type of rock are they

  made of? If not, what is your evidence?

  _____

Figure 7.22 (*opposite*) Part of the map, Geology of the Southeast Quarter of the Cross Mountain Quadrangle, California. The map explanation and symbols key follow on page 134.

EXPLANATION

MAP SYMBOLS

Fault, showing dip
Dashed where approximately located
U, upthrown side; D, downthrown side

Contact
Dashed where approximately located

Axis of anticline showing plunge

Axis of syncline showing plunge

Strike and dip of beds

Strike and dip of overturned beds

Prospect

Mine   Quarry

Mill

Geology by Howard S. Samsel

Base from U.S.G.S. 7½' quadrangles

QUATERNARY

Alluvium — Qal
Arkosic sand and gravel, volcanic debris, and fine-grained clastic and chemical playa sediments.

Stream terraces — Qts
Arkosic stream sands and gravels. Mostly remnants of an older alluvial cover.

Pediment terraces — Qtp
Arkosic sands and gravels; locally cemented with caliche.

Playa deposits — Qp
Buff clay, silt, and arkosic stream sands and gravel. Locally contain saline layers.

UNCONFORMITY

Ricardo Formation — TQr
Reddish to buff, lenticular, calcareous, arkosic conglomerate and sandstone, clay and siltstone. (Members 6, 7, and 8 of Dibblee, 1952)

Hornblende andesite — Ta, Tai
Reddish to purple, prophyritic to aphanitic hornblende andesite. Ta = flows; Tai = intrusive

---

TERTIARY

UNCONFORMITY

Rhyolite — Tr, Tri
Tr — Buff, reddish to pale purple, massive, vesicular, porphyritic rhyolite in flows and welded tuff breccia.

Tri — White to buff, yellow and reddish-purple porphyritic to aphanitic dacite to rhyolite in plugs and dikes. Locally have glassy borders.

Basalt — Tbd
Dark gray, porphyritic, xenolithic hornblende augite basalt dikes.

Sandstone, tuff and basalt flows — Tsv, Tsb
Tsv — Red to buff arkosic sandstone and conglomerate; green to white tuff, silicified tuff and tuff breccia; green to pale purple rhyolite tuff breccia.

Tsb — Dark gray, hypocrystalline, trachytoidal, porphyritic, xenolithic, hornblende augite basalt flows.

MESOZOIC

UNCONFORMITY

Aplite — ga
Fine- to medium-grained, equigranular granite aplite stocks and dikes.

Granite — gr
Coarse-grained, rusty equigranular biotite granite.

Dacite — dd
Pale gray, porphyritic dacite dikes.

Granodiorite — gd, gdx
gd — Coarse- to medium-grained, equigranular to porphyritic biotite granodiorite.

gdx — Undivided biotite granodiorite and xenoliths of schistose and gneissose rocks.

PRE-MESOZOIC (?)

Metasedimentary rocks — ms
Undivided metasedimentary rocks. Consists of fine-grained, white to dark gray, thin-bedded limestone, dolomitic limestone, and fissile argillaceous layers; fine-grained, alternately layered, pale and dark gray quartz-sericite schist. Gray to white, banded, hornblende-andesine gneiss.

GEOLOGY OF THE SOUTHEAST QUARTER OF THE CROSS MOUNTAIN QUADRANGLE, CALIF.

From: California Division of Mines and Geology Map Sheet 2

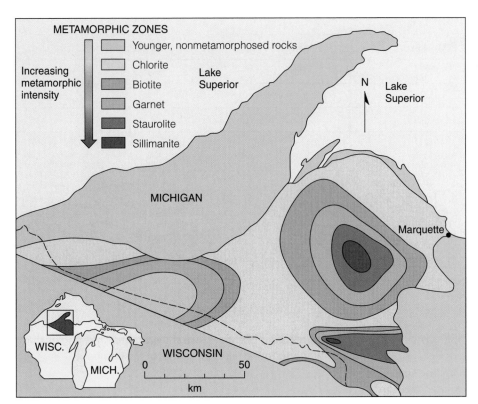

Figure 7.23 Map of the metamorphic zones in Upper Peninsular Michigan. Metamorphosed shales in this region best delineate the isograds, shown as black lines separating the colored metamorphic zones. Metamorphism here took place about 1.5 billion years ago.

METAMORPHIC ZONES

Increasing metamorphic intensity

- Younger, nonmetamorphosed rocks
- Chlorite
- Biotite
- Garnet
- Staurolite
- Sillimanite

Lake Superior

MICHIGAN

N

Lake Superior

Marquette

WISC.

MICH.

WISCONSIN

0   50

km

---

c. What type of rock is pJsc (colored green)?

_____

pJph? _____

Which formed at higher temperature? _____

d. Is pJsc or pJph more often found next to pre-Jurassic gneiss or Mesozoic granitic rocks? _____

Which is more often next to pre-Jurassic sedimentary rocks? _____

Does this make sense in terms of the temperatures of formation of all of these rocks? _____

Why or why not? _____

_____

16. Examine the metamorphic zone map of northwestern Michigan in Figure 7.23.

a. What do different colors on this map represent? (Generalize.) _____

b. Did heat from magma cause the metamorphism in this area—(contact metamorphism)? _____

What evidence supports this conclusion? _____

_____

c. Where would you expect to find garnets in this region? Refer to Table 6.2 to assist you with this answer.

_____

d. What is the line separating yellow from light brown?

_____

e. Where did the highest temperature of metamorphism occur? _____

Structures and Deformation

Deposition and erosion, and deformation, including folding and faulting, each produce structures in rocks. Some of these structures are planar features, others are not. Different styles and forces of deformation produce different structures.

Measurement of Orientation of Planar Geological Features

The most common form of geological structures is a plane. Beds or strata, dikes, sills, foliation (slaty cleavage in slate, schistosity in schist, gneissic banding in gneiss), and faults are all planar features.

17. Examine the rock samples in the structure display. Determine which samples have some sort of planar structure and enter this information in Table 7.1.

Planes may have many different orientations, such as vertical, horizontal, or anything in between. The orientation of a plane (**strike** and **dip**) is described by three quantities discussed here and shown in Figure 7.24:

- The **strike** is the direction of a horizontal line in the plane measured clockwise with respect to north, as indicated on

Table 7.1 Planar Features in Rocks

Rock Number and Name	Type of Planar Structure	Rock Number and Name	Type of Planar Structure

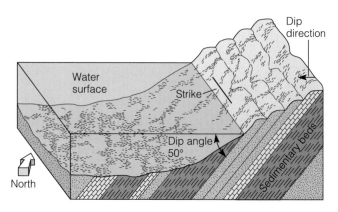

Figure 7.24 Strike and dip of tilted bedding. The strike is parallel to the waterline, and measured clockwise from north, which makes it 140° or 320°, depending on which end of the line you measure. The dip direction is perpendicular to the strike in the direction the beds tilt downward, in this case SW. The dip angle is measured from horizontal.

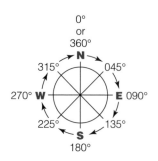

Figure 7.25 Compass rose showing the strike angles that correspond to the different points of the compass

the compass rose in Figure 7.25.[1] The strike is measured from either end of the line, so measurements with exactly 180° difference have the same strike. Here are some examples: Beds striking due north also strike due south; their strike is written 0°, 180°, or 360°. Strata striking due east and west have 90° or 270° strike. A northwest-southeast strike could be written either 135° or 315°.

- The **dip angle** is the angle of tilt of the plane measured with respect to the horizontal (Fig. 7.24).
- The general **dip direction** is the downward direction on the plane, perpendicular to the strike (Fig. 7.24) and expressed in general terms (such as NE, NW, etc.).

For the rock layers in Figure 7.24, the strike is measured as degrees from north from either end of the strike line, so it is 140° or 320°. The strike and dip are written with strike first, then dip angle, then dip direction: 320°, 50°SW or 140°, 50°SW.

You measure strike and dip with an instrument called a *compass clinometer*. Look at an inclined planar feature, and decide which direction is the steepest slope of the plane. Wa-

ter would tend to run down in this direction. This is the dip direction. Hold the compass clinometer level against the bed so the long edge of the compass is perpendicular to the dip direction. If you do this correctly, the long edge of the compass will be along a horizontal line on the plane, the strike. Your instructor will show you how to read the compass as the method varies with the compass. Write down the strike angle. While the compass is still in this position, note the dip direction (perpendicular and down the slope). Use the clinometer part of the compass to measure the dip. The procedure varies with the compass so follow the directions given by your instructor.

One of the chief benefits of being able to measure strike and dip is to be able to place information on a geologic map about the angle or tilting of planar geologic features such as sedimentary bedding, lava flows, volcanic ash beds, and faults. A geologist represents the strike and dip of bedding planes on a map using symbols such as those shown in Figure 7.26 or of foliation using symbols such as those shown in pCbs and pCv on the map in Figure 7.1a.

Measuring and Plotting Strikes and Dips

18. Models of dipping planes set up in the lab simulate tilted bedding planes in sedimentary rocks (or any other kind of planar feature) in the field. For these models, use a compass clinometer to determine the strike and dip of each model, and enter the measurements in Table 7.2. Be careful not to move the models as this will change the results.

....

[1]Another method divides the compass rose into four quadrants measured from north or south, so that the strike in Figure 7.24 would be written N40°W.

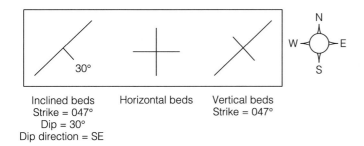

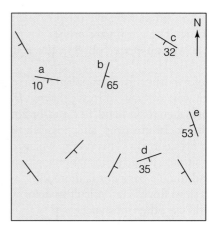

Figure 7.26 Strike and dip symbols. The strike is a long line positioned with the correct angle measured from north. The dip direction is a short tick perpendicular to it starting at the middle of the strike line and pointing in the dip direction. The dip angle is written next to the symbol. Horizontal beds are represented by a large plus sign, and vertical beds have a strike line with a shorter line perpendicular to it.

Figure 7.27 Map with strike and dip symbols

19. For each lettered strike and dip symbol on the map in Figure 7.27, write out the approximate numeric strike and dip.

   a. _____

   b. _____

   c. _____

   d. _____

   e. _____

20. For the four strikes and dips at the lettered locations on the map in Figure 7.28, draw the appropriate strike and dip symbol on the map. The crossing point for each symbol should be positioned at the point. The measurements are as follows:
   a. 59°, 90°
   b. 340°, 10°NE
   c. 127°, 80°SW
   d. 260°, 55°S

21. Imagine that while mapping the area in Figure 7.5 you measured five strikes and dips at the lettered locations on the map and recorded your measurements in your notebook. To continue making your geologic map in Figure 7.5, put the appropriate strike and dip symbol at each lettered location on the map.

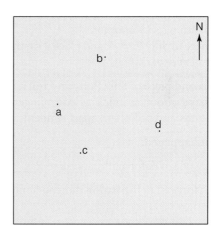

Figure 7.28 The points labeled a, b, c, and d on this map are outcrops where strike and dip were measured.

The measurements are as follows:
   a. 228°, 42°SE
   b. 52°, 28°SE
   c. horizontal
   d. 230°, 20°NW
   e. 49°, 32°NW

Deformation Experiments

Now you can measure strike and dip and read strike and dip symbols on a map, but what do they signify? From the principle of original horizontality, we would expect sedimentary bedding to be horizontal, yet clearly it is not always.

Table 7.2 Measurements of Strike and Dip Models

Model Number	Strike	Dip Direction	Dip Angle

Deformation can tilt and change the position and orientation of bedding. The reason we bother to measure strike and dip is so we can better observe the results of these movements.

Rocks deform when force is applied to them in two basically different ways or types of rock behavior. **Ductile** deformation occurs when rocks bend, flex, and/or flow and generally when the rocks are deep and possibly warm. **Brittle** deformation breaks rocks. This occurs when rocks are cold and near the surface. Rapid deformation is also more likely to be brittle, and ductile behavior is more likely when the deformation is slow. Rocks are difficult to deform in a laboratory, so we will use other substances.

Three different forces result in deformation. **Compression** squeezes things together or has forces moving toward each other. **Tension** pulls them apart or has forces moving away from each other. **Shear** is a scissorlike motion causing one rock mass to pass by another.

Take a piece of Silly Putty® and place it on the table while you do your next experiment.

22. Take a second piece of Silly Putty and roll it into an elongated shape about as thick as your finger. Hold onto the two ends and pull them apart suddenly.

  a. What happened? _____

  b. What style of deformation occurred? _____

  c. What force of deformation did you apply?

    _____

23. Put the pieces back together again. Warm the Silly Putty slightly in your hands and roll it into the same shape as before. Hold on to the two ends and pull them apart gradually.

  a. What happened? _____

  b. What style of deformation occurred? _____

  c. What force of deformation did you apply?

    _____

24. Observe the Silly Putty you left on the table.

  a. What happened to it? _____

    _____

  b. What style of deformation occurred? _____

25. Explain what circumstances or situations lead to the results you observed in these Silly Putty experiments.

  _____

  _____

  _____

Take a piece of wax and hold it in your hand to warm it for about 2 minutes while you do the next experiment.

26. Place another piece of wax in a vise, and crank the vise together until you see significant deformation in the wax.

  a. What happened? _____

  b. What style of deformation occurred? _____

  c. What force of deformation did you apply?

    _____

  d. Sketch the deformed wax in the space below. Draw an approximate north arrow on the top of each piece of wax, and notice the orientation position and arrangement of your deformed wax so you can reposition it after removing it from the vise. Keep these for a later exercise.

27. Place the warmed piece of wax in the vise, and crank the vise together until you see significant deformation in the wax.

  a. What happened? _____

  b. What style of deformation occurred? _____

  c. What force of deformation did you apply?

    _____

  d. Before you remove your piece of deformed wax, mark a north arrow on the top so you know the orientation it had in the vise for some later questions.

Folded Layers: Anticlines and Synclines

Folds are formed whenever layered rocks are *compressed* parallel to their layering, and the beds are buried deeply enough that they behave in a *ductile* manner (i.e., they bend rather than break). The beds may fold upward and form an **anticline** (Fig. 7.29a) or bend downward and form a **syncline** (Fig. 7.29b). If multiple folds occur in an area and the beds are not broken (faulted), anticlines and synclines must alternate.

Often folds are too large to be seen at one locality; therefore, their presence has to be deduced from observation of strikes and dips of individual beds. If strikes and dips are plotted on a map, then the presence of folds becomes apparent. Beds *dipping away* from an axis define an anticline; beds *dipping toward* an axis define a syncline. Due to the principle of stratigraphic superposition, the beds *dip toward younger* sedimentary layers (except if overturned). This means that you can use the age of rock layers to determine which folds are anticlines and which are synclines, or you can use the dip of rock layers to tell the relative age of the beds.

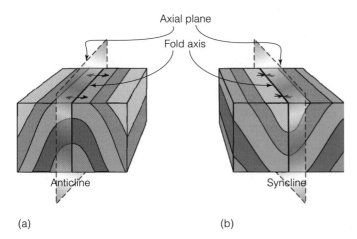

Axial plane

Fold axis

Anticline   Syncline

(a)   (b)

Figure 7.29 Ductile deformation may produce folds. (a) Upright anticline. (b) Upright syncline. The line around which bending takes place (similar to the crease in a folded sheet of paper) is called the **fold axis** (plural is *fold axes*). The fold axes are represented by symbols shown in Figure 7.30. Fold axes lie in a plane that cuts through the bent part of the fold, called the *axial plane*. The central folded part of the fold is known as the **core,** which is flanked on both sides by **limbs.**

Upright folds are anticlines and synclines that have horizontal fold axes and vertical axial planes. The folds shown in Figure 7.29 are upright folds. The strikes of sedimentary beds in an upright fold are parallel to each other, parallel to the fold axes, and the fold axes lie along horizontal beds.

28. Examine the deformation in the warmed wax experiment. What type of structure(s) did you make?

_____

Play-doh Folds   Make a series of anticlines and synclines in Play-doh as follows:

■ Take three (roughly) 1-in. balls of Play-doh, each a different color.
■ For each ball, place it between a folded sheet of plastic wrap and carefully roll it out with the rolling pin until it is about 2 1/2 in. × 4 in.
■ Carefully peel off the Play-doh layer and set it aside. Use the same sheet of plastic wrap for each ball.
■ Place one layer of Play-doh down on the plastic wrap; this simulates deposition of the first sedimentary layer. Place two more layers on top, representing continuing deposition.
■ Horizontally compress the Play-doh layers and plastic wrap lengthwise by squeezing the sides together parallel to the tabletop. You should get more than one anticline and syncline. If you didn't, then your layers should be thinner.
■ Carefully remove the plastic wrap without unbending the folds. The Play-doh represents some rock layers in a mountain range just after folding.

■ Erosion is likely to strip away rock faster from the highest hills than from the valleys. Simulate erosion by carefully cutting *horizontally* through your Play-doh folds using a cheese cutter. Cut deep enough so that you can just see the lowest of the three colors.
■ To see a cross section of your Play-doh folds, use the cheese cutter to slice vertically through your folds. Do this carefully so you do not flatten your folds as you cut.

29. a. Sketch a geologic map and a cross section of your Play-doh folds in the space provided below by drawing the contacts between different colors of Play-doh. Color both the map and the cross section.

b. What color is the oldest layer? _____ The youngest? _____

c. Draw fold axes on your map for each anticline and syncline using the symbols in Figure 7.30.

d. The layers in the core of the anticline(s) are older/younger than in the core of the syncline(s). Circle which.

e. Draw arrows on your map and cross section showing the direction of compression or squeezing that formed these folds.

f. What type of deformation did you produce in the Play-doh, brittle or ductile? _____

■ Wrap and save your folds for a later activity.

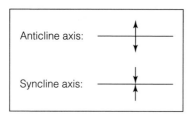

Figure 7.30 Fold axis symbols used on maps (not on cross sections)

Making a Model of an Area in a Geologic Map (Fig. 7.31) Sometimes students have a little trouble visualizing strike and dip and anticlines and synclines from symbols on a map. The next exercise is intended to help you do this.

30. Work together in a group. You will need a tray (or stream table) with sand and a quantity of streak plates.

a. Carefully study the map in Figure 7.31. Draw, in the sand, the outline of the map and the contacts between the shale, siltstone, and sandstone layers.

b. Next, examine the strike and dip symbols on the map. Starting with one strike and dip symbol, locate the appropriate position in your sand model, and shove a streak plate firmly into the sand so that it has the same strike and dip as represented by the symbol. Remember that the dip direction is down into the ground. Take turns placing streak plates for the majority of strike and dip symbols on the map. Help each other out if necessary, and ask for help from your instructor if you get stuck. Have your instructor check your model.

c. Carefully study your model and the map in Figure 7.31. Determine where all anticlines and synclines are on the map, and draw in the axes for these folds in the sand and on the map using the symbols in Figure 7.30.

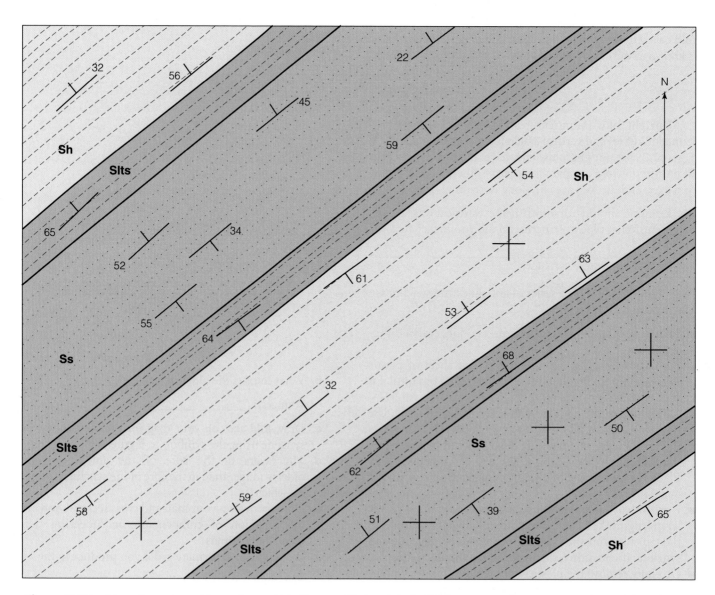

Figure 7.31 Map of a region with sandstone (Ss), siltstone (Slts) and shale (Sh) formations showing the strike and dip of bedding

d. Are the folds upright, with horizontal fold axes? That is, do the fold axes pass through horizontal beds (or dipping beds)? _____

e. Notice that for upright folds all the beds strike in the same direction. Is the strike of the beds parallel to or at an angle to the fold axes? _____

f. What was the direction (in general terms, N-S, NW-SE, etc.) of compression that affected the area? _____

g. List the age of the three rock layers in order from oldest to youngest: _____

_____

How do you know this? _____

_____

The relationships observed in Questions d and e will help you to recognize upright folds in the future. Upright folds show parallel stripes in map view and have beds that strike parallel to each other and parallel to the fold axes.

31. Examine the map you made in Figure 7.5. What structures do you detect? _____

Finish the map by drawing in the appropriate fold symbols.

Faults

Faults form by brittle deformation in the upper part of the Earth's crust (the upper 10 km) and are fractures across which displacement has taken place so that the two sides remain in contact. The displacement can be from a few centimeters to hundreds of kilometers. On geologic maps, faults are usually shown as very thick lines. Faults are easily recognized because formation contacts are abruptly displaced by faults. Faults are planes so they, too, have strikes and dips. If movement of the rocks on either side of the fault is parallel to the dip of the fault plane, the fault is said to be a **dip-slip fault** (Fig. 7.32a). **Strike-slip faults** are ones that do not have a component of up or down motion but have slipped horizontally—that is, parallel to the strike (Fig. 7.32b). An **oblique-slip fault** has movement between the strike and dip of the fault plane (Fig. 7.32c). For faults with a component of up or down motion, such as dip-slip

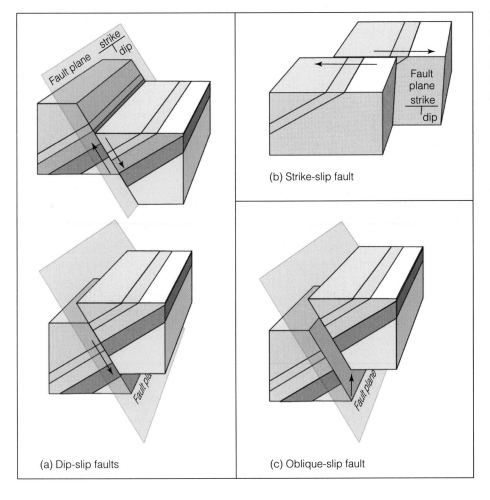

(a) Dip-slip faults

(b) Strike-slip fault

(c) Oblique-slip fault

Figure 7.32 Faults have strike and dip. The movement along the fault may be (a) parallel to the dip in dip-slip faults, (b) parallel to the strike in strike-slip faults, or (c) some direction in between in oblique-slip faults.

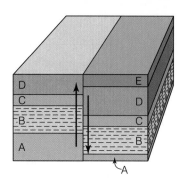

Figure 7.33 Age of rocks on either side of a vertical dip-slip fault

and oblique-slip faults (Fig. 7.32a and c), younger rocks on the side that moved down are commonly adjacent to older rocks across the fault, on the side that moved up. Confirm this using stratigraphic superposition by examining the diagram in Figure 7.33. Figures 7.32a and c also show this relationship.

In some places along faults, the slipping of rocks past each other leave smooth, slick striations on the rocks known as **slickensides** (Fig. 6.6f) that show whether the movement was strike slip, dip-slip, or oblique-slip.

32. Look at the deformed wax and your sketch from Exercise 26. What type of fault did you produce? _____

Also, observe the surface of the break you created and write your observations in the spaces below. Describe what it looks like. Is there any indication from this surface of the direction of movement of the two pieces of wax? If so, name the feature _____ and describe it. _____

_____

_____

_____

33. Examine the cross section A-A′ in Figure 7.1b.
   a. What is the youngest rock cut by the fault slightly left beneath Cheops Pyramid? _____

   b. What is the oldest rock that cuts this fault?

   _____

   c. What is the relationship between the age of the fault and the age of these rocks? _____

   _____

   d. What type of fault is this? _____

Unconformities

Another structural feature found in rocks is an unconformity. When an area is deformed, it is often uplifted and eroded. Later subsidence (sinking of the land) or submer-

gence due to sea level rise may take place, and then sediment is deposited on the eroded surface. The rocks above and below the erosional surface are quite different in age, and the long period of time between those ages produced no rocks. This time gap represented by the erosional surface is called an **unconformity** (Fig. 7.34). Thus, an unconformity is a substantial time gap in the rock record where rocks were either deposited then eroded or simply not deposited.

For all unconformities, the rocks younger than the unconformity are sedimentary and/or volcanic rocks and are approximately parallel to the unconformity surface. The older rocks may also be sedimentary and/or volcanic rocks and parallel to the unconformity surface making it a **disconformity**. They may be tilted or folded sedimentary/volcanic rocks and not parallel to the unconformity surface making it an **angular nonconformity**. Alternatively, they may be plutonic igneous or metamorphic rocks, making it an **unconformity**. One unconformity may change from one type to another as demonstrated in Figure 7.34.

Formation of an Unconformity with Play-doh Use the Play-doh folds that you set aside from Exercise 29. Roll out two or three more layers of Play-doh of different colors, and place these on top of the "erosional surface" of your folds. This represents further deposition.

34. Sketch all the Play-doh layers in the space below as viewed in the cross section. Label the unconformity on your cross section.

35. Summarize all the steps of formation of the angular unconformity you just made. To do this you will need to review the earlier steps that took place to form the folded layers.

_____

_____

_____

_____

_____

_____

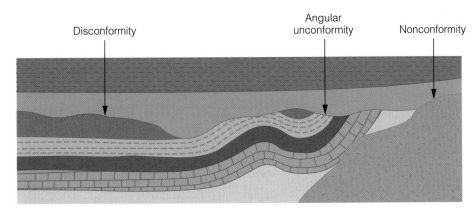

Figure 7.34 The nature of an unconformity may change with position. At the left, this is a disconformity between parallel strata. It becomes an angular unconformity in the central section with tilted and folded rocks below the unconformity. At the right, it is a nonconformity, with igneous rock below.

When you are done with the Play-doh, if the colors come apart easily, separate the colors. Put any unmixed colors away with the appropriate color. Put any mixed colors away as instructed. Keep the Play-doh covered so it doesn't dry out.

36. Look back at Figure 7.1.

a. What does the symbol €t stand for? _____

b. How about p€bs? _____

c. Examine the contact between these two rocks in cross section A-A′ in Figure 7.1b. What type of structure is this? _____

d. In the same cross section, what type of structure occurs between €t and p€d, p€s, or p€h?

_____

e. Use the geologic time scale in Figure 8.21 to decide whether this cross section has another unconformity.

An unconformity occurs between what pairs of rocks? (*Hint:* Look for missing time periods.)

_____

What type of unconformity is this? _____

Rocks in a Cross Section

37. Examine the cross section in Figure 7.35 and the corresponding rock samples. Which igneous intrusive rock sample matches its volcanic equivalent in composition?

_____

In the cross section we see rock layers that change from left to right. To help you match up the correct rocks to beds in the cross section, first decide which rocks are metamorphic. Then list, in Table 7.3, the parent rock samples that are sedimentary or igneous and pair them

NW   **SE**

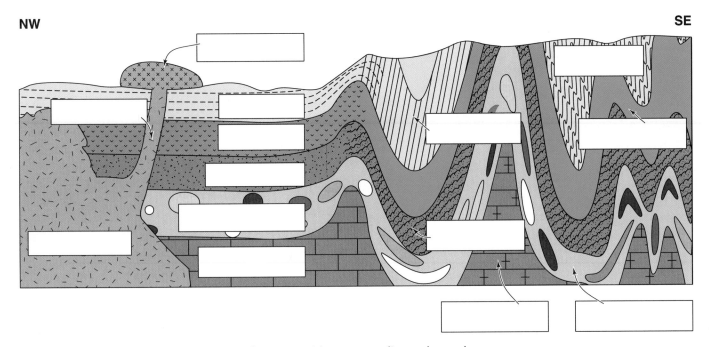

Figure 7.35 Hypothetical cross section of a region with corresponding rock samples

Table 7.3 Pairs of Parent and Metamorphic Rocks for the Cross Section in Figure 7.35

Parent Rock	Metamorphic Equivalent	

with the metamorphic rock they could become. One parent rock has two metamorphic equivalents, list these in Table 7.3 in the row that has three spaces.

38. Select which samples match which rocks in the cross section in Figure 7.35, and write in the sample name and number on the cross section. Also include the rock type for each sample (abbreviate: I = igneous, S = sedimentary, and M = metamorphic). You may want to refer back to Figure 7.2. Describe your reasoning here. What rock characteristics, cross section features, or other evidence made you decide the location for each rock?

_____

_____

_____

_____

_____

_____

_____

39. What structures are present in this cross section?

_____

Igneous features? _____

40. As a class or in smaller groups, discuss the geologic history of the area shown in this cross section. Summarize your discussion here:

_____

_____

_____

_____

_____

_____

_____

_____

_____

_____

_____

Geologic Time and Geologic History

One of the goals of the field of *historical geology* is understanding how the Earth has changed over eons of time. Reconstructing the Earth's history requires establishing both a sequence of events and various ways of measuring geologic time. In the 19th century, geologists established a detailed sequence of Earth's history using observations of the relationships between rocks to determine their relative age. **Relative age** means that the ages of rocks are established in comparison to each other using words such as *older* or *younger* rather than ages scaled in years or millions of years.

In this lab we will follow the procedures used by geologists, especially stratigraphers,[1] to establish the relative ages of rocks. The stratigraphic tools we use can be very powerful as they can reveal the past growth and subsequent erosion of entire mountain ranges, the development of past landscapes, and the evolution of life.

The discovery of radioactivity in 1898 by Henri Becquerel revolutionized the study of geologic time in the 20th century. The natural decay of radioactive isotopes gave geologists a way of measuring the age, called **isotopic** or **radiometric dating,** of certain geologic materials. **Absolute age** refers to the numerical age, with some degree of error. This term is a bit unfortunate because there is nothing absolute about it; however, it is widely used. We will use isotopic dat-

....

[1]A person who studies stratigraphy.

ing to ascertain the absolute, or numerical, ages of rocks. Applying isotopic dating indirectly, we will see how to establish the numerical ages of rocks that cannot be measured directly. The geologic time scale was established with this combination of numerical and relative ages.

Relative Age

Three principles help to determine the relative age of rocks: stratigraphic superposition, cross-cutting relationships, and fossil succession. The first two were introduced in Lab 7. Let's review and practice them before going on to fossil succession.

Recall that stratigraphic superposition states that younger sediments are deposited above older ones. Sediments, sedimentary rocks, and volcanic rocks obey this principle (Fig. 8.1a). Cross-cutting relationships indicate that younger features cut across older ones. Faults, erosion, and intrusions such as dikes, stocks, and batholiths can accomplish the cutting (Fig. 8.1b). Erosion is associated with unconformities and cuts through the dike and the fault in Figure 8.1b. Folds do not exactly cut across layers, but we can tell that the episode of folding is younger than the layers that are folded. Similarly, metamorphism is younger than the rocks it metamorphosed, and sills are younger than the layers they intrude.

1. Use stratigraphic superposition to list the order of formation of the sedimentary rocks in each column in Figure 8.2, from oldest to youngest. Use the rocks labeled with *letters* only at this stage.

   Column 1_____

   Column 2_____

   Column 3_____

2. Examine the columnar sections in Figure 8.2 and determine the relative age of the following dikes using the principle of cross-cutting relationships. Narrow down the age range as much as possible by choosing rocks that are closer in age rather than farther apart, as in the following example.

Example: Dike 1 is younger than D and G because it cuts through them. Since D is younger than G, Rock D is the closest

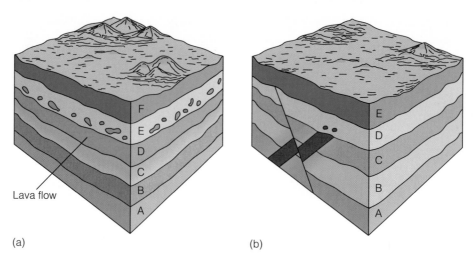

Figure 8.1 Relative dating. (a) Sedimentary layers and lava flows follow the principle of stratigraphic superposition so that the rocks shown formed in sequence, starting with A, then B, C, D, E, and F. Notice that inclusions of D in E also indicate that E is younger than D—the rock containing inclusions is younger than the rock from which the inclusions originated. (b) Cross-cutting relationships show that the relative age of the dike (gray) is younger than C and older than D. Erosion cut the top of the dike so D is younger than the dike. Similarly, the fault (diagonal black line) is younger than D and older than E. This block diagram has two unconformities, one between C and D and one between D and E.

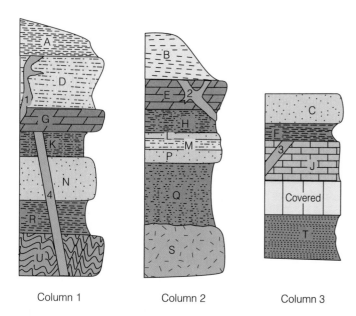

Figure 8.2 Columnar sections of three hypothetical regions. The lettered rocks obey the principle of stratigraphic superposition. Patterns indicate rock types as shown in Figure 7.2. Numbered rocks are dikes, which obey the principle of cross-cutting relationships. In Column 3, "covered" indicates the rocks there are not visible because they are buried under soil or sediment.

in age and helps narrow down the possible age range better than G does. Dike 1 is older than A because erosion cut through Dike 1 before A was deposited.

Dike 1 is younger than ____*D*____ and older than ____*A*____.

Dike 2 is younger than _____ and older than _____.

Dike 3 is younger than _____ and older than _____.

Dike 4 is younger than _____ and older than _____.

Fossil Succession

Recall from Lab 5 that a **fossil** is any evidence of preexisting life preserved in a rock. A clam shell in a rock can be a fossil; so can a footprint of a dinosaur. Fossils can be used in dating rocks because of the principle of **fossil succession** (also known as the law of *faunal succession*). This principle states that organisms evolve in a definite order (e.g., Fig. 8.3), that species evolve and become extinct, never to re-evolve. Thus, the evolution of a species and its extinction become time markers separating time into three units: one before the organism existed, one during the existence of the organism, and one after the organism went extinct. The presence of a fossil of a particular species indicates that the rock containing the fossil (assuming the fossil was not itself transported to the location at a later time) formed at some time between the evolution and extinction of that species. Fossils of organisms that existed for a short period of time, known as **index fossils** (also **guide fossils**), narrow the possible age of rocks containing them more than fossils that existed for a long time (Fig. 8.4). Thus, the fossils present in a rock are characteristic of the geologic period in which the rock formed and the preserved organisms lived. This is analogous to how fashion has changed over time. If you see a picture of men in tricorn hats and women in long dresses and bonnets, you can infer that the scene is from the 18th century. If, on the other hand, the scene contains men and women in jeans and running shoes, then you can infer that it represents the late 20th or early 21st century.

Rocks with matching sets of fossils from two different localities are probably of similar age (Fig. 8.5) and are said to be **correlative.** The process of matching them is called **correlation.**

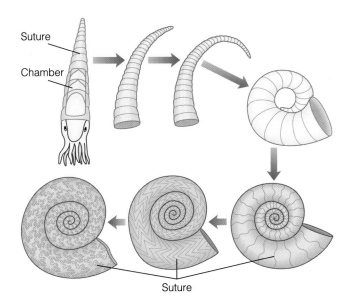

Figure 8.3 The evolution of ammonites. Each form shown is a species that evolved, gave rise to the next species, and went extinct. In the case of ammonites, the evolution produced a gradual progression to ammonites with more complicated sutures—the outline of the chamber wall on the outer shell wall.

3. Figure 8.6 shows the fossils in each sedimentary layer in Figure 8.2. Which rock layers have exactly matching *sets* of fossils? _____ _____ , _____ _____ ,

_____ _____ , _____ _____

What does this tell you about these rock layers?

_____

Sometimes a unique or distinctive layer, such as a volcanic ash bed, that forms from a widespread event of short duration, such as an explosive volcanic eruption, can be used as a time marker for correlation (Fig. 8.7).

Figure 8.6 has no such time-marking beds, but the layers with matching fossils can be correlated, as shown in Figure 8.8 for the first two columns.

Groups of fossils can also aid determination of relative age. In this next exercise, you will establish the relative age sequence of rocks in the three columns in Figures 8.2 and 8.6. Stratigraphic superposition provides evidence of the relative ages within each column but cannot do so for completely separate locations; fossils, however, can.

4. Work with a lab partner so you will have one whole uncut Figure 8.6 and one you can cut into pieces. Cut the layers into thin strips along the horizontal lines. Intersperse strips from Column 2 between strips from Column 1 by obeying the following rules but without looking at Figure 8.8.

■ Keep all rocks in each column in their original stratigraphic order. Notice that within one column the letters are in alphabetical order.

■ Place rocks with matching sets of fossils on top of each other, in a pile.

Cenozoic	Quaternary			
	Tertiary			
Mesozoic	Cretaceous	*Lingula*		
	Jurassic			
	Triassic			
Paleozoic	Permian			
	Pennsylvanian			
	Mississippian			
	Devonian	*Atrypa*		
	Silurian			
	Ordovician			
	Cambrian	*Paradoxides*		

Figure 8.4 The time ranges for three fossils. *Lingula* existed for a long range of time and still exists today and thus is not very useful for indicating the age of rocks in which it is found. *Atrypa* and *Paradoxides* both have short geologic ranges and so are good indicators of geologic age—known as **index** or **guide fossils.**

■ Group all rocks with the same individual fossils consecutively in a continuous sequence vertically.

Since Rocks U and S have no fossils, you will not be able to tell exactly where in the sequence these fit. Check your sequence in Figure 8.8.

5. Next, integrate Column 3 into the sequence with 1 and 2. Write in the rock labels C, F, J, and T in Figure 8.8 at the correct position within the time sequence. As for U and S, the exact position of the covered part of Column 3 is indeterminant. Use cross cutting relationships to determine where dike 3 should fit in the time line, and label it on the figure as well.

The particular columns in Figures 8.2 and 8.6 represent hundreds of millions of years of time during which many species of organisms evolved and went extinct. Notice that *Bellerophon* is present in layers from A to N, H to Q, and J and T. This means that *Bellerophon* lived while these layers were deposited and probably had not evolved before that (lower layers) and went extinct after that (higher layers). We can also see

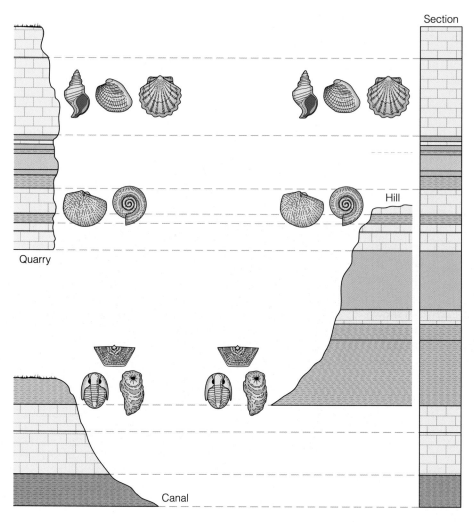

Section

Figure 8.5 Matching strata of the same age (correlation) using fossil succession. Rocks exposed in a quarry, on a hill, and in the walls of a canal have been correlated on the basis of equivalent sets of fossils. Once the correlations have been made, a complete section for the region, at the right, shows the relative ages of all of the beds.

Hill

Quarry

Canal

from Column 1 that *Turritella* (from Layer A) has not yet evolved at the time of D and G, but *Favosites, Mucrospirifer,* and *Pycnosaccus* (from Layer K) have gone extinct. In Column 2 the beds where these organisms evolve and become extinct is between Layer L and Layer H. Notice that *Calamites* and *Pseudoparalegoceras* are missing from Column 2. Missing fossils indicate an unconformity—between L and H, in this case.[2]

6. Make a time line in Figure 8.8 of each of the fossil organisms based on the information you have collected with Figure 8.6. Show when each organism evolved, when it lived, and when it went extinct. You can show this by simply drawing a vertical line for the time the organism lived and two short crossing lines at each end for

the evolution and extinction of the organism. Then write the organism's name sideways along the vertical line. Bellerophon has been done for you in Figure 8.8. Fossils that extend to the top of the time line may not have gone extinct, so leave off the crossing line at the top for them.

Unconformities

Unconformities can be obvious or subtle and can represent anywhere from modest to vast amounts of geologic time and history. Unconformities where the layers above and below are parallel (i.e., disconformities) are often quite difficult to recognize, yet they may represent much passage of time. For all unconformities, an episode of rock formation is interrupted by a period of erosion followed by a transition to deposition. Since erosion is most effective on steep mountainous slopes, many unconformities denote a period of mountain formation. Most people think of mountains as permanent stationary and unchanging features, and they seem to be from the perspective of a human lifespan. But over tens of millions of years, mountains form and erode away. This means that an unconformity is likely to represent the passage of a great deal of time, which becomes a gap in the rock record. In Figure 8.8,

....
[2]In these exercises we assume that the absence of a fossil in a column means the organism did not exist on Earth at that time, even though in reality its absence may be caused by a variety of reasons. For example, the organism might first appear in one area because it migrated there when the environment became favorable, or might disappear because of migration away from an unfavorable environment. Another possibility is the lack of preservation of remains of organisms that were present.

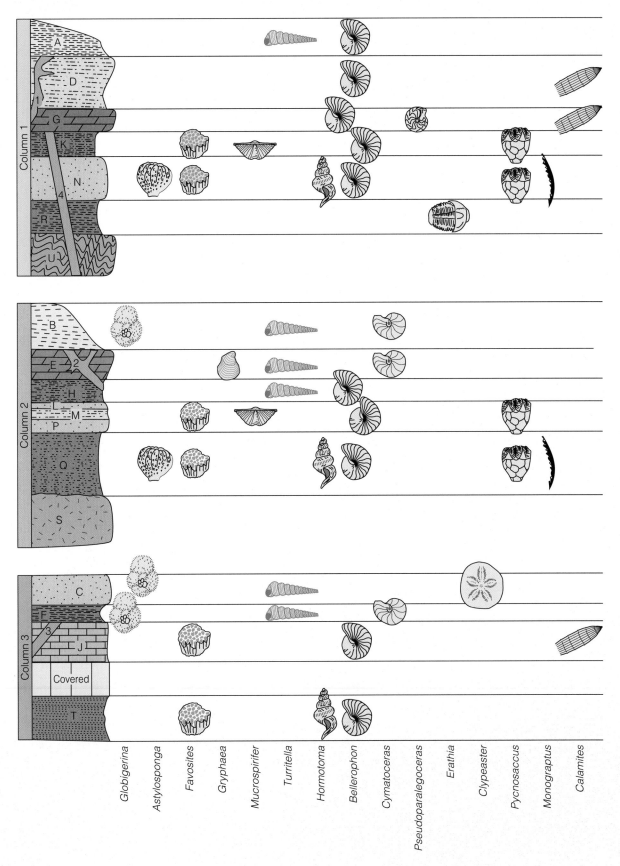

Figure 8.6 The stratigraphic columns from Figure 8.2 with fossils that occur in each layer. Tear out the page, and cut and arrange the layers as instructed in Exercises 4 and 5.

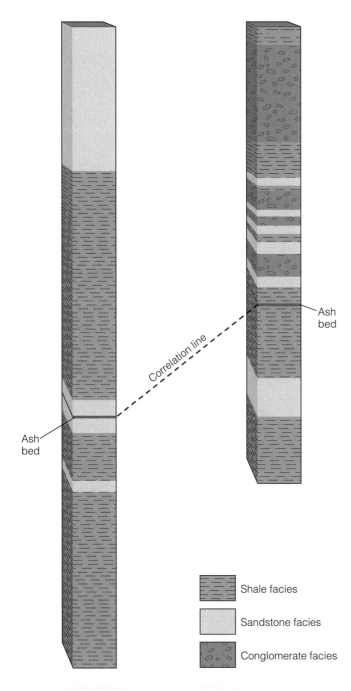

Figure 8.7 Two columns of strata formed in different environments. Both contain a volcanic ash bed that allows for correlation between them. The ash bed is a time marker.

Ash bed

Ash bed

Correlation line

Shale facies

Sandstone facies

Conglomerate facies

the rocks are arranged along a time line, showing gaps between the rocks where the unconformities occur.

7. Examine Figure 8.2 again:

a. In Column 1, what type of feature is the boundary between R and U? _____

b. In Column 2, what type of feature is the boundary between Q and S if the bottom of Q is not metamorphosed? _____

c. What evidence is there, if any, in Figure 8.2, that an unconformity exists between A and D? (If none, write "none.") _____

Between G and K? _____

Between B and E? _____

Between F and J? _____

Between N and R? _____

Label each of these unconformities on Figures 8.8 and 8.2.

d. What evidence in your sequence of strips from Figure 8.6 or the fossil time line you made in Figure 8.8 tells you there is an unconformity between H and L

in Column 2?_____

_____

Label this unconformity on Figures 8.2 and 8.8.

Notice that you could not identify some of the unconformities if it were not for the information provided by fossils in Figure 8.6.

Geologic History

The Earth is constantly changing. Earth movements, volcanic eruptions, intrusions, rivers, glaciers, waves, and other geologic processes, are continually altering the interior and surface of our planet. Although many geologic processes are very slow, they are proceeding now and have modified the Earth extensively in its past. Geologic history is a study of what geologic processes or events occurred in the past that shaped the Earth. Geologic history can be deciphered using the principle of original horizontality, stratigraphic superposition, cross-cutting relationships, and fossil succession. Recall that the principle of original horizontality states that sedimentary layers are deposited approximately horizontally so that inclined strata (not counting cross bedding) indicate tilting of the layers after deposition. Tilted layers, the superposition of strata, and cross-cutting relationships of faults, unconformities, and igneous intrusions can be seen on geologic maps, cross sections (see also Lab 16), and in the field. From these, the history of past geological events can be determined. The relative dating techniques that we already used apply not only to rocks and fossils but also to geologic events so that a historical sequence of geologic events can be deduced. The following step-by-step explanation of an example geologic cross section demonstrates this technique.

Carefully examine the cross section in Figure 8.9. What geologic events occurred to produce the rocks and their geometry in this cross section? We will use the principles of stratigraphic superposition, cross-cutting relationships, and original horizontality to determine what happened here. Figure 8.10 shows a series of cross sections **a–g** that illustrate the sequence of events that resulted in the final cross section. The

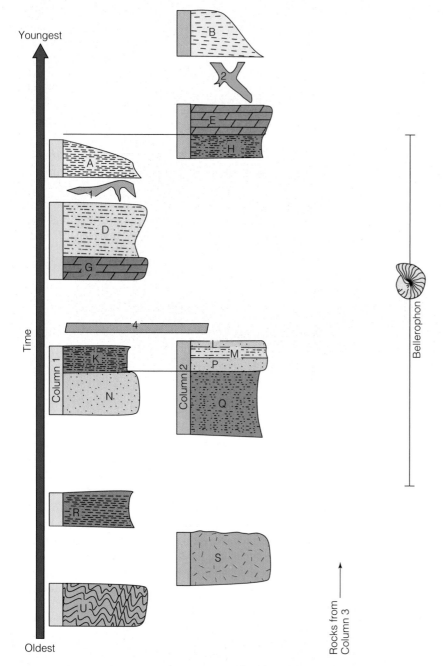

Figure 8.8 The stratigraphic columns from Figure 8.2 are arranged on a time scale with each rock positioned at the time it formed. Time proceeds vertically from oldest at the bottom to youngest at the top. Each dike is also shown at its time of formation. Unconformities occur where there is space (a time gap) between the rock layers. Insert rocks from column 3 (Exercise 5) and add fossils from Figure 8.6 to the timeline (Exercise 6), using the space at the right.

following geologic events occurred and are also summarized in Table 8.1:

a. First stage: *deposition* of beds in order from **1** to **6** followed by *lithification* (Fig. 8.10a).

b. *Folding and uplift* of Beds **1–6** by compression, possibly at a convergent plate boundary. Dike **7** is not folded and must come later (Fig. 8.10b).

c. *Intrusion* of Dike **7** (cuts across beds **1–6**, but not the uncon- formity of beds **8–11**) and volcanic eruption of Lava Flow **7**

(which are not seen in the final Cross Section **g**, because they have been eroded away) (Fig. 8.10c).

d. *Erosion.* The erosional surface (the angular unconfor- mity) in Figure 8.9 cuts across Dike **7** (Fig. 8.10d).

e. *Subsidence* allows the transition from erosion to deposi- tion. *Submergence or sea-level rise* could accomplish the same result (Fig. 8.10e).

f. Then *deposition* of **8–13** in order above the unconformity. *Lithification* follows. Beds **12** and **13** are not seen in the fi-

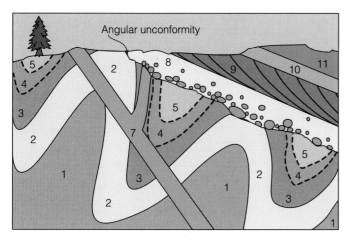

Figure 8.9 Deciphering geologic history from a cross section. The sequence of events that formed this cross section is illustrated step by step in Figure 8.10.

nal Cross Section **g,** because they have been eroded away (Fig. 8.10f).

g. Last stage: *tilting* and *erosion.* The principle of original horizontality tells us that Beds **8–11** were originally horizontal, so they must have been tilted subsequent to deposition (Fig. 8.10g).

Notice how each statement, and almost the complete history, can be deduced from examination of Cross Section g (or Fig. 8.9). Especially important clues in g are (1) cross-cutting relationships—where the dike, the unconformity, and the land surface (an erosional surface) cut across other features indicating that the latter must be older—and (2) stratigraphic superposition—the oldest sediments or volcanic rocks are on the bottom. It is this "detective" type of reasoning that has enabled geologists to deduce so much about the sequence of geologic events in a small region or even the geologic history of the entire Earth.

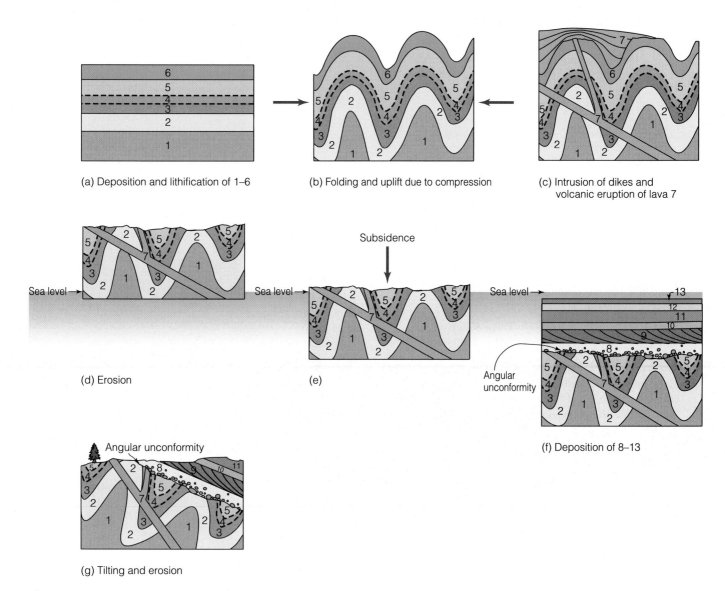

Figure 8.10 Step-by-step sequence of cross sections illustrating the geologic history for the cross section in Figure 8.9

Table 8.1 Sequence of Events That Formed the Cross Section in Figure 8.9

Event Sequence	Geologic Events	Description
Last (youngest)	Erosion	removed 12 and 13 and parts of other layers
9th	Tilting	caused by plate movements
8th	Lithification	of 8–11 and possibly 12 and 13
7th	Deposition	of 8–11 and possibly 12 and 13
6th	Subsidence	or submergence. These allowed for the transition from erosion to deposition.
5th	Erosion	removed 6 and parts of other layers, wearing away mountains to a nearly flat surface.
4th	Intrusion	of 7. Magma intruded along two cracks and volcanic eruptions extruded lava flows. Both then solidified.
3rd	Folding and uplift	compression and mountain building at a convergent plate boundary
2nd	Lithification	of 1–5 and possibly 6
1st (oldest)	Deposition	of 1–5 and possibly 6

Note: Since layers form from the bottom up, the sequence of events is listed the same way.

Table 8.2

Events for Problem Number _____

Geologic History Computer Exercise

Use the computer to practice deciphering geologic history:

- Start **In-TERRA-Active.**
- Click **Materials** then **Geologic Time: Relative.** Click **APPLY.**

8. Do Problem 1 or Problem 2, following the instructions. As you work the problem, fill in the correct sequence of events in Table 8.2. If you do not understand the choices, click EXPLORE, go through the exercise, then go back to APPLY.

- Click **Quit** and **yes.**

9. Deduce the sequence of geologic events in the cross section in Figure 8.11, and enter them in order from oldest to youngest in Table 8.3. Beware: there is a trick in the sequence; carefully examine the graded bedding and the cross bedding in the cross section. Add some additional details about each event in the column labeled "Description." Use Table 8.1 as an example.

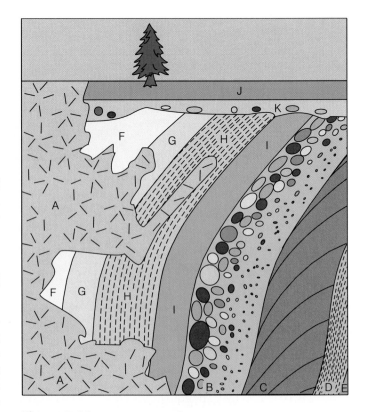

Figure 8.11 Cross section for Exercise 9

Table 8.3 Sequence of Events that Formed the Cross Section in Figure 8.11

Event Sequence	Geologic Events	Description
Last (youngest)		
...		
8th		
7th		
6th		
5th		
4th		
3rd		
2nd		
1st (oldest)		

Cross Section of Devils Fence Quadrangle, Montana

Use the techniques you have acquired to interpret real geologic cross sections of Devils Fence Quadrangle, Montana.

10. Examine the Structure Sections B-B′, C-C′, D-D′, and the map of the Devils Fence Quadrangle, Montana (Fig. 8.12). Unless otherwise directed, use all three sections to answer the following questions.

    a. What rocks are cut by the fault near Єw on section B-B′? _____

    b. Is faulting more likely to have occurred during deposition or after deposition stopped?_____

    c. Did the folding occur before or after the intrusion of gd?_____

    d. Look up ad in the Explanation on the map. What is its rock name and type?_____ _____

    e. What type of rock mass is it?_____

    f. Did the folding occur before or after the process forming ad?_____

    g. Where does ad fit in the sequence of events? _____

    h. In Table 8.4 , list the sequence of events that occurred in the Devils Fence Quadrangle from oldest to youngest. Use your answers to Questions **a–g** to help determine the sequence and the details. Next to each event, explain the event in more detail.

Numerical Dating

Numerical dating (or absolute dating) of a sample determines the age in years of a rock, fossil, or artifact using decaying isotopes in the specimens either directly or indirectly. These dates are given in millions of years (Myr) for some dating techniques and thousands of years for others. Notice *numerical* or *absolute* here does not mean the same as *precise* or *accurate,* so error may be associated with the numerical date, and is usually denoted by a plus-or-minus sign. Whereas relative dating relies on stratigraphic superposition and cross-cutting relationships, numerical dating uses *isotopic dating* and *indirect dating* to build the history of an area.

Isotopic Dating

In the Earth, the nuclei of some atoms are unstable and decay in a process called **radioactivity.** That means they spontaneously convert to other elements by gaining or losing particles from the nucleus (Fig. 8.13). Such atoms are called radioactive or **radiogenic isotopes.**[3] Isotopic dating uses this natural radioactive decay to measure the age of rocks. Isotopes that are radioactive have a natural rate of decay that is known as the *half-life.* Half-life is the length of time it takes for half of parent atoms to decay into daughter atoms. Radioactive decay in a crystal is shown in Figure 8.14. Each half-life reduces the number of parent atoms (green) by half and increases the number of daughter atoms (red) by the same amount. The sum of parent and daughter atoms remains

·············
[3]**Isotopes** are atoms having the same number of protons but different numbers of neutrons, which means that they are the same element and chemically similar but have different atomic mass.

Figure 8.12 (*see next page*) Geologic map and structure sections of part of Devils Fence Quadrangle, Montana

GEOLOGIC MAP OF DEVILS FENCE QUADRANGLE, MT.

EXPLANATION

Contour interval 40 feet
Datum is mean sea level

QUATERNARY

Recent

Qal

Alluvium

Pliocene (?)

Tfg

Fan gravel and pediment gravel
Unconsolidated or weakly consolidated fan deposits and gravel veneer on pediments

TERTIARY

Miocene (?)

Ttg

Tuff and gravel
Light-colored rhyolitic tuff with a few interbeds of stream gravel

UNCONFORMITY

CRETACEOUS

Lower and Upper Cretaceous

Kcm / Kcl

Colorado formation
Upper black shale unit not mapped. Middle siliceous mudstone and sandstone unit, Kcm, is intertongued marine and nonmarine sandstone, mudstone, and siliceous mudstone, in part of volcanic origin. Lower black shale unit, Kcl, is a basal tan-weathering quartz sandstone, drab and olive-gray blocky siltstone, very dark gray to black shale, and an upper dark-gray carbonaceous, limonitic speckled to mottled sandstone

EROSIONAL UNCONFORMITY

Lower Cretaceous

Kk

Kootenai formation
Comprises three units. Upper unit is 10 to 25 ft of gastropod-bearing limestone overlain by a few feet to 80 ft of drab mudstone; middle unit is red and green mudstone and shale with concretions and lentils of limestone; lower unit is crossbedded "pepper-and-salt" sandstone and interbedded shale and mudstone

JURASSIC

Upper Jurassic

Ju

Morrison and Swift formations
Morrison formation: varicolored nonmarine shale, mudstone, and siltstone, with thin beds of limestone and sandstone, and near the top a unit of black shale. Upper hundred feet locally contains thick, lenticular "pepper-and-salt" sandstone and grades into overlying Kootenai formation
Swift formation: grayish-brown punky calcareous marine sandstone, 20 to 35 ft thick, with a basal chert-pebble conglomerate

EROSIONAL UNCONFORMITY

PERMIAN

Pp / PPq

Phosphoria formation
Brown and gray chert and sandstone, in part phosphatic; may locally contain one or two thin beds of phosphate rock, Pp. In places mapped with the Quadrant formation, PPq

PENNSYLVANIAN

Pq

Quadrant formation
Light-colored quartzitic sandstone and interbedded light-gray sugary-textured sandy dolomite

PMa

Amsden formation
Red to grayish-red mudstone, shale, and subordinate amounts of carbonate rock with interbeds of gray, brown, or yellow argillaceous sandstone in upper and lower parts; middle part of medium- to dark-gray thick-bedded dolomite

EROSIONAL UNCONFORMITY (?)

MISSISSIPPIAN

Madison group

br / Mmc

Mission Canyon limestone
Medium-gray to light-gray medium-grained thickly and indistinctly bedded limestone, with a few thin siliceous layers in lower 200 ft and sparse gray chert nodules and lentils in upper half. A breccia unit, br, about 200 ft below top of formation has been mapped locally

Ml

Lodgepole limestone
Upper part of medium-gray fine- to medium-grained limestone in distinct beds as much as 3 ft thick alternating with zones of much thinner beds containing rare mudstone partings; lower part of medium-gray limestone in beds 1 in. to 1 ft thick with partings and interbeds of yellow to red calcareous mudstone; grades into Mission Canyon limestone through a 150- to 200-ft zone

CARBONIFEROUS

46° 10'

B'

D'

STRUCTURE SECTIONS

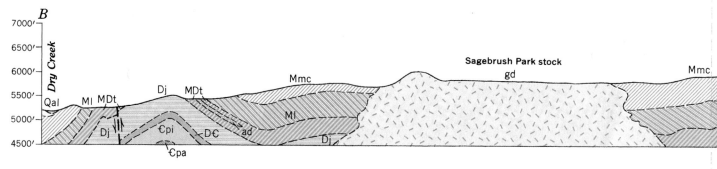

EXPLANATION Continued

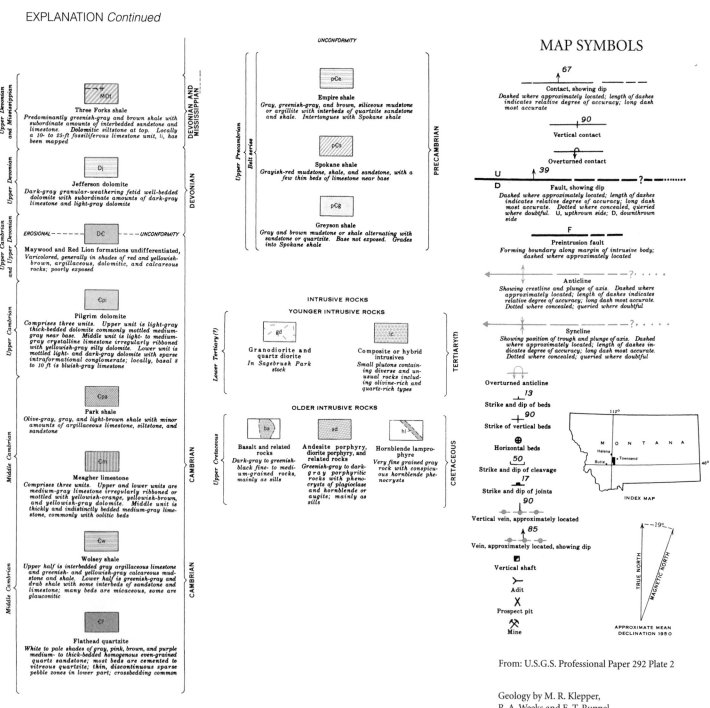

Upper Devonian and Mississippian

Three Forks shale

Predominantly greenish-gray and brown shale with subordinate amounts of interbedded sandstone and limestone. Dolomitic siltstone at top. Locally a 10- to 25-ft fossiliferous limestone unit, li, has been mapped

Upper Devonian

Jefferson dolomite

Dark-gray granular-weathering fetid well-bedded dolomite with subordinate amounts of dark-gray limestone and light-gray dolomite

EROSIONAL — UNCONFORMITY

Upper Cambrian and Upper Devonian

Maywood and Red Lion formations undifferentiated,
Varicolored, generally in shades of red and yellowish-brown, argillaceous, dolomitic, and calcareous rocks; poorly exposed

Upper Cambrian

Pilgrim dolomite

Comprises three units. Upper unit is light-gray thick-bedded dolomite commonly mottled medium-gray near base. Middle unit is light- to medium-gray crystalline limestone irregularly ribboned with yellowish-gray silty dolomite. Lower unit is mottled light- and dark-gray dolomite with sparse intraformational conglomerate; locally, basal 8 to 10 ft is bluish-gray limestone

Park shale

Olive-gray, gray, and light-brown shale with minor amounts of argillaceous limestone, siltstone, and sandstone

Middle Cambrian

Meagher limestone

Comprises three units. Upper and lower units are medium-gray limestone irregularly ribboned or mottled with yellowish-orange, yellowish-brown, and yellowish-gray dolomite. Middle unit is thickly and indistinctly bedded medium-gray limestone, commonly with oolitic beds

Wolsey shale

Upper half is interbedded gray argillaceous limestone and greenish- and yellowish-gray calcareous mudstone and shale. Lower half is greenish-gray and drab shale with some interbeds of sandstone and limestone; many beds are micaceous, some are glauconitic

Flathead quartzite

White to pale shades of gray, pink, brown, and purple medium- to thick-bedded homogenous even-grained quartz sandstone; most beds are cemented to vitreous quartzite; thin, discontinuous sparse pebble zones in lower part; crossbedding common

---

UNCONFORMITY

Upper Precambrian — Belt series

Empire shale

Gray, greenish-gray, and brown, siliceous mudstone or argillite with interbeds of quartzite sandstone and shale. Intertongues with Spokane shale

Spokane shale

Grayish-red mudstone, shale, and sandstone, with a few thin beds of limestone near base

Greyson shale

Gray and brown mudstone or shale alternating with sandstone and quartzite. Base not exposed. Grades into Spokane shale

PRECAMBRIAN

INTRUSIVE ROCKS

YOUNGER INTRUSIVE ROCKS

Granodiorite and quartz diorite
In Sagebrush Park stock

Lower Tertiary (?)

Composite or hybrid intrusives
Small plutons containing diverse and unusual rocks including olivine-rich and quartz-rich types

TERTIARY (?)

OLDER INTRUSIVE ROCKS

Basalt and related rocks
Dark-gray to greenish-black fine- to medium-grained rocks, mainly as sills

Andesite porphyry, diorite porphyry, and related rocks
Greenish-gray to dark-gray porphyritic rocks with phenocrysts of plagioclase and hornblende or augite; mainly as sills

Hornblende lamprophyre
Very fine grained gray rock with conspicuous hornblende phenocrysts

Upper Cretaceous — *CRETACEOUS*

MAP SYMBOLS

67
Contact, showing dip
Dashed where approximately located; length of dashes indicates relative degree of accuracy; long dash most accurate

90
Vertical contact

Overturned contact

U   **39**   ?
D
Fault, showing dip
Dashed where approximately located; length of dashes indicates relative degree of accuracy; long dash most accurate. Dotted where concealed; queried where doubtful. U, upthrown side; D, downthrown side

F
Preintrusion fault
Forming boundary along margin of intrusive body; dashed where approximately located

Anticline
Showing crestline and plunge of axis. Dashed where approximately located; length of dashes indicates relative degree of accuracy; long dash most accurate. Dotted where concealed; queried where doubtful

Syncline
Showing position of trough and plunge of axis. Dashed where approximately located; length of dashes indicates relative degree of accuracy; long dash most accurate. Dotted where concealed; queried where doubtful

Overturned anticline

13
Strike and dip of beds

90
Strike of vertical beds

Horizontal beds

50
Strike and dip of cleavage

17
Strike and dip of joints

90
Vertical vein, approximately located

85
Vein, approximately located, showing dip

Vertical shaft

Adit

Prospect pit

Mine

INDEX MAP

From: U.S.G.S. Professional Paper 292 Plate 2

Geology by M. R. Klepper,
R. A. Weeks and E. T. Ruppel

Base from U. S. Geological Survey Form Map of Devils Fence Quadrangle, Montana.

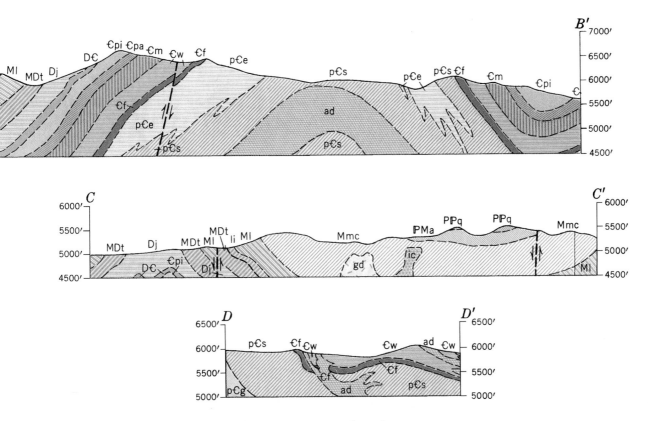

Table 8.4 Sequence of Events for Devils Fence Quadrangle

Event Sequence	Geologic Events	Description
Last (youngest)		
8th		
7th		
6th		
5th		
4th		
3rd		
2nd		
1st (oldest)		

constant, because as the parents decrease, the daughters increase. Each isotope has a fixed half-life, which can be measured by present-time observations made in a laboratory of the decay of the radioactive isotope. The rate of decay of the radioactive isotope is constant and specific for each isotope; some examples are given in Table 8.5.

Parent and daughter atoms are separated from each other by chemical reactions such as crystallization from magma or recrystallization during metamorphism (Figs. 8.15 and 8.16). So at the time of the chemical reactions that form minerals in a rock, when radioactive material becomes incorporated into a rock, the clock starts ticking (Figs. 8.15b

and 8.16a and c). At this starting time, no daughter isotopes should be present,[4] because they are chemically different from their parent atoms and incompatible with the crystal or chemical substance forming. Then, as time passes, the parent atoms decay and daughter atoms begin to appear in the mineral at a predictable rate (based on the half-life, Figs. 8.15c and 8.16b and d). Every time a parent decays, a daugh-

··········

[4]In practice, daughter atoms can be inherited from previous minerals, so this is not always perfect. If this happens, the measured age of the rock is older than the actual age.

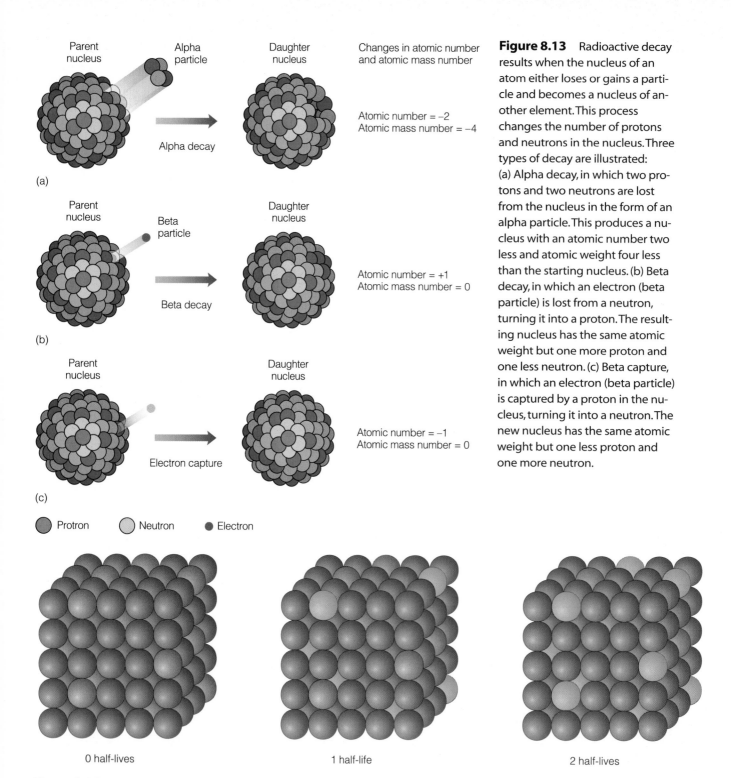

Parent nucleus — Alpha particle — Daughter nucleus

Alpha decay

(a)

Changes in atomic number and atomic mass number

Atomic number = –2
Atomic mass number = –4

Parent nucleus — Beta particle — Daughter nucleus

Beta decay

(b)

Atomic number = +1
Atomic mass number = 0

Parent nucleus — Daughter nucleus

Electron capture

(c)

Atomic number = –1
Atomic mass number = 0

Figure 8.13 Radioactive decay results when the nucleus of an atom either loses or gains a particle and becomes a nucleus of another element. This process changes the number of protons and neutrons in the nucleus. Three types of decay are illustrated: (a) Alpha decay, in which two protons and two neutrons are lost from the nucleus in the form of an alpha particle. This produces a nucleus with an atomic number two less and atomic weight four less than the starting nucleus. (b) Beta decay, in which an electron (beta particle) is lost from a neutron, turning it into a proton. The resulting nucleus has the same atomic weight but one more proton and one less neutron. (c) Beta capture, in which an electron (beta particle) is captured by a proton in the nucleus, turning it into a neutron. The new nucleus has the same atomic weight but one less proton and one more neutron.

⬤ Protron ⬤ Neutron • Electron

0 half-lives   1 half-life   2 half-lives

Figure 8.14 A crystal starts at time equal to zero half-lives, with blue atoms representing stable isotopes and green ones radioactive isotopes of the same element. At first there are no daughter isotopes because they are chemically incompatible with the blue and green element. After one half-life has passed, half of the green atoms have decayed to their daughter isotope, the red atoms. After two half-lives, another half of the green atoms has decayed, leaving one-fourth of the original amount and three-fourths of the red daughter atoms. Continued decay reduces the number of parents by half again and again.

ter forms from it. The daughter atoms become trapped in the crystal and cannot subsequently escape,[5] and its isotopic

....

[5]Again, in practice, daughter atoms may sometimes leak out, especially if they are gases (e.g., ^{40}Ar). If this happens, the measured age of the rock is younger than the actual age.

date gives the time since this crystallization occurred. To date a rock, we need three pieces of information: the half-life of an appropriate radioactive isotope present in the rock, the relative proportion of parent atoms (the radioactive isotope) and of daughter atoms present now. Instead of amounts, ratios or proportions of atoms are sufficient.

Table 8.5 Some Radioactive Isotopes Used for Dating Rocks

Name of Dating System	Radioactive Isotope (parent) Used for Dating Rocks	Symbol of Parent Isotope	Daughter Isotope	Symbol of Daughter Isotope	Half-Life	Age Range	Rock Type
Carbon-14 dating	Carbon-14	^{14}C	Nitrogen-14	^{14}N	5730 yr ±30	100 and 70,000 yr	Sedimentary
Potassium-argon dating	Potassium-40	^{40}K	Argon-40	^{40}Ar	1.3 Byr	50,000 yr to 4.6 Byr	Igneous/ metamorphic
Uranium-lead dating	Uranium-238	^{238}U	Lead -206	^{206}Pb	4.56 Byr	10 M to 4.6 Byr	Igneous/ metamorphic
Uranium-235 dating	Uranium-235	^{235}U	Lead-207	^{207}Pb	704 Myr	10 M to 4.6 Byr	Igneous/ metamorphic
Rubidium-strontium dating	Rubidium-87	^{87}R	Strontium-87	^{87}Sr	48.8 Byr	10 M to 4.6 Byr	Igneous/ metamorphic

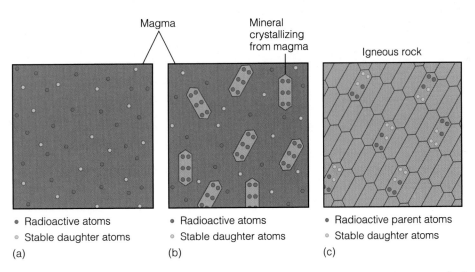

Magma   Mineral crystallizing from magma   Igneous rock

- Radioactive atoms
- Stable daughter atoms

(a)

- Radioactive atoms
- Stable daughter atoms

(b)

- Radioactive parent atoms
- Stable daughter atoms

(c)

Figure 8.15 Chemical reactions, such as the formation of crystals from a magma, start the radiometric clock ticking. (a) Both radioactive parent and stable daughter isotopes swim around in the magma. (b) As the magma cools, some minerals crystallize with radioactive parent atoms. The daughter atoms are chemically different and therefore incompatible with and excluded from crystals containing the parent atoms. The inclusion of parents and exclusion of daughters chemically sets the radiometric clock to 0. (c) Within the crystals, as time passes, the parent atoms gradually convert to daughter atoms. Notice that one half-life has passed in this diagram, because there are equal numbers of parent and daughter atoms in the crystals.

Half-Life Activity   The following activity is intended to help you understand the half-life and isotopic dating. Some groups will have to start as observers (see observation later). Start with 128 playing cards (the decks may look different). Your group should secretly decide how long the half-life in this exercise will be—some value between 20 seconds and 3 minutes—and for how many half-lives you will run your experiment—a number between 2 and 6. Cards face down will represent the radioactive isotope and face up will be the daughter isotope. Make piles of eight cards, so it will be faster to flip the correct number over. Practice your experiment as follows until you get it right. Start with all the cards face down. After one half-life has gone by, flip over half of the cards. After another half-life has passed, flip over half of the remaining face-down cards. Repeat this process until you have reached the number of half-lives you chose. Record the following information:

11. How many cards were left face down at the end of your experiment? _____

How many face up? _____

How long is the half-life you selected? _____

How many half-lives passed before you stopped the experiment? _____

How long did your experiment run? _____
Show any calculations here.

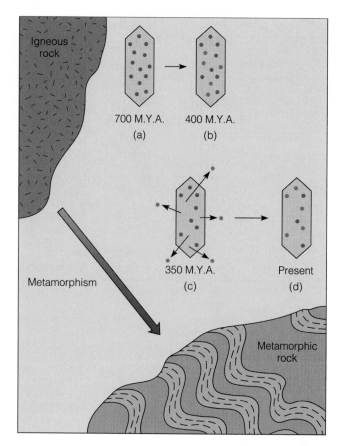

Figure 8.16 Chemical reactions, such as the recrystallization of minerals during metamorphism, resets the radiometric clock within the crystals to 0. (a) Crystallization from magma produces a crystal with only parent atoms (blue). (b) Passage of time converts parents in the crystal to daughter atoms (red). (c) Metamorphism triggers chemical reactions that expel the chemically incompatible daughter atoms from the crystal. (d) Passage of more time converts more parents to daughters. If this rock remained in chemical isolation, so the daughters stayed in the rock during metamorphism (although not in the crystal with the parents), dating the crystal would give the time of metamorphism.

When you have your timing worked out ask another group to observe. Run your experiment and tell the other group precisely when it is completed. Let the other group figure out for themselves the information you recorded.

Observation: Now find a group that has practiced demonstrating their half-life exercise and is ready for your group to observe. Observe them to determine and record the following information:

12. How long is their half-life? _____

How many cards were left face down at the end of their experiment? _____

How many cards were face up? _____

How many half-lives passed before they stopped the experiment? _____

How long did their experiment run? _____

This is similar to determining the age of a rock using radioactive isotopes. Show your calculations below. Compare your answers to their answers.

Are you close? _____ If so, how far off are your answers, and what were the sources of any differences?

 Isotopic Dating Computer Exercise

- Start **Earth Systems Today.**
- Go to **Topic: Earth's Processes.**
- Go to **Module: Geologic Time.**
- Explore activities in this module to discover the following for yourself.
- Use the hour-glass animation to explore the decay rate of a radioactive system.

13. How does the line plotted relate to the age?

_____

_____

_____

14. Use the parent-daughter cross-plot graph to answer these questions:

a. How does the amount of daughter isotope change as the amount of parent changes? _____

_____

b. How do the decay rates of different age samples of the same parent compare? _____

_____

15. Complete the following:
a. Fill in Table 8.6 with of the number of parent and daughter isotopes there would be after each number of half-lives if you started with 100 quadrillion atoms of the parent isotope (or you can think of this as 100%).
b. Graph your results on the graph paper in Figure 8.17.

- Place time in half-lives on the horizontal axis and percentage or number of atoms on the vertical axis.

Table 8.6 Radioactive decay of 100 Quadrillion Atoms of a Radiogenic Isotope

Time in Half-Lives	Number of Parent Atoms in Quadrillions (or %)	Number of Daughter Atoms in Quadrillions (or %)
0	100	
1		
2		
3		
4		
5		
6		
7		
8		

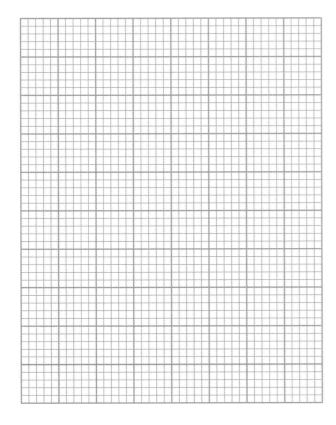

Figure 8.17 Linear graph paper

- Label the horizontal axis with numbers in units of 1 and the vertical axis in units of 10.
- Plot parent atoms in one color or with one symbol and daughter atoms in another.
- Make a key showing which is which.
- Give your graph a title that describes what it shows. Do not use "Time versus Parent Atoms" as the title; this information is already on the graph. Instead, choose a title that gives additional information.

- Draw two smooth curves through the points on your graph, one for parent atoms and one for daughter atoms.

16. Reading from your graph, how many half-lives would be equivalent to 75% parent atoms? _____ How certain are you of this number? _____ If you had drawn your curve slightly differently, would you get the same answer? _____

Example Problem 1 Let's look at an example rock and try to determine its age.

A granite contains crystals with 150 ppma (parts per million by atomic proportions) of ^{235}U and 1050 ppma of ^{207}Pb. The half-life for the decay of ^{235}U to ^{207}Pb is 704 Myr. How old is the rock?

Calculations:

Add up the parents and daughters. This equals the original number of parents present at the time of the chemical reactions that formed the minerals in the rock:

$$\begin{array}{r} 150 \\ +1050 \\ \hline 1200 \end{array}$$

Therefore, when the rock formed, 1200 ppma radioactive parent atoms were present. We know from the definition of the half-life that the uranium parents will decay by half in 704 Myr. Keep on dividing the parents in half, as shown in Table 8.7, until you get the number found today. Count how many times you divided by 2. We divided the parents in half three times to get to the present level of uranium. Thus, three half-lives have passed since the rock formed.

3 half-lives × 704 Myr/(half-life) =

Answer:

2112 Myr = 2.11 billion years (Byr)

Table 8.7 (Example 1)

Time (in half-lives)	Parents (ppma)
0	1200
1	600
2	300
3	150

$$\frac{1200}{2^h} = 150,$$

where h is the number of half-lives

$$\frac{1200}{2^3} = 150$$

← Stop here—this is how many parents are present today; it took 3 half-lives of time to get this many parent atoms.

17. Determine the age of shale with leaf fossils containing 21 ppma of ^{14}C and 651 ppma of ^{14}N. Look up the half-life in Table 8.5. Write out the complete step-by-step procedure used to determine your answer:

When determining the age of a rock by isotopic dating, if the number of half-lives is a whole number, you need only divide by 2 the right number of times to determine the answer. But to get an accurate age with partial half-lives, you need to use a mathematical expression or a graph. The following equation will give the age in half-lives since Time 0. (Time zero is the time of the chemical reactions that formed the substance being measured):

$$h = \text{age in half-lives} = 3.322 \times \log[(p + d)/p],$$

where p is the number of parent atoms and d is the number of daughter atoms. Parent and daughter atom figures need to be in the same units. Try this equation with the numbers used earlier to see whether you get the same result.

18. If you graph the values in Table 8.6 on semilogarithmic graph paper, they should lie on a straight line. Try this by plotting the values for Time 0 and for Time 8 on the graph paper in Figure 8.18. Draw a straight line between these two points, and see whether your other points would fall on the same line. Do they? _____

Reading from this graph, how many half-lives would be equivalent to 75% parent atoms? _____

Since with this type of graph you could draw a straight line through the points, your answer for 75% is more dependable than for your previous graph.

Clastic Sedimentary Rocks and Isotopic Dating

Example Problem 2 In this next example, the number of half-lives does not come out as a whole number, so use the

equation. Also, notice that this is a clastic sedimentary rock, which will be treated differently (discussed shortly).

A sandstone contains grains with 2.03 ppma ^{238}U and 0.17 ppma ^{206}Pb. How old is the rock?

Use the equation to determine the age of the grains.

$$p = 2.03 \text{ ppma}; d = 0.17 \text{ ppma}$$

$$3.322 \times \log[(2.03 + 0.17)/2.03] = 0.116$$

This means that 0.116 half-lives are needed to bring this number of parents down to the present level at 2.03 ppma. Look up the half-life for ^{238}U in Table 8.5.

0.116 half-lives $\times$ 4.56 Byr/(half-life) = 529 Myr

Answer:

529 Myr or 0.529 Byr

19. Another way to do this same problem is to use your graph in Figure 8.18. Calculate what percentage 2.03 is of the total original number of parents (2.03 + 0.17), then read the half-lives off the graph: _____. How old is this in millions of years? _____ How does this value compare to the calculated value?

_____

Now let's get back to the fact that this is a clastic sedimentary rock. From the calculations, the minerals in this rock are 529 Myr old. Time 0 was 529 Myr ago. But how does Time 0 relate to the age of this clastic sedimentary rock? This is *not* the age of the rock. Time 0 is the age of the clastic grains that came from a former igneous or metamorphic rock that became weathered and eroded to make the sediment in this sedimentary rock. This rock must be younger than 529 Myr, and that is all you can say. The dating technique does not tell you how much younger—just younger.

Carbon-14 dating can give ages of sedimentary rocks and fossils since both of these can have chemical reactions involving carbon close to the time they form. Carbon-14 has a short half-life compared to the age of many rocks and cannot be used if the sample is over 70,000 yr (Table 8.5). For substances older than 70,000 yr, isotopic dating generally applies to igneous and metamorphic rocks. We can, however, combine isotopic and relative dating. Using what we call *indirect dating* (discussed later), we can determine ranges of ages for materials that do not have isotopic age measurements, especially sedimentary rocks and fossils.

20. Determine the age of a conglomerate that has andesite clasts containing crystals with a $^{87}Rb/^{86}Sr$ ratio of 63.90 and a $^{87}Sr/^{86}Sr$ ratio of 0.73. (In this problem, the units are the same for the parent and daughter atoms, so you can use the values as in the example even though they are ratios.) Table 8.5 provides additional information you may need. Use the equation to determine the age of the andesite in half-lives: _____, and in millions of years: _____. Compare these answers with

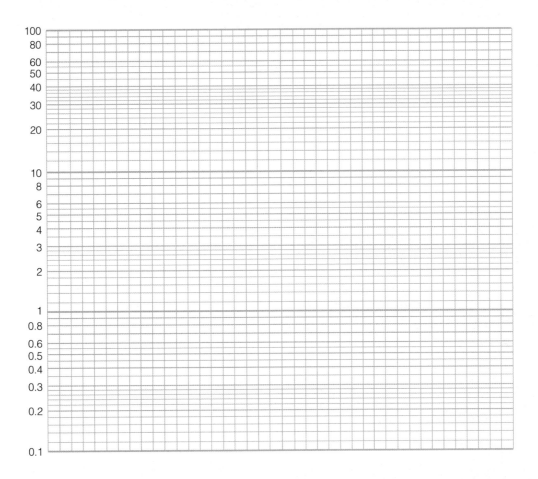

Figure 8.18 Semilogarithmic graph paper

answers using your graph in Figure 8.18. How old is the conglomerate? _____

Indirect Dating

We saw earlier that isotopic dating does not generally work on old sedimentary rocks. On the other hand, indirect methods can produce dates in numerical terms (Fig. 8.19). Isotopic dating can give the age of volcanic rocks in layered sequences with sedimentary rocks, plutonic rocks that cut across sedimentary rocks (Fig. 8.19a), and even igneous (Fig. 8.19a) or metamorphic rocks below a nonconformity (Fig. 8.19b), or where the sedimentary rock has been metamorphosed. Then the relative dating techniques of stratigraphic superposition and cross-cutting relationships can give the range of possible ages for the sedimentary rocks.

Example Problem 3 Use the cross section in Figure 8.20 and the isotopic ages of the igneous rocks labeled in the cross section to determine the age of the Dakota Sandstone.

Answer:

The Dakota Sandstone is below and therefore older than the Mancos Shale, which is older than the igneous dike dated at 66 Myr. The Dakota Sandstone is above and

therefore younger than the volcanic ash dated at 160 Myr. So the age of the Dakota is between 66 and 160 Myr.

21. How old are the Morrison Formation and the Wasatch Formation in Figure 8.20? Explain your reasoning.

_____

_____

_____

_____

22. If Layer I in Figure 8.11 is a basalt lava flow dated at 240 Myr and Intrusion A is a gabbro dated at 184 Myr, how old is Conglomerate B? _____ How old is Shale H? _____ How old is Conglomerate K?

_____

23. Study the dikes in Figure 7.21. If the isotopic age of P is 120 Myr and of W is 52 Myr, how old are the undated rocks, F and N?

F _____ N _____

24. The ages of the dikes, other igneous rocks, and metamorphic rocks from Figure 8.2 have been determined us-

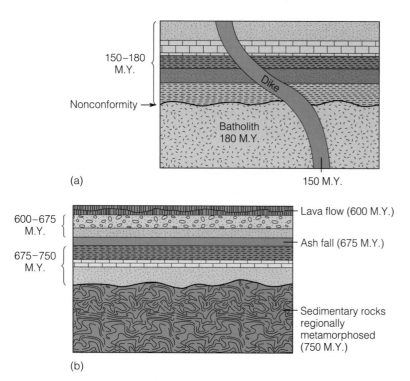

(a)

150–180
M.Y.

Nonconformity →

Dike

Batholith
180 M.Y.

150 M.Y.

(b)

600–675
M.Y.

675–750
M.Y.

Lava flow (600 M.Y.)

Ash fall (675 M.Y.)

Sedimentary rocks
regionally
metamorphosed
(750 M.Y.)

Figure 8.19 Indirect dating. Relative dating gives approximate or ranges of numerical ages for sedimentary rocks when used in conjunction with isotopic dating of associated igneous and metamorphic rocks. (a) The sedimentary rocks were deposited on the batholith and are therefore younger than the batholith. They were intruded by the dike and are therefore older than the dike. Thus, they formed between 180 and 150 Myr ago. (b) Old sedimentary rocks that were later metamorphosed formed before the metamorphism 750 Myr ago. The sedimentary layers just above the metamorphic rocks formed between the metamorphism (750 Myr) and the ash fall (675 Myr). The upper sedimentary rocks formed between the ash fall (675 Myr) and the lava flow (600 Myr).

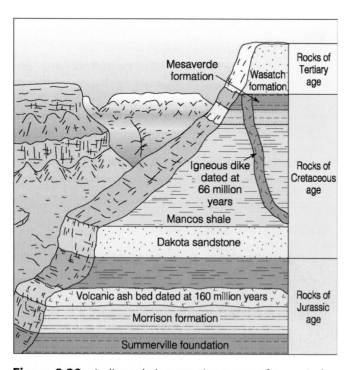

Mesaverde
formation
Wasatch
formation

Rocks of
Tertiary
age

Igneous dike
dated at
66 million
years

Rocks of
Cretaceous
age

Mancos shale

Dakota sandstone

Volcanic ash bed dated at 160 million years

Morrison formation

Summerville foundation

Rocks of
Jurassic
age

Figure 8.20 Indirect dating can give ranges of numerical ages for the sedimentary rocks. See text for details.

Table 8.8 Isotopic Ages of Rocks in Columnar Sections in Figure 8.2

Rock	Isotopic Age (yr)
1	240 Million
2	140 Million
3	70 Million
4	370 Million
S	560 Million
U	640 Million

By now you probably have some respect for the talents needed to collect and process the information to complete the geologic history for an area. You may have experienced some frustration with these numerical ages because you know that *Elrathia*, for example, is older that *Mucrospirifer*, but the numerical ages do not show the difference. The frustrations over apparent conflicts and missing data are usually overcome by the inquisitiveness of the investigators and their desire to complete the "puzzle" of the area. It is only through careful work with more indirect dating that the age ranges can be narrowed. Nevertheless, you have now completed a geologic time scale for the rocks in Figures 8.2 and 8.8.

ing isotopic dating. The values are given in Table 8.8. Use this information and the time line in Figure 8.8 that you developed using principles of superposition and cross-cutting relationships to help you determine the possible range of ages for each sedimentary rock layer and fossil. Enter these in Table 8.9.

25. If you found a rock containing the fossil *Gryphaea*, how old would the rock be? _____

Table 8.9 Age Ranges for Sedimentary Layers and Fossils in Figure 8.6 Based on Indirect Dating

Rock Layer or Fossil	Possible Age Range (Myr)	Geologic Period(s)
B C F *Globigerina* *Clypeaster*		
Turritella *Cymatoceras*		
A E H		
D G J *Pseudoparaleoceras* *Calamites*		
K L M N P Q R T *Astylospongia* *Mucrospirifer* *Hormotoma* *Elrathia* *Pycnosaccus* *Monograptus*		
Favosites		
Bellerophon		

Geologic Time Scale

The geologic time scale (Fig. 8.21) was assembled over many years using data from many regions of the Earth in a similar manner to the way you worked with Figures 8.2, 8.6, and 8.8. At first, the time scale delineated age that was purely relative without any specific times or dates. It was only later, when isotopic dating techniques became available, that the geologic time scale acquired its numerical ages, the numbers along its edge (Fig. 8.21). For example, rocks of the Devonian period were known to be younger than Cambrian rocks and older than Permian rocks, but not how much younger and older, and even an approximate age for Devonian was unknown. The numbers on the geologic time scale were determined by indirect dating techniques in many places all over the world. They are still being revised as new rocks and fossils are found and dates are measured with improved isotopic techniques.

26. What are the corresponding geologic period(s) (Fig. 8.21) for each rock and fossil in Table 8.9? Enter these in the table.

27. Discuss the following question in a group and summarize your conclusions here: If you were geologists trying to narrow down the numerical age of fossils defining the Cretaceous-Tertiary boundary (KT boundary), how would you go about it?

_____

_____

_____

The Earth is approximately 4600 million, or 4.6 billion, years old. Life on Earth began sometime around 3.6 to 3.8 Byr ago. However, *Homo sapiens* such as ourselves have only been around 100,000 yr, which is a tiny fraction of the age of the Earth or the existence of life on Earth. If we compare the time life has existed on Earth to 1 day (24 h), modern humans have only been around for about 2 seconds. Let's see whether we can gain a better understanding of the scale of the vastness of geologic time.

28. Find items such as paperclips in the lab that are small, numerous, and fairly uniform. Use your imagination.

    a. Measure a modest quantity of these things so that you can estimate how much would make 100,000 of them. You may want to count 20 or so and measure in some way how much, how big, or how heavy those are. Then calculate how much, big, or heavy 100,000 of them would be. Make conversions as needed so the measurements are intuitively meaningful to you. Write out what you did and the measurements and calculations you made.

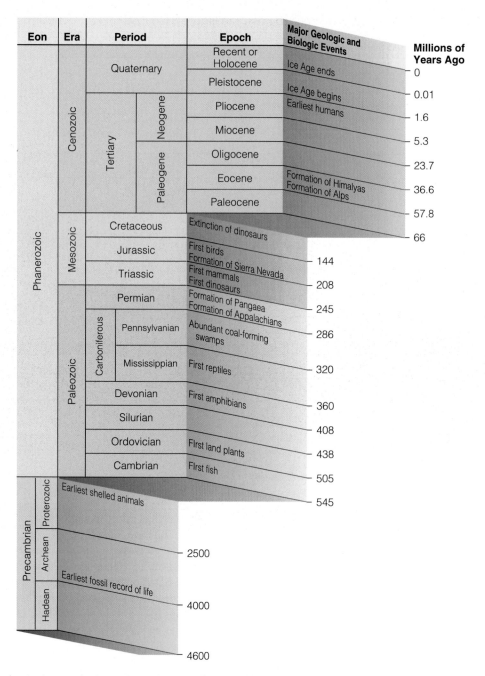

Eon	Era	Period		Epoch	Major Geologic and Biologic Events	Millions of Years Ago
Phanerozoic	Cenozoic	Quaternary		Recent or Holocene	Ice Age ends	0
				Pleistocene	Ice Age begins	0.01
		Tertiary	Neogene	Pliocene	Earliest humans	1.6
				Miocene		5.3
			Paleogene	Oligocene		23.7
				Eocene	Formation of Himalyas Formation of Alps	36.6
				Paleocene		57.8
	Mesozoic	Cretaceous			Extinction of dinosaurs	66
		Jurassic			First birds Formation of Sierra Nevada	144
		Triassic			First mammals First dinosaurs	208
	Paleozoic	Permian			Formation of Pangaea Formation of Appalachians	245
		Carboniferous	Pennsylvanian		Abundant coal-forming swamps	286
			Mississippian		First reptiles	320
		Devonian			First amphibians	360
		Silurian				408
		Ordovician			First land plants	438
		Cambrian			First fish	505
Precambrian	Proterozoic				Earliest shelled animals	545
	Archean					2500
	Hadean				Earliest fossil record of life	4000
						4600

Figure 8.21 Geologic time scale. Some important geologic and biological milestones are listed at the right.

b. Next estimate how much would make 4.6 billion of them. Again, make conversions so the measurements are intuitively meaningful to you. Write out what you did and the measurements and calculations you made. Compare your results and choice of objects with others in the class. Write down your impressions or reactions to the results.

Earth's Structure and Plate Tectonics

The internal structure of Earth and its surface structure interact to produce the ocean basins and the continents, the deepest abyss and the highest mountains. To understand this interaction, we start with a look at Earth's interior layers. We study its surface layer and its division into pieces called *plates,* especially the edges of the plates and how they move, a field known as **plate tectonics.** Plate boundaries strongly influence the shape of the land surface and the configuration of the ocean floor. To understand these influences, we will study the differences among the types of plate boundaries: divergent, convergent, and transform. Later in this lab, we will explore how different types of plate boundaries and the movement of plates influence the shapes of continents and oceans and the formation of mountain ranges and rocks.

The Structure of Earth

Earth's Layers

Earth is a layered body, with the densest layers in the center and lower **density** (mass/volume) layers toward the outside. The chemically distinct layers can be divided into the crust (Fig. 9.1), the mantle, and the core (Fig. 9.2). Another way of looking at Earth characterizes its different layers physically—that is, strong or weak solids or liquid. In this system the lithosphere is the uppermost layer. Figure 9.3 shows the layers according to both chemical and physical distinctions. Table 9.1 gives some vital statistics on the layers.

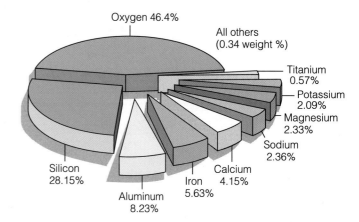

Figure 9.1 Average chemical composition of Earth's continental crust

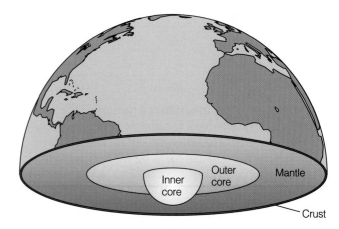

Figure 9.2 Earth's chemical layers. Crust, mantle, and core are shown here with the thickness of the crust exaggerated.

1. Use Figure 9.3 and Table 9.1 to make the following comparisons:
   a. Compare the lithosphere and the asthenosphere.

   _____

   _____

   b. Compare the lithosphere and the crust. _____

   _____

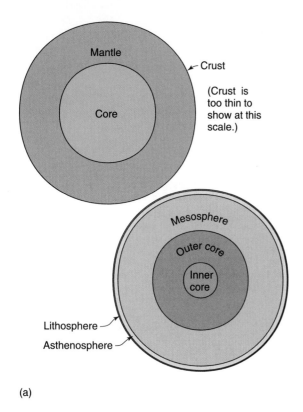

(a)

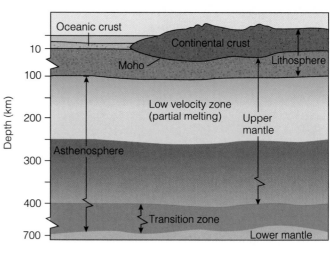

(b)

Figure 9.3 (a) Scale drawings of the interior layers of Earth according to the chemical (top) and physical (bottom) classification systems. (b) Sketch of the uppermost layers of the Earth. The chemical layers, crust and upper mantle, overlap with the lithosphere and asthenosphere. The lower mantle corresponds to the mesosphere in (a). (Notice that the scale in (b) varies in different parts of the diagram.)

2. Why do you think the core is chemically uniform but has distinct physical layers separated into an outer and inner core?

_____

_____

_____

_____

The Lithosphere The **lithosphere** is the surface layer defined by its physical characteristics as composed of strong rigid rock. It reaches about 100 km below the surface and sits above the asthenosphere (Fig. 9.3a and b). As we will see later in this lab, the lithosphere is divided into segments called **plates** that move across the surface at speeds about as fast as your finger-

nails grow. The upper part of the lithosphere, the crust, is chemically different than its lower part, the mantle (Fig. 9.3b).

The Crust

Earth's crust is of two types: the layer we live on, **continental crust,** and the layer that directly underlies the oceans, **oceanic crust.** Seventy-one percent of Earth's surface is covered with oceans. If the crust were perfectly flat, Earth's whole surface would be covered with water to a depth of more than 2500 m. Differences in elevation of the crust of substantially greater than 2500 m are necessary for any continents and living space to exist. As we will see later in the lab, large elevation differences are produced by plate tectonics.

One important difference between oceanic and continental crust is their elevations. The highest point on the con-

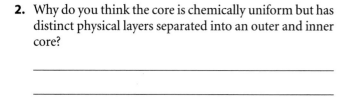

Table 9.1 Layers of Earth

CLASSIFICATION SYSTEM	LAYERS	AVERAGE DEPTH TO BOTTOM OF THE LAYER (KM)
Chemically Distinct Layers	**Crust:** mostly silicates	20
	Mantle: mostly magnesium silicates	2883
	Core: mostly iron	6371
Physically Distinct Layers	**Lithosphere:** strong rigid solid	100
	Asthenosphere: weak, mostly solid	350
	Mesosphere: stronger solid	2883
	Outer core: liquid	5140
	Inner core: solid metal	6371

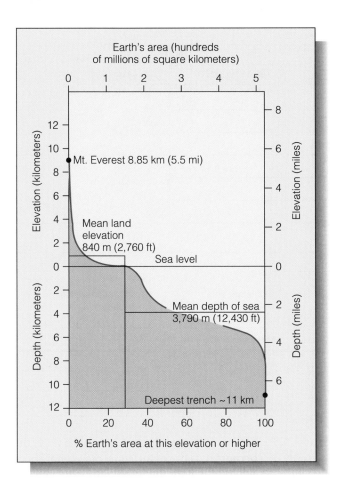

Figure 9.4 Plot of distribution of elevations of the land and depths of the sea. This is not a profile or cross section but a graph of the proportion of the area of the Earth's surface at each elevation.

tinental crust is Mount Everest, with an elevation of 29,028 ft, or 8847 m. The deepest part of the oceans, on oceanic crust, is the Mariana Trench at a depth of 36,198 ft, or 11,033 m. The average depth of the ocean is about 3790 m; the average elevation of the continents is 840 m above sea level (Fig. 9.4).

Oceanic crust and continental crust are also different in thickness: oceanic crust averages 8 km thick, and continental crust averages about 45 km thick.

Why Are the Oceans Where They Are? Have you ever wondered why the oceans are where they are? If we remember that water flows downhill, we immediately realize that the oceans occur where the elevation of the crust is low (observed in a data gathering stage of the scientific method). This brings up a new question based on this observation of

crustal elevation: *Why* is the crust under the oceans lower in elevation than the continental crust? The answer to this question would be a hypothesis that explains *what causes* the oceanic crust to be lower than the continental crust. Let's hypothesize that the answer has something to do with the rocks of the oceanic and continental crust. In the ensuing exercises, we will follow the scientific method (refer to the Introduction to this lab manual) to test and improve the hypothesis. What characteristic(s) of these rocks could cause oceanic crust to be lower than continental crust? We will look at the characteristics of mass, volume, and density.

- **Mass:** a measure of a body's resistance to acceleration. On Earth, the mass of an object is proportional to its weight.
- **Volume:** size or extent in three dimensions
- **Density:** mass divided by volume, or the amount of mass in a unit of volume

3. Examine the oceanic crustal samples and the continental crustal samples with special attention to their volume and mass. Use the balance to measure the mass of one sample of each type of crust and include that information in Table 9.2. Each student in your group should measure a different pair of samples. Make sure the units are grams (g). Then place a sample on the balance and take down the reading.

4. Your lab instructor will give you the volume of each sample (the communication phase of the scientific method). Enter the volumes in Table 9.2. Calculate the density for each sample, and enter that information in Table 9.2.

5. Discuss with your lab partners what hypothesis you can devise that explains why oceanic crust is lower than continental crust and write it out here.

_____

_____

_____

_____

6. Based on your hypothesis, what would you predict about the other samples of oceanic and continental crust?

_____

_____

_____

7. Communicate with members of your group to get density data for the other samples, and enter all the densities in Table 9.3. Calculate the average for each crust type.

Table 9.2 Mass, Volume, and Density of Rocks from Oceanic and Continental Crust (Exercises 3 and 4)

Type of Crust	Sample Number	Rock Name	Mass (g)	Volume (cm³)	Density (g/cm³)
Oceanic					
Continental					

Table 9.3 Density of Rocks from Oceanic and Continental Crust (Exercise 7)

Type of Crust	Sample Number	Rock Name	Density (g/cm³)	Average Density for Crust Type (g/cm³)
Oceanic				
Continental				

8. Is there any pattern to the data? Write down your observations.

_____

_____

_____

_____

9. Do the results in Table 9.3 match your hypothesis in Exercise 5? Yes / No

10. Does your hypothesis need to be modified? Can you refine or improve your hypothesis? If so, write the new hypothesis in the space here.

_____

_____

_____

_____

_____

_____

Scientists over the years have discovered, in fact, that rocks from the oceanic crust are consistently different in density and chemical composition from continental rocks. Oceanic crust is made of basalt and chemically similar rocks, such as gabbro, and amphibolite. Continental crust is largely made up of granite or chemically similar rocks, such as quartz-rich diorite and gneiss. If the oceanic and continental crust were the same thickness, the base and the top of the oceanic crust would sit lower than those of the continental crust because of its greater density. But like the full container ship in Figure 9.5, the continental crust displaces more of the mantle than the oceanic crust does (Fig. 9.3b) because it is heavier due to its thickness. Although its surface is higher, its base is lower than the oceanic crust.

The Theory of Plate Tectonics

Remember that a *theory* is a well-tested hypothesis that accounts for a large body of data collected over a long time. Oc-

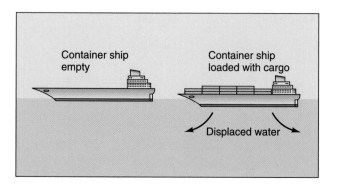

Figure 9.5 A filled container ship sits lower in the water than an empty one because it displaces a volume of water of equivalent weight—the more weight, the more water displaced.

casionally, in science, a theory explains such a wide variety of data, phenomena, and observations that it provides a **unifying theory** or guiding principle for its area of science. The *theory of plate tectonics* provides such a guiding principle or unifying theme for geology. It helps to explain and predict a wide variety of geologic processes, including earthquakes, volcanoes, mountain building, and even the locations where some rocks and fossils are found. The body of evidence that supports this theory is as large and varied as the phenomena it helps to explain.

The **theory of plate tectonics,** simply put, proposes that Earth's lithosphere is divided into separate **plates,** each of which moves as a unit over the asthenosphere relative to the other plates. The plates and their movement are another structural aspect of Earth. These plates can be oceanic, continental, or a combination of both. Three types of boundaries between the plates can be defined on the basis of relative movement of the plates on either side of the boundary (Fig. 9.6):

■ **Divergent,** where plates move away from each other (Figs. 9.6 and 9.7). This type can be further subdivided into oceanic and continental divergent plate boundaries. Divergence creates oceanic crust (Fig. 9.8) and breaks up or rifts apart continents (Fig. 9.9).

■ **Convergent,** where plates move toward each other (Figs. 9.6 and 9.7) and one plate may descend under the other. This type can be further subdivided into ocean-ocean con-

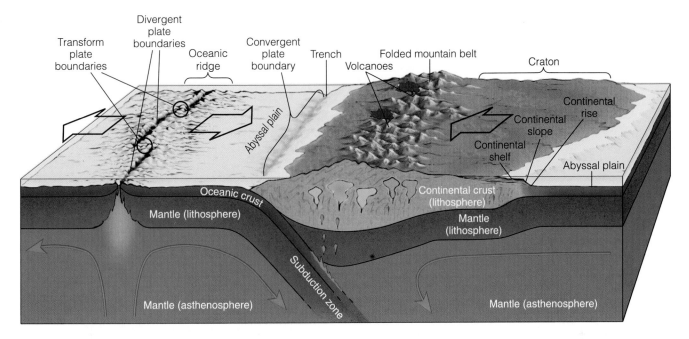

Figure 9.6 Block diagram showing different types of plate boundaries

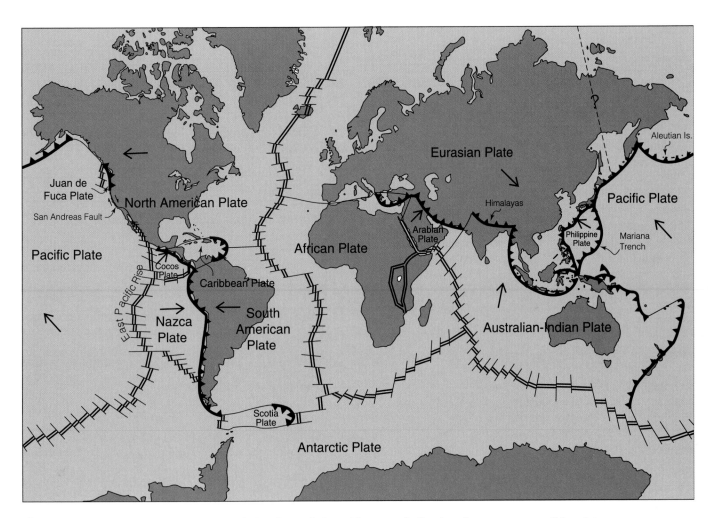

Figure 9.7 World map of the plates and plate boundaries with arrows indicating the movements of the plates

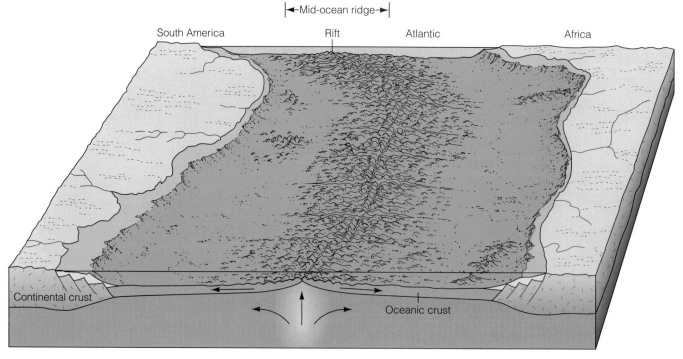

Figure 9.8 The divergent margin and Atlantic Ocean basin between South America and Africa, showing the Mid-Atlantic Ridge. At divergent plate boundaries, oceanic crust forms. If divergence continues for tens of millions of years an ocean basin such as this may result.

vergent (Fig. 9.10), ocean-continent convergent plate boundaries (Fig. 9.6), and continental collisions (continent-continent convergent plate boundaries; Fig. 9.11). Convergence creates or thickens continental crust, creates mountains, builds continents, and destroys oceanic crust.

■ **Transform fault,** where plates slide past each other (Figs. 9.6 and 9.7). This type can be further subdivided into oceanic (Fig. 9.12) and continental transform-fault plate boundaries (Fig. 9.13). Oceanic transform faults are generally found as "offsets" in oceanic divergent plate boundaries (Fig. 9.12) and are much more common than continental transform-fault plate boundaries. Transform movement conserves crust, neither creating nor destroying it.

 ## Plate Tectonics Computer Exercise

■ Start **TASA Plate Tectonics.**
■ Click the advance (▶) button.
■ Click the **10 Plate Boundaries** button.
■ Click the advance (▶) button.
■ Continue by clicking the advance (▶) button and answering the following questions:

11. Divergent plate boundaries create _____

_____.

12. What fills the gap as plates spread apart?

_____

13. Write down the answers to the review questions in the Divergent Boundaries section:

  a. Plates are moving apart at _____

  _____.

  b. Most divergent boundaries are located _____

  _____.

  c. New oceanic crust is created at divergent boundaries

  at a rate of about _____.

14. At what type of plate boundaries do you find oceanic

trenches? _____

15. What types of convergent plates result in the formation

of magma and volcanoes? _____

_____

16. What types of convergent plates result in the formation of mountains? Keep in mind that islands are mountains

in the sea. _____

_____

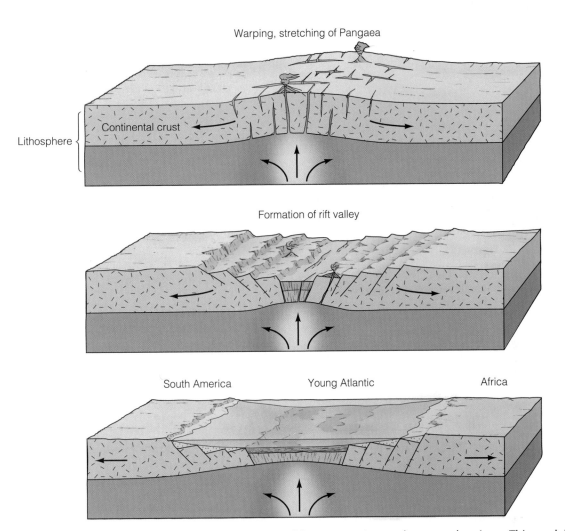

Warping, stretching of Pangaea

Lithosphere { Continental crust

Formation of rift valley

South America   Young Atlantic   Africa

Figure 9.9 Continental divergence rifts the continent apart and forms oceanic crust between the pieces. This model shows the formation of the early Atlantic Ocean. After 150 to 200 million years of divergence, the area looks like Figure 9.8.

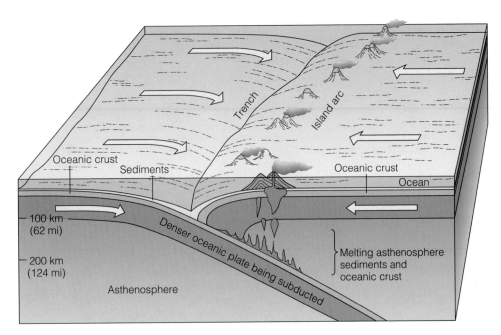

Trench
Island arc
Oceanic crust
Sediments
Oceanic crust
Ocean
100 km (62 mi)
Denser oceanic plate being subducted
200 km (124 mi)
Melting asthenosphere sediments and oceanic crust
Asthenosphere

Figure 9.10 Block diagram of an ocean-ocean convergent plate boundary showing subduction of the older, denser oceanic plate

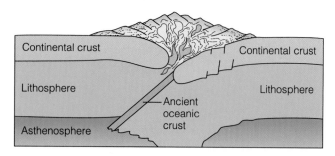

Figure 9.11 In a continental collision, neither continental plate subducts; instead, both plates crumple, undergoing folding and faulting. The old oceanic part of one of the plates that originally separated the two continents may continue to subduct.

17. Examine the sequence that shows the movements of the plates since the breakup of the supercontinent, Pangaea.

    a. What happened to India?_____

    _____

    _____

    b. What is the geographic feature found at the present plate boundary north of India? _____

    c. What type of plate boundary is this? _____

    _____

18. Write down the answers to the review questions in the Convergent Boundaries section of the TASA CD:

    a. The boundaries along which older oceanic crust is being destroyed are called _____

    _____.

    b. The region where an oceanic plate descends into the asthenosphere is called _____

    _____.

    c. As an oceanic plate slides beneath the overriding plate, it bends and produces a(n) _____

    _____.

    d. When two plates converge, _____

    _____.

    e. Match the following with choices from the program:

        ■ Oceanic-oceanic plate boundary: _____

        ■ Oceanic-continental plate boundary: _____

        ■ Continental-continental plate boundary: _____

    _____

19. Write down the answers to this review question in the Transform Fault Boundaries section:

    a. Which of these are true about transform faults?

    _____

    _____

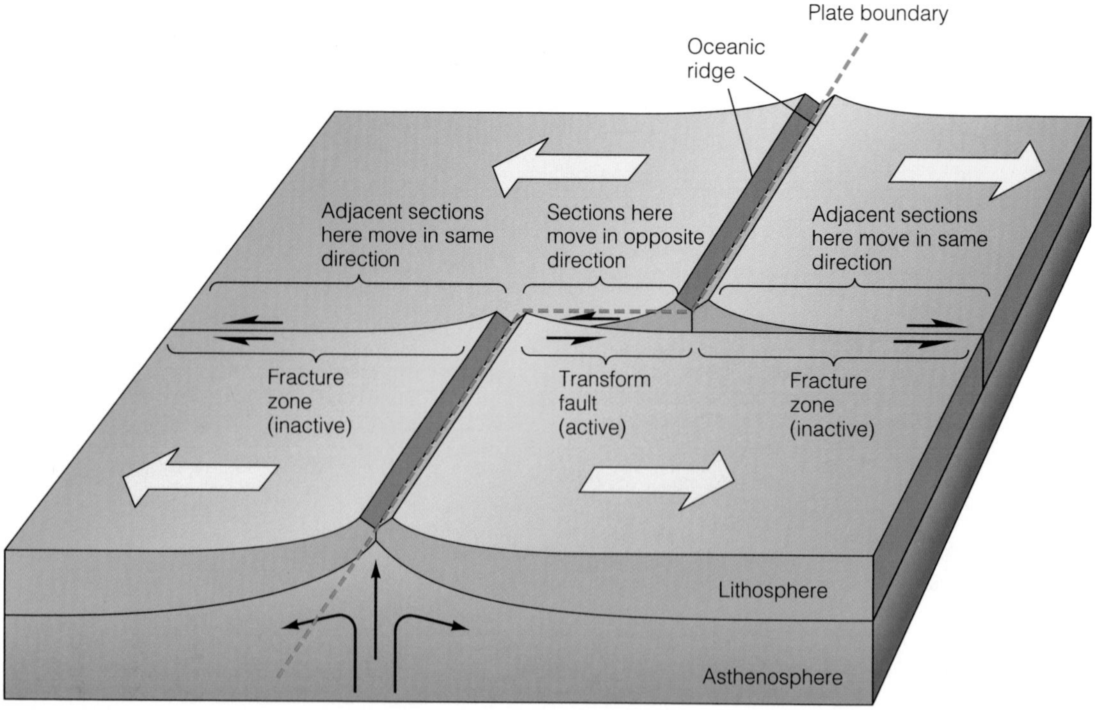

Figure 9.12 Block diagram of two segments of a mid-ocean ridge divergent plate boundary separated by a transform fault

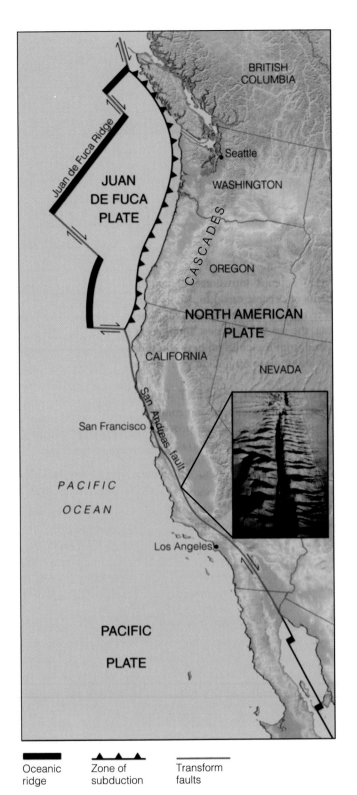

Oceanic
ridge

Zone of
subduction

Transform
faults

Figure 9.13 The San Andreas Fault is a continental transform fault. Notice also some oceanic transform faults bordering the Juan de Fuca Plate. Subduction of the Juan de Fuca Plate produces the Cascade Mountain Range.

- Click the stop (■) button.

■

Plate Map and Globe

20. Examining the plate map in Figure 9.7, determine which symbol shown in Table 9.4 corresponds to which type of plate boundary. Write the types of plate boundary next to the appropriate symbol in Table 9.4.

21. On the globe and in Figures 9.7 and 1.2, find the features listed in Table 9.5, write them in on Figure 9.7, and fill in Table 9.5.

22. Based on these observations, what hypotheses can you make about the relationships of plate boundaries and each of the following types of physiographic features? Find an example of each feature in Table 9.5 and its type of plate boundary and generalize.

    Example: *Long narrow seas* such as the Red Sea and the Gulf of California: Long narrow seas occur at continent-continent divergent plate boundaries.

    a. *Oceanic trenches* such as the Peru Chile Trench and the Aleutian Trench:

    _____

    _____

    b. *Volcanic island arcs* such as the Aleutians:

    _____

    _____

    c. *Continental volcanic arcs* such as the Andes:

    _____

    _____

    d. *Mid-ocean ridges* such as the Mid-Atlantic Ridge and the East Pacific Rise:

    _____

    _____

Table 9.4 Symbols for the Three Types of Plate Boundaries (Exercise 20)

Symbol	Type of Plate Boundary
▼▼▼▼▼▼	
═══	
───	

Table 9.5 Physiographic Features at Different Plate Boundaries (Exercise 21)

Physiographic Feature	Plate Boundary Type	Crust Type*	What Plates Are on Either Side of the Boundary?
Peru Chile Trench			
Andes Mountains			
Aleutian Trench			
Aleutian Islands			
Himalayas			
Mid-Atlantic Ridge			
East Pacific Rise			
San Andreas Fault			
Red Sea			

*Ocean-ocean, ocean-continent, or continent-continent.

23. Examine the map in Figure 9.14, showing the seismicity of the world. At what type(s) of plate boundaries are earthquakes found? _____ At what type of plate boundaries are deep earthquakes found? _____

_____

Divergent Plate Boundaries and Transform Faults

Remember that plates move away from each other at divergent plate boundaries, also called **spreading centers.** The spreading creates oceanic crust, which is denser than continental crust (Fig. 9.8). When a continent consistently diverges, oceanic crust forms, separating the continent into two smaller pieces (Fig. 9.9). This is known as a **continental rift.** As new ocean floor is added in the center, the age of the oceanic crust becomes progressively older away from the spreading center (Fig. 9.15). Eventually the oceanic crust becomes wide enough to form an ocean basin (Fig. 9.8). The plate boundary remains in the middle of the ocean basin and gives rise to a mid-ocean ridge at that location. Neither the boundary nor the ridge are straight and continuous but have repeated discontinuities where transform faults occur. In the following group of exercises, you will make a model of a continent that rifts at a divergent plate boundary and is further cut by two transform faults.

24. Work in groups of two or more.

- One person should cut out Plate A and another Plate B from Figures 9.16 and 9.17.
- A third person should separate the page containing the base of the model (Fig. 9.18) from the manual and make the folds and cuts indicated on the figure.
- Use a paper clip to join the triangles on Plate A and Plate B together *face to face*, the end of each paper flush with the other. Likewise, match and clip flowers and stars. When assembled, the pages should not lie flat but should be arranged as shown in Figure 9.19.
- Insert the paper-clipped triangle end of the plates from Figures 9.16 and 9.17 into the slot marked with a triangle in the model base (Fig. 9.19). Insert flower and star ends into their respective slots.
- Gently pull the three tabs through the slots so they are as far as they will go—do not crease or flatten.

You have a single continent made up of two plates. You are now ready to watch divergence in action. Fifteen million years ago the continent began to rift.

25. One person should hold the model base horizontally at the two ends where it says "hold here." A second person should pull Plate A and Plate B horizontally away from each other, keeping them at about the same level as the model base. The numbers on the plates are the age of the rocks.

a. Starting 15 Myr ago, pull the plates apart until you reach the configuration for 13 Myr ago. The age of the rocks is marked in millions of years on the oceanic crust. In the space here, sketch a map of the two plates. Indicate where the plates were oceanic and where they were continental. The parts that were oceanic were probably filled with water.

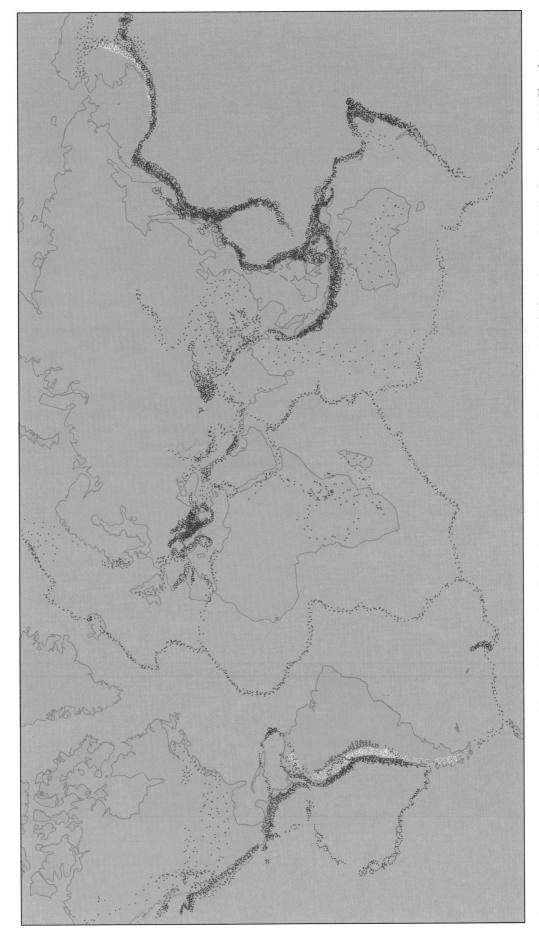

Figure 9.14 Map of the seismicity of the world; shows the distribution of about 10,000 earthquakes worldwide from January 1977 to December 1986. Blue dots represent shallow earthquakes at 0–70 km depths, yellow represent intermediate depth earthquakes from 70–300 km, and red represent earthquakes deeper than 300km.

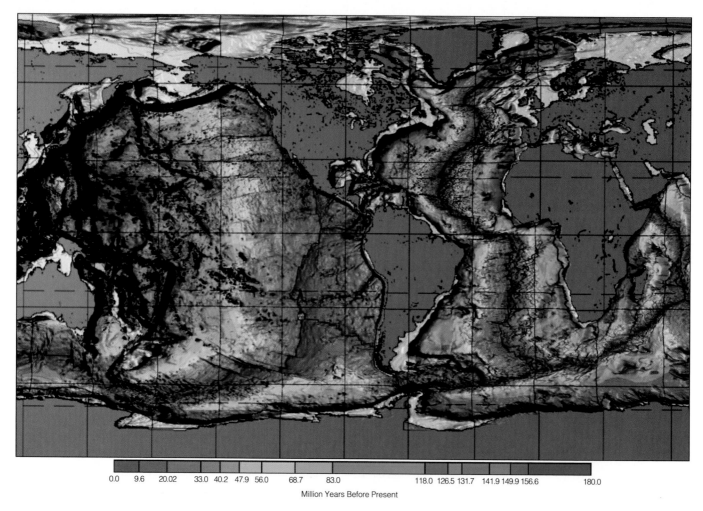

Figure 9.15 Age of the ocean floor determined from paleomagnetism. Colors represent different-aged rocks on the sea floor, with red the youngest and purple the oldest.

b. In your model, how many continents were present

13 Myr ago? _____

How many plates were there? _____

Remember that one continent can be made up of more than one plate and that one plate can have both continental and oceanic parts.

c. At each time step in Table 9.6 as you proceed with the lettered exercises, fill in a description of the land masses and oceans from the following list:

- Two continents with a narrow land bridge
- One continent with narrow sea inlets and a land-locked basin
- Two continents separated by an ocean
- Two continents separated by a narrow ocean
- One continent with no seas

Table 9.6 Configuration of Model Oceans and Continents of the Past (Exercise 25c)

Time (Myr ago)	Description of Model
15	
13	
10	
0	

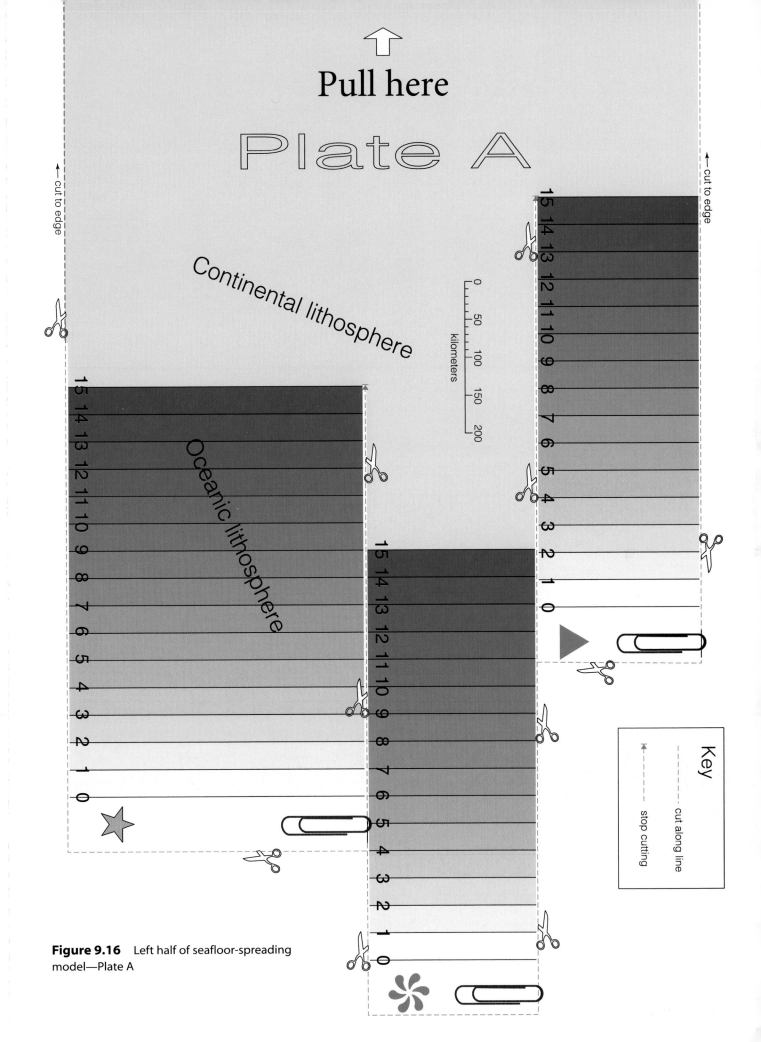

Figure 9.16 Left half of seafloor-spreading model—Plate A

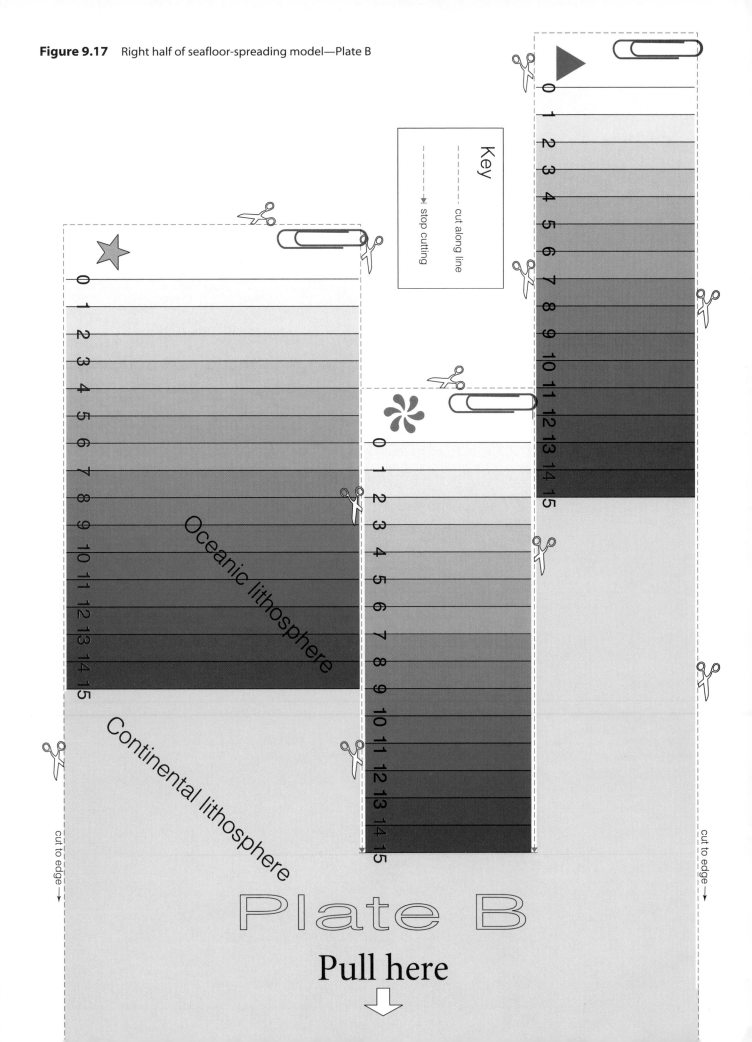

Figure 9.17 Right half of seafloor-spreading model—Plate B

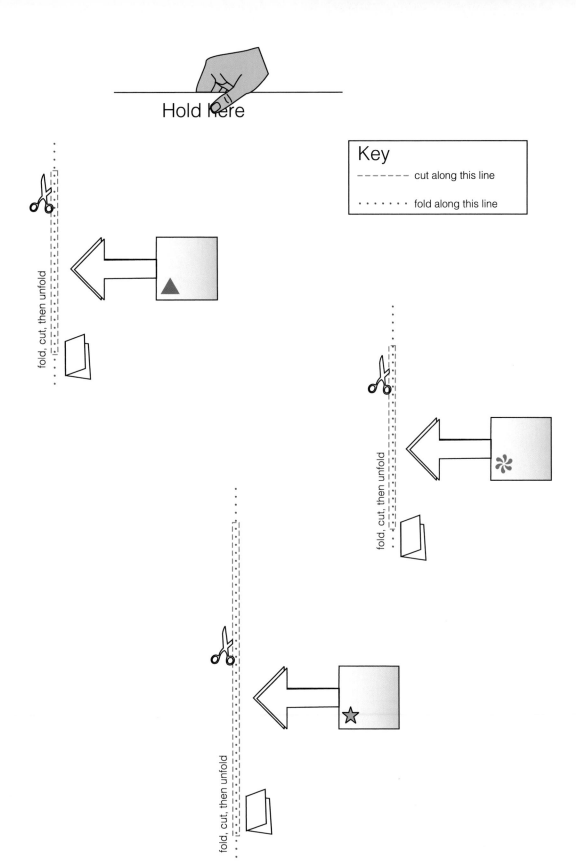

Figure 9.18 Base for seafloor-spreading model

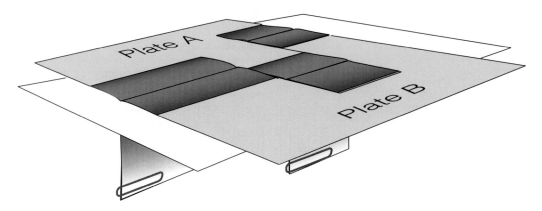

Figure 9.19 Sketch showing how to set up the seafloor-spreading model in Figures 9.16, 9.17 and 9.18

d. Continue spreading your model apart until you reach 10 Myr ago. Sketch the new configuration.

e. Spread your model until you reach the present (0 Myr). Sketch the new configuration.

Land organisms can migrate from one continent to another as long as there is some sort of connection or land bridge between the continents.

f. On your model diverging continent, when did species of organisms roaming the continent become isolated

into two separate populations? _____

26. Use the age of the rocks on your model (the numbers) and the map scale on Plate A to determine how fast Plate A was moving relative to Plate B in kilometers/million years or millimeters/year.[1] This number is your full spreading rate. Show your calculations.

Hint: Measure the distance from 0 to 15 in centimeters; multiply by 2 for the full spreading rate (i.e., the measurement from 15 to 15 when the model is open all the way to 0). Then convert centimeters on the map to kilometers on the ground using the conversion factor 1 cm = 50 km, measured off the map scale (i.e., multiply centimeters by 50).

Full spreading rate = _____

27. Next calculate how fast Plate A or Plate B was moving relative to the spreading center in kilometers/million years or millimeters/year. This number is the one-half spreading rate.

Rate of Plate A (or B) = _____

28. The slowest one-half spreading rates are about 20 mm/yr, and the fastest ones are about 120 mm/yr. Was your spreading rate reasonable (i.e., did it fall within this range)? Yes / No?

29. Study the relationship between the age of rocks and their distance from the spreading center. What general pattern do you see?

_____

_____

Transform faults in mid-ocean ridge systems may be a bit puzzling. Seeing a snapshot of a series of segments of a mid-ocean ridge, the movement along transform faults, at first glance, appears backwards. Before seafloor spreading was understood in the 1960s, geologists interpreted transform faults as displacing the mid-ocean ridges. They thought that the mid-ocean ridges were once continuous and had become

·············
[1]Note that kilometers/million years = millimeters/year since there are 1,000,000 mm in a kilometer.

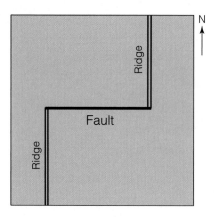

Figure 9.20 Map of a transform fault between two segments of mid-ocean ridge.

offset by the transform faults. For example, they would have interpreted the far block of the transform fault in Figure 9.12 to have moved to the right (right-lateral strike-slip fault; see Lab 16). However, after studying actual earthquake movements they realized their predicted motion was wrong. The far block of the fault in Figure 9.12 happens to have moved left (left-lateral strike-slip fault) instead. That means that seafloor spreading must be occurring at each segment of the mid-ocean ridge (instead of off-set of the ridge by the fault) to account for the actual motion along the transform faults.

30. Study the map in Fig. 9.20, which depicts a portion of a mid-ocean ridge spreading center. Until the 1960s, geologists would have hypothesized that area below (south of) the fault on the map moved left relative to the area above (north of) the fault. Study your model of Plate A and Plate B.

    a. Do you agree with their assessment of the fault?

    _____

    b. What motion do you think occurred along the fault?

    _____

    _____

    c. What type of faults are each of the transform faults on your model? Label them with arrows showing the motion on one of your sketches in Exercise 25.

31. What incorrect assumption led geologists to the wrong conclusion?

    _____

    _____

    _____

    _____

 Divergence Computer Exercise

■ Start **INTERACTIVE PLATE TECTONICS.**
■ Click **BASICS.**
■ Click **Divergent Boundaries.**
■ Click the button at the top of the right-hand part of the screen.

32. a. Play (▶) the video sequence. What allows partial melting of the asthenosphere? _____

   b. What type of lava forms? _____

   c. Recalling what you learned earlier in the lab, what property of the lava causes the new material at the spreading center to become oceanic rather than continental crust (i.e., what causes it to sink to a lower-level than the continental crust)?

   _____

■ Computer exercise continues below.

■

Convergence, Subduction, and Mountain Building

When a slab of lithosphere moves downward into the asthenosphere, the process is called **subduction.**

 Convergence Computer Exercise

■ Now click **Convergent Boundaries,** and, if necessary, review convergent plate boundaries by reading the screen.
■ Advance to the next screen using the down (▼) button, then click **OCEAN-OCEAN CONVERGENCE.**
■ Click the button at the top of the screen, and play (▶) the video sequence.

33. At what depth does melting start? _____

   This melting may be triggered by _____

   _____

   as it descends.

■ Click **Q** to quit.

■

At ocean-continent convergent plate boundaries, the oceanic plate is subducted beneath the continental plate. Earthquakes occur along this subduction zone, caused by break up and possibly by metamorphism of the descending plate. As the cold subducting slab descends deeper into the Earth, it is warmed by rocks on either side that may have temperatures of as much as 1500°C at a depth of about 100 km. Eventually the cold slab heats up enough to release water and to partially melt. The water aids melting of the overlying mantle. The magma from both the melting slab and the overlying mantle rises, some to be trapped beneath the crust, some to become intrusions (such as granite and diorite plutons) in the crust, and some to erupt as lava (especially andesite) or volcanic ash. In the area around the magmatic/volcanic arc, regional metamorphism generally occurs. Convergence creates directed pressure, and the heat from the magma helps warm the rocks forming low-, moderate-, and high-grade foliated rocks such as slate, phyllite, schist, and gneiss. The convergence, plutonic activity, and volcanism create a broad band or strip of mountains with the volcanic peaks of this continental volcanic arc system situated on top.

At the surface, the volcanoes are eroded, shedding volcanic sediments toward the oceanic trench at the plate boundary. Older volcanoes may be entirely eroded away to reveal the plutonic rocks beneath. These in turn may shed sediment toward the sea. The sediment is deposited on the continental shelf and in the oceanic trench and may eventually become rocks such as graywacke (in this case, a volcanic-rich sandstone) and shale.

34. Examine Figure 9.21, which shows profiles across segments of the Aleutian and Tonga Island arcs. The stars on the diagrams represent the locations of earthquakes and are a guide to the location of the subduction zone. How does the dip of the subduction zone at Amchitka compare with that for the Tonga Island arc?

_____

_____

35. Use the locations of the volcanic islands and the fact that most magma rises vertically to help you determine the minimum depth at which melt-generating processes start in the subduction zone at Amchitka in Figure 9.21.

    a. What is it? _____ km.

    b. At Tonga Islands? _____ km.

36. Using the same approximate minimum depth of melting, the positions of the trenches, and the position of the volcanic arcs (at the small volcano), sketch in where you expect the subduction zone to be for Cook Inlet and for Skwentna in Figure 9.21.

37. Based on your observations, which of the following statements do you think is true?
    a. All subduction zones dip at the same angle.
    b. Subduction zones dip 62° on average.
    c. Subduction zones have different dips in different places, and the distance from the trench to the arc is related to the dip.

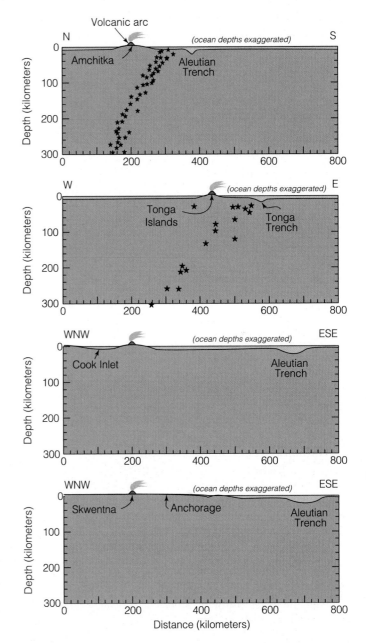

Figure 9.21 Profiles across segments of the Aleutian and Tonga arcs. Stars show the locations of earthquakes.

    d. Subduction zones have different dips and the depth to the subduction zone at the volcanic arc is dependent on that dip.

The *Cascade Mountain Range* spans the distance between northern California and northern Washington (Fig. 9.13). It makes a north-south sweep through the center of the part of Washington shown on the map in Figure 1.3.

38. Examine the geologic map of Washington (Figure 1.3). What are the two most common types of rocks?

_____ _____

39. In what way do you think plate tectonics has influenced this area? What is your evidence?

_____

_____

_____

_____

_____

_____

The Relationship Among Density, Crust Type, and Plate Tectonics

In this lab, you have studied cross sections through the Earth, determined the density of rocks, looked at maps, and created models of oceans, continents, and plates. Do these disparate subjects have anything in common? The following exercise will demonstrate how they are related.

At a convergent plate boundary, two plates move toward each other. Each plate is a slab of lithosphere, which is made up of crust at the top and a bit of upper mantle below. As the plates come together, either they collide and crumple, or one plate goes down under the other. What determines which action will occur depends on the type of crust that makes up part of the lithosphere of the two plates.

40. Based on your data in Table 9.3, which plate will pass under the other, if any, at an ocean-continent convergent plate boundary? Why?

_____

_____

_____

_____

_____

Subduction creates oceanic trenches and adjacent volcanic mountains. At an ocean-ocean convergent plate boundary, usually the older oceanic crust is cooler and thus slightly denser than the younger crust.

41. At an ocean-ocean convergent plate boundary, which one of the two oceanic plates do you think will subduct

beneath the other? _____

How did you arrive at this answer? _____

_____

42. What do you think happened at the ancient plate boundary shown in Figure 9.22? _____

_____

_____

Locate the resulting feature on the map of North America in Figure 1.1. What is its name?

_____

43. Neither continent is subducted at a continent-continent convergent plate boundary (e.g., the Himalayas). Instead, the plates collide and crumple, creating mountains but neither volcanoes nor a subduction zone. Why do you think these things are true?

_____

_____

_____

_____

So subduction is closely related to the density of the rocks on either side of the convergent plate boundary. The denser plate is the one that subducts, provided it has oceanic crust. Continental crust, with its low density, does not subduct readily. Subducting continental crust would be a bit like trying to sink a cork in a bucket of water—it is just too light to go down.

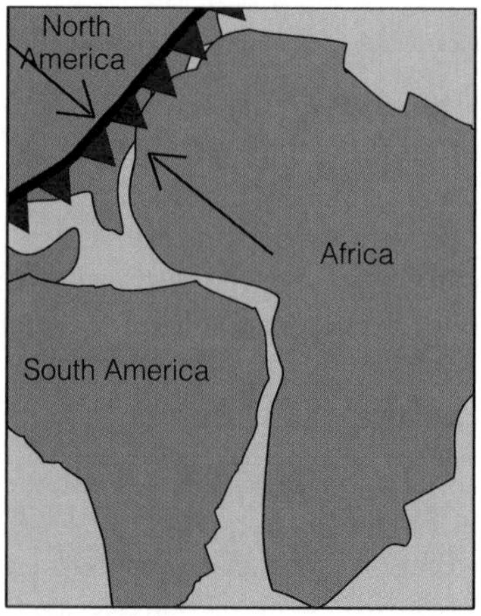

Figure 9.22 Plate map of the Americas and Africa about 300 Myr ago

Geometric Fit of Continents

From the time of the first maps of South America and Africa, people have noticed the geometric fit of these two continents as if they were giant pieces of a jigsaw puzzle (Fig. 9.23a). Sir Francis Bacon observed this in 1620 and suggested that the two continents had once been joined. Alfred Wegener in 1915 explored this hypothesis, providing data to indicate that the fit is more than geometric; it is also geologic, with matching rocks, evidence of glaciation, and fossils (Fig. 9.23b–e). Since that time, geologists and geophysicists have gathered abundant data to indicate that the continents and the sea floor are and have been moving. With these data, they have established possible positions of the continents in the past. At one time, we now believe, all of the continents were assembled into one large supercontinent known as **Pangaea** (Fig. 9.24). At other times, the northern continents of North America, Europe, and Asia were assembled into a supercontinent called **Laurasia,** and the southern continents were assembled into **Gondwana** or **Gondwanaland** (Fig. 9.23d and e).

44. Ask your lab instructor to set up the Continental Drift Globe for about 250 Myr ago during the Permian period (refer to the geologic time scale in Fig. 8.21).

   a. Describe the positions of the continents at that time.

   _____

   _____

   _____

   b. Did the Atlantic Ocean exist at that time? _____

   If so, how was it different than today?

   _____

   _____

   c. Did the Pacific Ocean exist at that time? _____

   If so, how was it different than today?

   _____

   _____

   d. How was the Mediterranean Sea different at that time? _____

   _____

   _____

   _____

   e. Locate what is now the southeastern United States. Was it in the Northern or Southern Hemisphere?

   _____

What present-day continent was its neighbor then?

   _____

   f. Two hundred fifty million years ago, the Earth's climate was cool enough to have glaciation at 60°N and S latitude. Where would continental glaciers have been? Name the present-day continents and the approximate locations of the glaciers on those continents.

   _____

   _____

   g. What do we call the supercontinent of 250 Myr ago?

   _____

45. Now have your instructor reassemble the continents at positions for roughly 175 Myr ago, during the Jurassic (Fig. 8.21).

   a. What has changed? What type of plate boundary separates the continents?

   _____

   _____

   _____

   _____

   b. What do we call the northern supercontinent?

   _____

   c. What do we call the southern supercontinent?

   _____

   What continents or parts of continents were present in this supercontinent?

   _____

   _____

   _____

46. Have your instructor reassemble the continents at positions that represent roughly 130 Myr ago, during the Cretaceous (Fig. 8.21).

   a. What has changed? What new feature has developed that was not present in the Permian or the Jurassic periods?

   _____

   _____

   _____

   _____

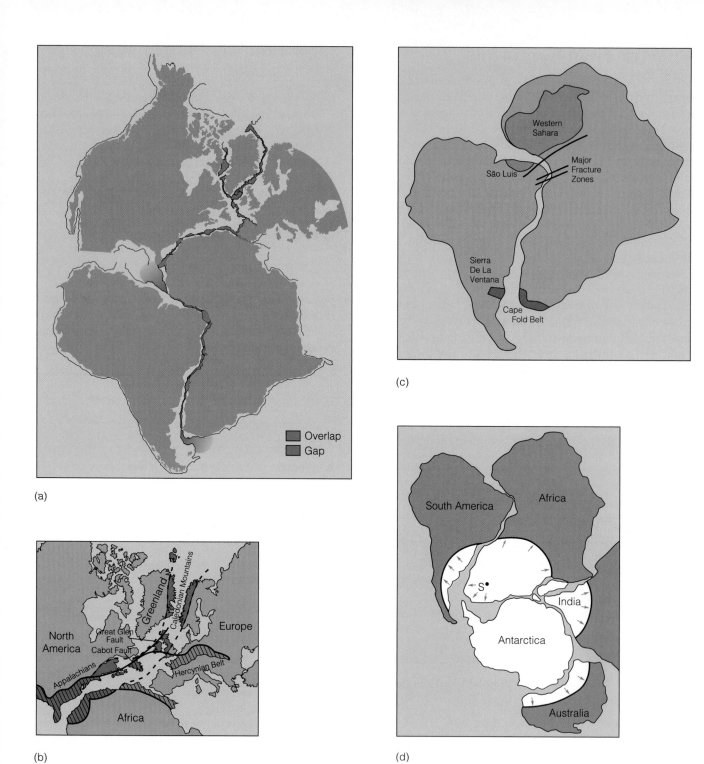

(a)

(c)

(b)

(d)

Figure 9.23 Jigsaw puzzle fit of the continents. (a) Matching shapes of continents. (b) Matching rocks among North America, Africa, and Europe. (c) Matching rocks between South America and Africa. (d) Matching distribution of evidence of glaciation in the southern continents reassembled as Gondwanaland. (e) (*opposite*) Matching fossils of Gondwanaland. These fossils are now found on widely separated continents.

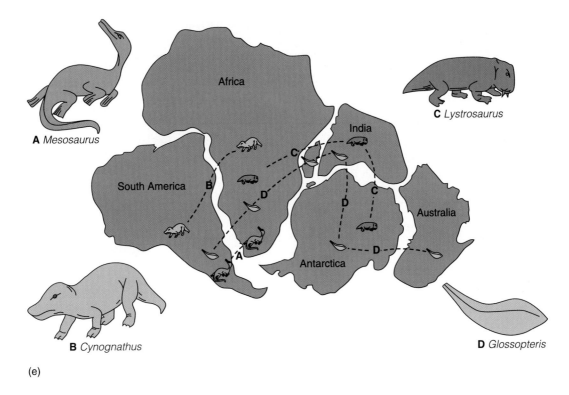

A *Mesosaurus*

C *Lystrosaurus*

Africa

India

South America

Antarctica

Australia

B *Cynognathus*

D *Glossopteris*

(e)

Figure 9.23 Jigsaw puzzle fit of the continents *Continued*

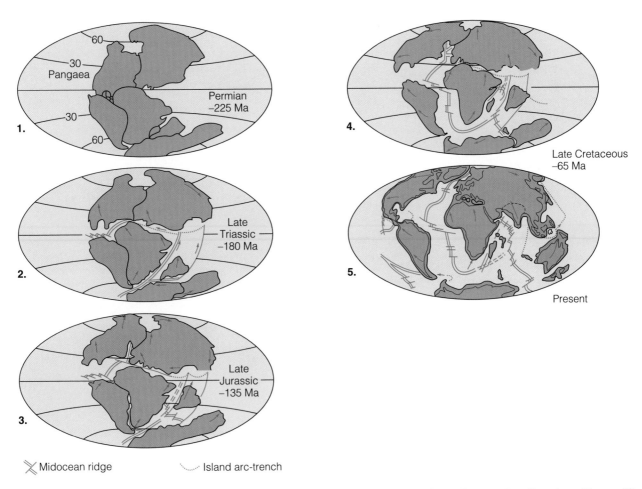

1. Pangaea — 60 — 30 — 30 — 60 — Permian −225 Ma

2. Late Triassic −180 Ma

3. Late Jurassic −135 Ma

4. Late Cretaceous −65 Ma

5. Present

✕ Midocean ridge   ⋯⋯ Island arc-trench

Figure 9.24 Reconstruction of the breakup of Pangaea in five steps. Red arrows show plate motion directions. Ma = million years ago.

b. Summarize how the Atlantic Ocean basin formed.

_____

_____

_____

_____

c. During the formation of the Atlantic Ocean basin, what was happening to the Pacific Ocean basin? What type of plate boundary would accomplish this?

_____

_____

_____

_____

47. Look back at the Geologic Map of Washington in Figure 1.3 and the Cascade Mountains (Fig. 9.13). Do the rock types and plate boundaries seen there fit with your hypothesis in Exercise 46c?

_____

_____

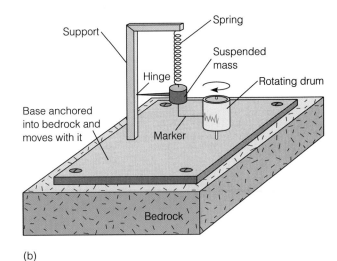

Earthquakes and Seismology

One of the most devastating of natural disasters is a great earthquake.[1] It vibrates and shakes the ground, sometimes so strongly that people, buildings, and highway overpasses cannot stand. The result is chaos and destruction. You may have experienced an earthquake yourself. If you haven't experienced one, you may not realize that a single earthquake has more than one vibration. An earthquake is actually a series of waves called **seismic waves** that pass through and vibrate the ground. If you have experienced an earthquake of magnitude 4, say, how would a magnitude 5 earthquake compare? How do scientists detect earthquakes from remote locations? These are things we will explore in this lab.

- An **earthquake** is the vibration of the ground caused by natural geologic forces.
- **Seismology** is the study of earthquakes and their waves.
- **Seismologists,** scientists who study earthquakes, use seismographs to detect earthquakes and seismograms to measure them.
- A **seismograph** is an instrument (Fig. 10.1) that inscribes the Earth's motion on a record called a **seismogram** (p. 206).
- The **focus** is the place beneath the surface where faulting generates an earthquake; it is where the earthquake starts (Fig. 10.2).
- The **epicenter** is the point on the land surface directly above the focus (Fig. 10.2). A map can only show the epicenter, not the focus of an earthquake, so people in the media use the term *epicenter* more often, and it is more familiar to most people.

Figure 10.1 The seismographs pictured here work on the principle of inertia. The suspended mass remains stationary, while the other parts of the instrument and the Earth around it move during an earthquake. (a) A horizontal-motion seismograph records the horizontal vibrations that occur during an earthquake. (b) A vertical-motion seismograph records vertical vibrations.

Earthquake Hazards

One of the largest earthquakes to occur historically within North America was the Anchorage, Alaska, earthquake in 1964. A thrust fault (Lab 16) displaced a vast area of rock,

----

[1]A great earthquake is one with a magnitude greater than 7.5 on the Richter scale.

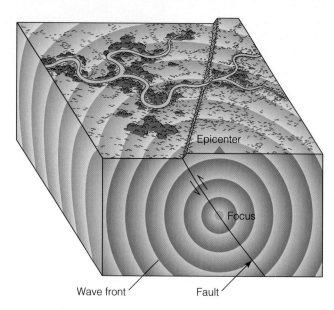

Figure 10.2 When an earthquake occurs, the movement starts at the **focus**, generally because of rupture along a fault, and radiates outward in all directions as seismic wave fronts (**body waves**). Directly above the focus, on the ground surface, is the **epicenter** of the earthquake. This is the mapped location of the earthquake. **Surface waves** radiate outward from the approximate location of the epicenter.

resulting in an earthquake of magnitude 8.4. Almost 3 km of coastal bluffs in Anchorage gave way due to a combination of liquefaction and landslides. About 1000 aftershocks ensued during the months after the main earthquake, some of them large earthquakes in their own right. Many of the approximately 130 deaths resulted from the tsunami generated by the sudden upward shift of the seafloor. The resulting tsunami hit coastal areas, drowning people as far south as Crescent City, California, where the third wave washed 500 m inland. In a matter of hours, a tsunami can move across the ocean from where it was generated (Fig. 10.3).

In highly populated areas the death toll and destruction of property can be much higher. The famous 1906 earthquake in San Francisco, with an estimated magnitude of 8.2, released substantially less energy than the Anchorage earthquake. Near San Francisco, a northern 430-km-long portion of the San Andreas Fault (Fig. 9.13) ruptured and moved up to 6 m. In San Francisco, the death toll from the earthquake may have reached 2500 people. Structures built on former marshland were especially hard hit due to liquefaction. A fire that raged out of control for 3 days afterward caused ten times more damage than the earthquake.

These earthquakes illustrate the major hazards associated with most large earthquakes. The following explains these hazards in more detail:

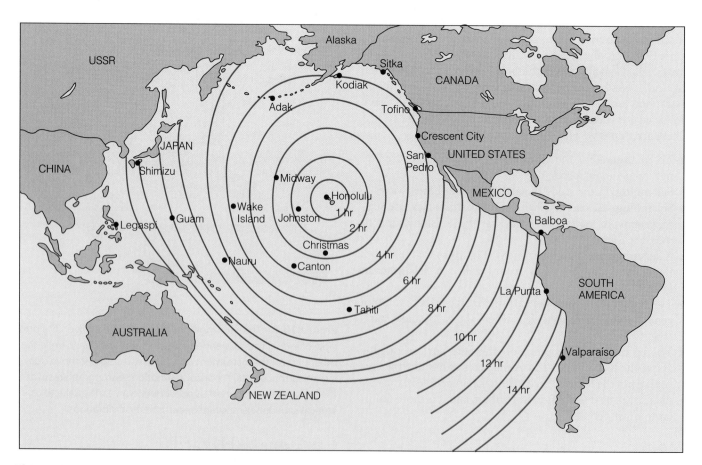

Figure 10.3 Tsunami Travel Times to Honolulu, Hawaii. If an earthquake struck somewhere in the Pacific Ocean basin, the time it would take for the resulting tsunami to reach Honolulu is indicated at the earthquake location.

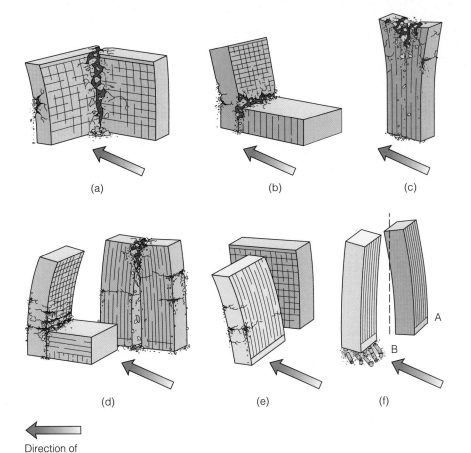

Figure 10.4 Effects of ground shaking on buildings of different shapes. (a) Separate wings of a building tend to have different motions, resulting in damage at the junction of the wings. (b) Different-height buildings respond differently to seismic waves, resulting in damage at the point where the height changes. (c) Vibrations increase with height of the building so that more damage occurs at the top. (d) Closely spaced high-rises may sway into each other. (e) Buildings with their long axis parallel to the seismic wave direction are more stable than if the long axis is perpendicular. (f) Buildings with more consistent and crossing supports, such as A, are less prone to damage than buildings with a "soft story" such as parking structures or large lobbies supported only by columns, such as B.

(a) (b) (c)

(d) (e) (f)

Direction of
seismic wave

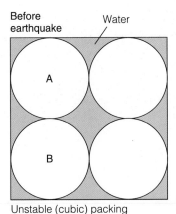

Before earthquake   Water

A

B

Unstable (cubic) packing

After earthquake   Excess pore water in liquefied soil carries load of overburden.

Water

A

B

Stable (hexagonal) packing

Figure 10.5 During liquefaction, loose, recently deposited sediment, or artificial fill becomes more compacted. Circles in the diagram represent spherical grains of sand in cross section. The spaces (pores) between the sediment decrease because of compaction, and water in these pores is forced upward. As a result, material and structures above have only water-laden sediment for support and collapse.

■ **Ground shaking** is the first hazard from earthquakes. Collapsing buildings and highway structures are the most direct and immediate result of this hazard (Fig. 10.4). **Aftershocks** are smaller earthquakes that occur after an earthquake and cause further damage to already weakened structures. The size and frequency of aftershocks diminish with time after the main shock. Ground shaking is not always the most destructive earthquake hazard.

■ In areas with steep slopes, earthquakes may trigger **landslides,** on land or under the sea, where rock or debris moves rapidly down the slope in response to gravity.

■ **Liquefaction** occurs when an earthquake vigorously shakes and compacts water-saturated sediment, which behaves like quicksand. Compaction of the sediment results

in displacement of pore water (Fig. 10.5) which flows upward, carrying sand with it. Buildings and structures where liquefaction occurs are much more likely to be damaged or collapse than in areas of bedrock, which have much less ground motion (Fig. 10.6).

■ Although a **tsunami** has nothing to do with tides, it is popularly called a *tidal wave*. It is a series of large ocean waves generated by a sudden disturbance of the seafloor (Fig. 10.7) that travel through the ocean in all directions from the epicenter (Fig. 10.8). In the open ocean, even large tsunamis are not noticed; but near shore, they become steep and very hazardous (Fig. 10.9). Tsunamis can also be generated in lakes such as Lake Tahoe, California and Nevada.

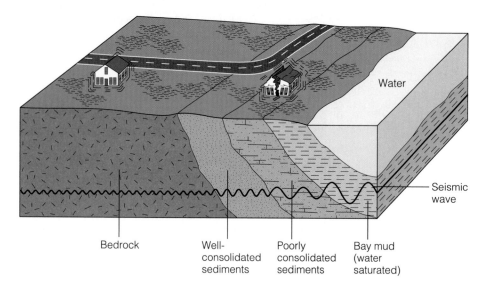

Figure 10.6 Ground motion increases, both in strength and duration, on water saturated and unconsolidated sediments. Seismic wave amplitudes are shown in bedrock and in progressively less consolidated and more saturated sediments. Buildings and other construction on weaker sediment are more prone to structural damage.

Bedrock

Well-consolidated sediments

Poorly consolidated sediments

Bay mud (water saturated)

Water

Seismic wave

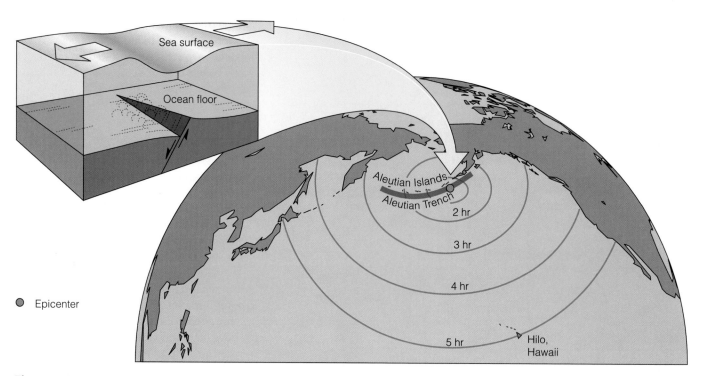

Epicenter

Figure 10.7 A tsunami generated by subduction in the Aleutian Trench in 1946 and traveling 472 mph hit the coasts of Japan and Hawaii 5 hours later and killed 150 people.

■ **Fire** is commonly associated with earthquakes, because during the earthquake flammable materials such as wood, paper, and gas come in contact with energy sources that can start fires, such as electric lines or flames from gas stoves. Utilities and transportation are often disrupted, making firefighting more difficult.

Earthquake Hazard Experiment

Read through the whole experiment before you start.

Materials needed:

- Two pans with sand >4 in. deep, one wet, one dry
- An earthquake simulation system
- Bricks to simulate buildings
- A clock, two stopwatches, or two watches

If previous students have used the same pans, stir the sand before you start. Your lab instructor will show you how to create vibrations to simulate an earthquake. Put identical "buildings" in each pan. A brick standing on its end makes a

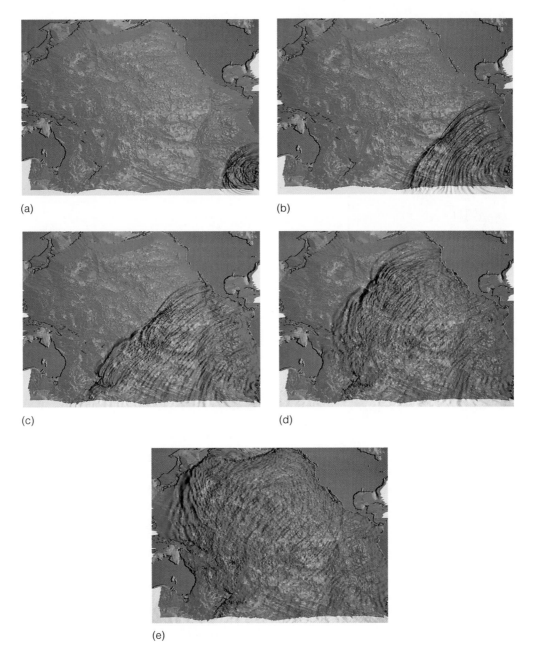

Figure 10.8 Computer-simulated tsunami in the Pacific Ocean. The waves were modeled after a 1960 tsunami generated in the Peru-Chile Trench off the coast of Chile. (a) 2.5 hours after the earthquake. (b) 7.5 h. (c) 12.5 h. Notice the waves hitting New Zealand. (d) 17.5 h—tsunami focusing on Hawaii. (e) 22.5 h—tsunami hitting Japan

good building. If both pans are vibrated at the same time, have two people observe, one for each pan. Time how long it takes from the start of the shaking for a building to topple.

1. Write the shorter time in Table 10.1 for Earthquake 1.
   a. Which pan of sand had more catastrophic failure or had buildings topple sooner? _____
   b. What happened to or in the sand during the vibrations? _____
   _____

Table 10.1 Relationship between Earthquake Repetitions and the Time for a Building to Collapse during an Earthquake

"Earthquake" Number	Time for a Building to Collapse (seconds)
1	
2	
3	
4	
5	

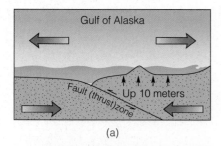

(a)

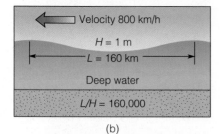

(b)

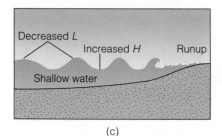

(c)

Figure 10.9 Changes in a tsunami as it approaches shore. (a) Faulting caused uplift of the seafloor in the Gulf of Alaska in 1964, causing a tsunami to move outward in all directions. (b) Tsunamis in deep water have low height and long wavelength and high speed. (c) As the waves move into shallow water, the wavelength decreases, wave height increases, and the wave eventually breaks. The runup is the area inundated by the wave above sea level.

c. How does the behavior of wet and dry sand differ?

_____

_____

d. What hazards discussed previously occurred in one or both of the pans? _____

_____

Very young bay mud behaves differently from older sediments during an earthquake. Test how the "age" of the sediment changes the results in the following experiment.

2. Repeat the prior experiment with the more unstable configuration four more times, without stirring or disturbing the sand between experiments, except to replace the "buildings." Tabulate and graph your results in Table 10.1 and Figure 10.10. Generalize what you observed.

_____

_____

_____

Figure 10.10   Graph paper for a bar graph

Earthquake Hazards Computer Exercise

Use the computer to view past earthquake disasters:

- Start **_Interactive Plate Tectonics._** Click the title screen to get to the menu.
- Click **CASE STUDIES,** and then click **EARTHQUAKES** under "**Case Studies and Brief Profiles.**"
- Click the flashing button in California and then the **STORY** button. Explore the information given about the Loma Prieta earthquake, and answer the following questions. (*note:* To return to a previous part of the program use the ⬅ button).

3. a. What was the magnitude of the Loma Prieta earthquake? _____

- Be sure to click the first button in the first row. Play (▶) the video of the earthquake during the World Series and watch the seismograph (in the inset) as the earthquake occurred. (You can rewind the video segment using the (◀◀) button to view it again.)

   b. What happened to the needle on the seismograph just before and during the earthquake?

   _____

   _____

   c. Why was there more damage in places 60 mi from the quake than close to the epicenter? What were most of the collapsed structures (buildings and viaducts) built on?

   _____

   _____

   _____

- Click the first and third buttons in the second row.

   d. What man-made structure caused most of the deaths during this earthquake? _____

- Explore the other buttons; in particular, click the third button in the first row, the red down (▼) button followed by the second button.

   e. What are the hazards people experienced during and after this particular earthquake?

   _____

   _____

   _____

- Next click TECTONIC FORCES and explore the information given. When you have read the first screen, click the red (▼) button to see more, especially the third button on the second screen. Answer the following questions:

4. a. The Loma Prieta earthquake occurred along _____ _____ fault, which is also a _____ fault.

   b. What type of plate boundary is found here?

   _____

   c. Which two plates caused the movement along the fault? _____

   _____

⊙ Tsunami Hazards Computer Exercise

- Start **Earth Systems Today.**
- Go to **Topic: Earth's Disasters.**
- Go to **Module: Earthquakes/Tsunamis.**

5. Use the activities in this module to discover:

   a. What regions of Earth are known for generating tsunamis? _____

   _____

   b. In what way are these regions related to or associated with plate tectonics? _____

   _____

   c. Where are tsunamis experienced? _____

   _____

   d. What is the connection between where tsunamis are generated and where they are experienced?

   _____

   _____

The Origin of Earthquakes

Each of the earthquakes discussed here occurred near a plate boundary. Although not all earthquakes occur at plate boundaries, it is the motions of the plates that usually provide the energy to cause earthquakes. Rocks start to bend or flex (usually) near the plate boundaries as the plates move. This continues until the forces causing the bending exceed the strength of the rocks or the friction keeping the rocks in place. At that point, the rocks break and slip along a fault. The earthquake is generated by the snapping-back action as the bent rocks return to their original shape (Fig. 10.11). This snapping back is known as **elastic rebound.** Waves of vibrational energy emanate from the point where the break starts; these waves are the earthquake or the shaking of the ground and are called *seismic waves*. You can observe elastic rebound in many objects such as a stick or a plastic fork.

6. Your instructor will provide you with an object to break. Read through this whole exercise before you start. Start by bending or flexing the object slightly with steady pressure. Listen as you do this—you may hear some slight crackling or popping sounds depending on the object you are bending.

   a. What do you hear, if anything, before it breaks?

   _____

   b. Observe the object carefully as you bend it but before you apply sufficient pressure to break it. If you accidentally break it before you make your observations, get another one and try again using slightly less force at first. Before it breaks, but while you are applying pressure, have your lab partner sketch the shape of the object. Are all areas equally deformed? _____

   c. Now bend the object until it breaks. What do you hear as it breaks? _____

   _____

   d. Sketch the parts of the object after it is broken—both the areas of breakage and the unbroken areas. Clearly label any areas of deformation.

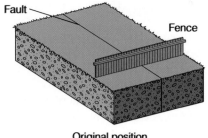

Fault

Fence

Original position

(a)

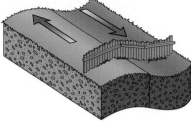

Deformation

(b)

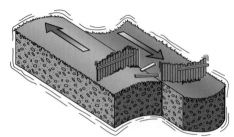

Rupture and release of energy

(c)

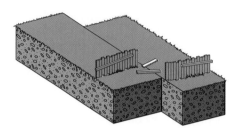

Rocks rebound to original undeformed shape

(d)

Figure 10.11 Elastic rebound theory states that earthquakes occur when accumulated stress is suddenly released and rocks snap back to their original shape elastically. (a) Before deformation. (b) Moving blocks have deformed the rocks near a fault so the rocks are bent out of shape. (c) When the rocks finally break, the earthquake occurs as a result of the rocks snapping back to their original shape. This is a snapshot while the rocks are in the process of rebounding. (d) After the earthquake, the rocks have returned to their undeformed shape, except for the area of rupture of the fault.

e. Were all areas equally deformed? _____

How do these areas compare to areas of deformation you sketched in **b?**

_____

_____

_____

f. How does the shape of these pieces compare to the shape of the object just before it broke and before you started to bend it?

_____

_____

_____

_____

_____

The sounds you heard as the object broke were a type of seismic wave called a *P wave*. Other waves also moved through the object when it broke but occurred too quickly to observe them.

Seismic Waves

Seismic Wave Computer Exercise

■ Start **In-TERRA-Active.**
■ Click **Interior** then **Earthquake Location.**
■ Read the material and advance through this section by clicking (▶) to discover the answer to the following:

7. You are visiting a seismograph station in Arizona when waves begin to register on the instruments. This seismogram shows three clear peaks in energy, separated by periods of reduced activity. Your geology instructor tells you that the reading comes from a single earthquake reported nearby. How can one earthquake yield a seismogram documenting different periods of high energy?

_____

_____

_____

_____

■ Click **Quit** and **yes.**

■

When an earthquake occurs, energy is released in the form of seismic waves, which spread out in all directions from the focus (Fig. 10.2), similar to ripples from a stone dropped in a pond. There are important differences, however. In the case of the stone, the waves you see move outward in a circle away from the center where the stone fell; in contrast, the waves from an earthquake's focus move outward in three dimensions in a spherical pattern. These are known as *body waves* (Figs. 10.12 and 10.13).

Body waves, which move through the Earth and are generated at the focus, are of two types, **P waves** and **S waves.** *P* stands for primary or push-pull waves and *S* stands for secondary or shear waves (Figs. 10.12b and c and Fig. 10.13a and b). P waves are called primary because they are the first to arrive, since they are faster than all other seismic waves. The P wave is a compressional wave with the vibration direction parallel to the direction it travels. This means that the wave

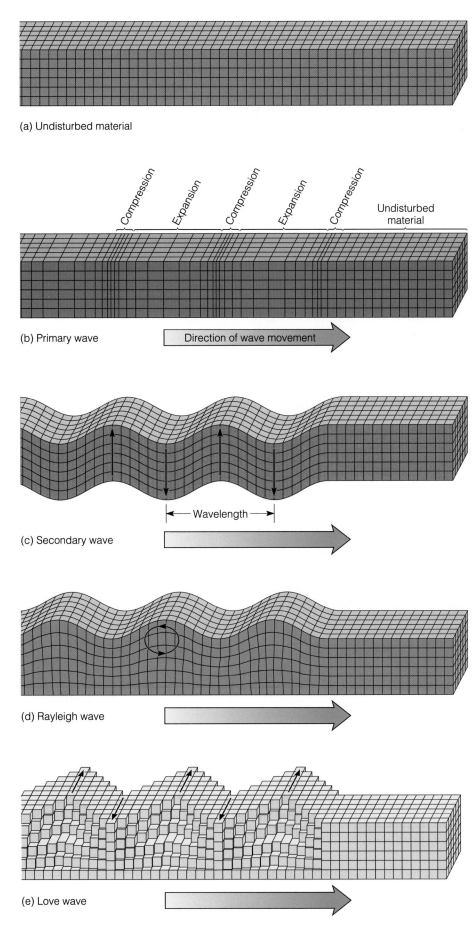

(a) Undisturbed material

Compression Expansion Compression Expansion Compression Undisturbed material

(b) Primary wave

Direction of wave movement

(c) Secondary wave

←— Wavelength —→

(d) Rayleigh wave

(e) Love wave

Figure 10.12 Elastic deformation of material as different seismic waves travel through it. (a) The material without any waves moving through it. (b) Primary waves (P waves), a type of body wave, compress and expand the material, causing particles of the material to vibrate parallel to the direction that the wave is traveling. (c) Secondary waves (S waves), another type of body wave, shear the material perpendicular to the wave travel direction. (d) Rayleigh waves are surface waves that produce a rolling motion like ocean waves, only retrograde—that is, backward from the rolling of ocean waves. These waves diminish with depth. (e) Love waves are surface waves that shear the material in a horizontal plane perpendicular to the wave travel direction and diminish with depth.

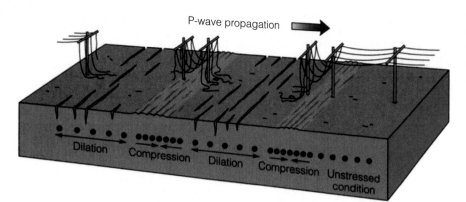

P-wave propagation

Dilation Compression Dilation Compression Unstressed condition

(a)

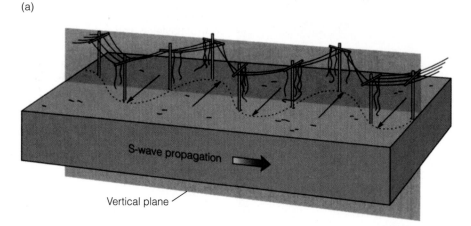

S-wave propagation

Vertical plane

(b)

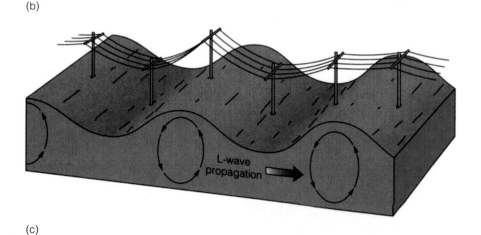

L-wave propagation

(c)

Figure 10.13 Some of the effects of seismic waves as they pass. (a) P waves. (b) S waves. (c) Surface waves, both Love and Rayleigh waves

causes material alternately to compress and expand as it moves (Figs. 10.12b and 10.13a). Sound waves are also compressional waves. The S wave vibrates perpendicular to its travel direction like the waves in a vibrating guitar string (Figs. 10.12c and 10.13b). Ponds and the liquid outer core have no S waves because S waves do not travel through liquids or gases. The way energy is transmitted by these waves may be demonstrated with a Slinky®.

8. Spread a Slinky across a tabletop. Have your lab partner hold one end of the Slinky while you hold the other. One of you produce a P wave by stretching out the Slinky and giving it a sharp push-pull movement at one end, parallel to its length. In the gap in Figure 10.14, sketch a snap-

shot of what the Slinky looked like as the wave moved through it.

9. Produce an S wave by sharply "waving" one end of the stretched Slinky perpendicular to its length. Sketch, in the gap in Figure 10.15, a snapshot of what the Slinky looked like as the S wave moved through it.

Another kind of seismic wave is a surface wave (Figs. 10.12d, e, and 10.13c). **Surface waves** only travel along Earth's exterior and are generated by P and S waves where they reach the surface. The waves you see in the pond are an example of surface waves. Surface waves are even slower than S waves and are last to arrive. Just as S and P waves have different types of motion, surface waves vibrate in other ways

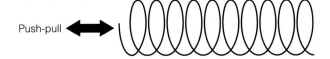 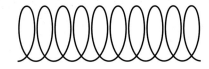

Figure 10.14 Sketch the movement of a P wave in the gap where no coils are drawn.

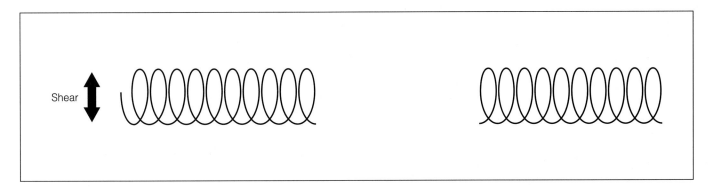

Figure 10.15 Sketch the movement of an S wave in the gap where no coils are drawn.

(Fig. 10.12d and e). One type moves a bit like a snake (Fig. 10.12e); another type rolls somewhat like water waves (Figs. 10.12d and 10.13c). People have reported seeing parking lots that looked like the ocean or feeling as though they are on a boat during an earthquake; these are surface waves. These waves are especially destructive because they cause different parts of a building or structure to move in different directions at the same time (Figs. 10.4 and 10.13c).

10. Review the sequence of waves, and write them down from fastest to slowest.

_____

_____

Imagine what it would be like to be in an earthquake at different locations near and far from the epicenter. Let's assume you are in a relatively safe location such as an open field with no structures or steep hills nearby, not too close to the sea, and where the underlying material is fairly firm so that liquefaction does not take place. Assume the location far from the epicenter is still close enough that you can feel each of the waves. Put the following types of motion in the order that you would experience them for each location: rolling motion, up-and-down vibration, and side-to-side vibration. Which is first, second, and third at these locations?

■ Near the epicenter? _____

_____

■ Far from the epicenter? _____

_____

(*Hint:* For P and S waves, think about the travel direction of the waves as they move toward the location. Then think about the kind of vibration each wave has with respect to that travel direction.)

■ What else would be different about the two locations? (Cite at least two differences.)

_____

_____

_____

_____

When all of the waves are recorded, the resulting seismogram looks something like Figure 10.16. Time progresses from left to right with a sequence of waves: P, S, then surface waves. The first arrival of each wave is labeled.

Magnitude

Magnitude, often called *Richter magnitude*, is one way that seismologists measure and compare the size of different earthquakes. The magnitude was originally based on a measure of the maximum amount of ground motion, the maximum amplitude (half the height of the squiggles in Figs. 10.16 and 10.17) of the seismic waves, at a 100-km distance from the epicenter (Fig. 10.17). The largest earthquakes have thousands of times more amplitude than small earthquakes, so the magnitude scale is compressed using logarithms. This means that the amplitude increases by a factor of 10 for each single-digit increase in the magnitude. The original scale developed by Charles Richter was dependent on the local area, the instrument used, and the size of the earthquake, so it has been modified considerably since its invention. Whichever method of calculation is used, the magnitude is related to the energy released during an earthquake.

Imagine you just visited (or live in) earthquake country and experienced an earthquake. Maybe it was a magnitude 4. You say, "That wasn't bad. What's the big deal?" It is important to understand how much bigger a 5 would be, or a 6 or 7. When comparing two earthquakes, such as Earthquake B and

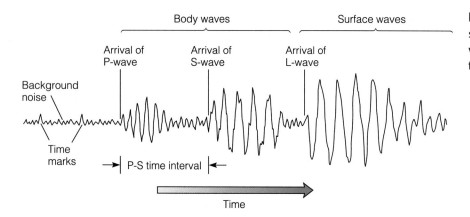

Body waves

Surface waves

Arrival of P-wave

Arrival of S-wave

Arrival of L-wave

Background noise

Time marks

P-S time interval

Time

Figure 10.16 A seismogram showing P wave arrival time, the S wave arrival, and the arrival of the first surface wave (L)

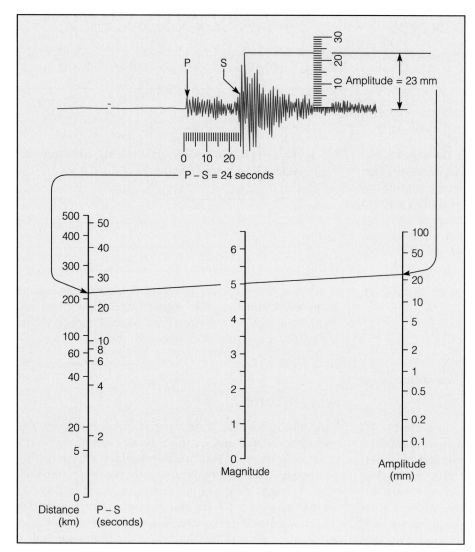

P

S

Amplitude = 23 mm

0 10 20

P – S = 24 seconds

500
400
300
200

100
60
40

20

5

0
Distance (km)

50
40
30
20

10
8
6
4

2

P – S (seconds)

6

5

4

3

2

1

0
Magnitude

100
50
20
10
5
2
1
0.5
0.2
0.1

Amplitude (mm)

Figure 10.17 Richter magnitude can be determined graphically using the maximum amplitude of the seismic waves, measured as shown by the amplitude scale, and the distance of the instrument from the earthquake's epicenter as determined from the time between the arrivals of the P and S waves (P - S in the diagram). This diagram is based on P and S wave behavior in southern California and is not appropriate for other areas.

Earthquake A, with magnitudes M_B and M_A respectively, the energy released by Earthquake B (E_B) is expressed in terms of the energy released by Earthquake A (E_A) as follows:

$$E_B \cong [30^{(M_B - M_A)}] \times E_A$$

This means that the ratio of energy released is

$$\frac{E_B}{E_A} \cong 30^{(M_B - M_A)}$$

The energy released by an earthquake is important because it is directly proportional to the destructiveness of the earthquake. Let's compare a magnitude 4 earthquake with an earthquake that is generally felt by people at rest indoors—a magnitude 3. The energy released by a 4 is approximately $30^{(4-3)} = 30^1 = 30$ times greater than the energy released by a magnitude 3 quake. A magnitude 5 releases about 900 times the energy of a magnitude 3.

Table 10.2 Energy of Different Earthquakes

Magnitude Difference, $M_B - M_A$	Ratio of Energy Released, E_B/E_A
0	
1	
2	
3	
4	
5	
6	

11. Fill in Table 10.2 using the equation.[2]

12. Draw a graph on the linear graph paper in Figure 10.18 showing the relationship between the difference in magnitudes versus the ratio of the energy released by two earthquakes of different magnitudes. That is, plot a graph with $(M_B - M_A)$ on the horizontal axis and (E_B/E_A) on the vertical axis. Remember to label your graph correctly.[3]

13. Draw another graph with the same data using the semi-log graph paper in Figure 10.19.

14. Describe the difference in the shape of the two graphs. Why do the graphs look different? Why is this important?

_____

_____

_____

_____

15. Read from your graph in Figure 10.19 to determine:
    a. How many times more energy was released by the Anchorage, Alaska, earthquake of 1964 (magnitude 8.4) compared with the San Francisco 1906 quake (estimated magnitude 8.2)? _____
    b. Compare the New Guinea 1998 earthquake (7.1) with the San Francisco quake. _____

16. Think of what it would be like for such an earthquake to hit your area today.
    a. Make a rough-sketch map of your area showing the locations of major highways, public transportation systems, residential areas, water, gas, and sewer plants, power generation plants, and dams. Also include any geographic features such as lakes, oceans, hills mountains, or slopes.

. . . .

[2]If your calculator does not handle very large numbers, use the following equivalent equation, or work the problem by hand:

$$\frac{E_B}{E_A} \cong 3^{(M_B - M_A)} \times 10^{(M_B - M_A)}$$

Calculate the last expression by hand, remembering that 10 to any power x is 1 followed by x zeros; for example, $10^4 = 10,000$.

[3]Correct labeling of a graph includes placing numbers along each axis, an axis description, the units of the numbers (if any), and a title or caption. A ratio has no units, and neither does Richter magnitude, but remember to use units on your graphs in other labs.

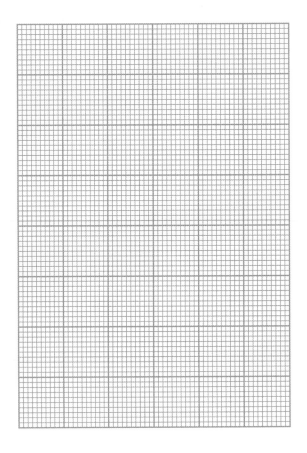

Figure 10.18 Linear graph paper

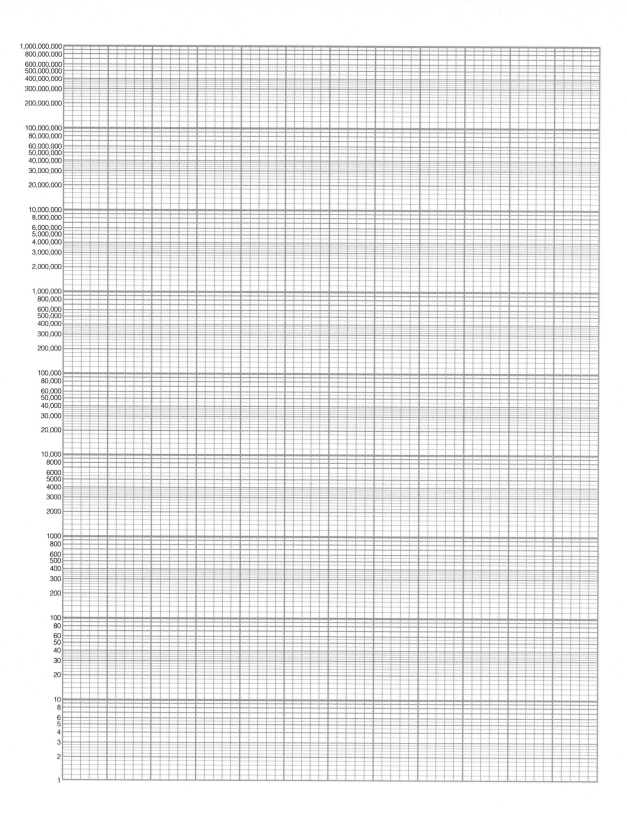

Figure 10.19 Nine cycle semilogarithmic graph paper

b. Considering all these systems and their interactions with the daily lives of those in your community, write an essay on a separate piece of paper or on p. 214 describing what might happen if a great earthquake hit your area today.

Locating an Earthquake

A seismograph detects seismic waves and records the results on a seismogram. To determine the distance from the seismograph to the earthquake focus, we can use a property of these seismic waves that we have already seen. The P and S waves travel at different speeds. The farther from the earthquake, the farther apart the waves will be in time of arrival at a particular location. If a seismograph were located at an epicenter, all of the waves would arrive at nearly the same time because the earthquake was so close. On the other hand, if the seismograph were some distance from the epicenter, the difference between the arrival times of the fast P waves and the slower S waves would be greater, because the faster waves would pull ahead of the slower waves.

Imagine an auto race in which one car, let's say a yellow one, is consistently faster than another, a blue one. Let's assume both cars start at the same place and time, like P and S waves leaving the earthquake's focus, and maintain a constant speed. After two laps around the racetrack, the yellow car will be twice as far ahead of the blue car as it was after the first lap. The farther they go, the farther ahead the yellow car gets. With a bit of calculation you could even tell how many laps the cars had gone by the difference in the time they pass you in the grandstand. That is what we will do with P and S waves, tell how far they have traveled based on the difference in time they arrive at a seismograph.

The proportionality is not quite correct for P and S waves because their speeds are not constant (Fig. 10.20), but if we use their average speeds, it will give us an approximation. The distance from the focus determines the difference in arrival times, so the reverse is also true; the arrival-time difference can be used to measure the approximate distance to the focus. Seismograms such as the one in Figure 10.16 are a record of the arrival times for different seismic waves. A seismologist measures the time between the P and S wave arrivals (S − P) from the seismogram and then calculates the approximate distance to the focus using the speed of the P and S waves. Since most earthquakes occur within shallow levels of the crust, distances to the focus are approximately distances to the epicenter as well. With at least three seismograms for the same earthquake from three different seismograph locations, the seismologist can determine the location of the epicenter of the earthquake by drawing circles on a map around each seismograph location (see later discussion).

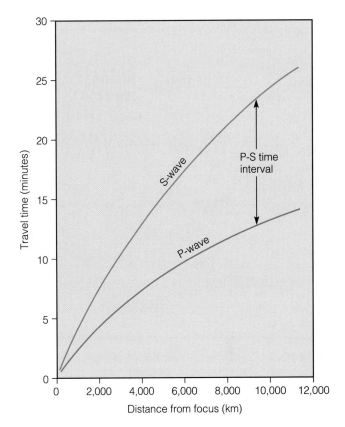

Figure 10.20 Travel time curves for P and S waves. If the velocities of these waves were constant, the travel time curves would be straight lines. However, notice that the curves have gentler slopes at greater distances—the waves speed up with distance. This is because waves that travel farther move through deeper parts of the Earth where the waves can travel faster in the denser material.

_____

_____

_____

_____

_____

Earthquake Location Computer Exercise

- Start **Earth Systems Today.**
- Go to **Topic: Earth's Disasters.**
- Go to **Module: Earthquakes/Tsunamis.**
- Use the activities in this module to discover the following for yourself.

17. Find an earthquake epicenter and write a step-by-step description of how you did this.

_____

_____

Distance to the Earthquake

Let's locate an earthquake, starting with the time lag between the P and S waves.

18. Examine Figure 10.21. Locate the P and S wave arrivals for each seismogram as illustrated in Figure 10.16. Use the scale on the seismograms to determine how much time passed between their arrival. This is called the time lag, S − P. The time lag is a time difference. You do not have to subtract to obtain this difference because you already have a difference when you measure the times between the P and S waves on the diagram. Fill in Table 10.3 with the time lags for each seismograph.

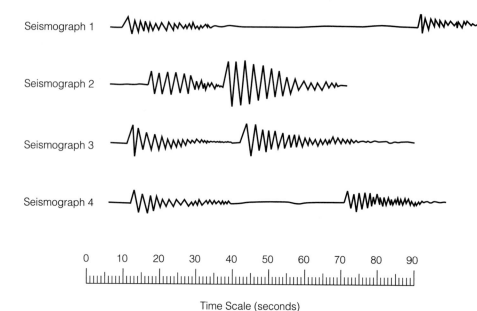

Seismograph 1

Seismograph 2

Seismograph 3

Seismograph 4

0 10 20 30 40 50 60 70 80 90

Time Scale (seconds)

Figure 10.21 Short segments of seismograms for an earthquake from four seismographs. Only the P and S wave parts of the seismograms are reproduced here.

Table 10.3 S-P Time Lag for Four Seismograms

Seismograph	$(S - P)_D$ Time Lag (s)
1	
2	
3	
4	

19. In your own words, how do you recognize the P wave arrival on the seismograms? S wave arrival?

_____

_____

_____

_____

_____

So far all we have is a difference in time for each seismograph. As discussed earlier, we can use this measurement to calculate the distance to the focus for each seismograph. We noticed that the time lag between race cars at the track is proportional to distance they traveled. If we knew the average speed of each car and the length of one lap on the racetrack, we could then use the time between the two cars to tell how many laps they had gone. Let's do the same with the earthquake waves by choosing a "lap" equal to 1 km. If you know how fast a P wave travels, you can calculate how long it will take to travel 1 km; the same holds for the S waves. If you subtract, you have the time lag for 1 km. This just means that after the waves have traveled 1 km, the S wave will be behind the P wave by this number

of seconds. The calculation is simple; you probably already know it:

$$speed = \frac{distance}{time}, so$$

$$time = \frac{distance}{speed}$$

The average P-wave and S-wave speeds in the crust are 6.1 km/s and 4.1 km/s, respectively. The speed of the waves would have been measured previously by setting off explosions or vibrations at known times and detecting the waves some distance away.

20. What is the time required for each wave to travel 1 km?

P waves travel 1 km in _____ s. S waves travel

1 km in _____ s. What is the difference in arrival times? Time lag, at 1 km is _____ s.

(*Hint:* Subtract the S and P travel times.) Let's call this time lag $(S - P)_1$. The subscript 1 stands for 1 km distance.

The longer the travel distance, the longer the time lag. We will assume they are proportional since we are using average speeds. So set up a proportionality equation. The distance we are trying to determine is D—the distance from the earthquake's focus to the seismograph—and the time lag we see on the seismogram is $(S - P)_D$. D is to $(S - P)_D$ as 1 km is to $(S - P)_1$. Or the distance is to its time lag as 1 km is to the time lag for 1 km.

$$\frac{D}{(S - P)_D} = \frac{1 \text{ km}}{(S - P)_1}$$

where

D = unknown distance in kilometers

$(S - P)_D$ = time lag for distance D

$(S - P)_1$ = time lag for 1 km

Then solve the equation for *D*, which is what we are trying to calculate.

$$D = \frac{1 \text{ km}}{(S - P)_1} (S - P)_D$$

We know $(S - P)_1$ from the question above, but what about $(S - P)_D$? That is just the symbol for the time lag you measured off the seismograms that you entered in Table 10.3. Now we can calculate *D* for each seismograph.

21. Calculate the distance from each seismograph to the earthquake's focus using the equation for *D* and the measurements in Table 10.3. Record the answers in Table 10.4.

Even with a number of distances, we still do not know where the earthquake started or exactly what time it hit.

Locating the Earthquake from Three Seismographs

Once you know the distance to an earthquake's focus from three seismographs, you can approximately locate the earthquake's epicenter on a map. Set a compass so that the distance between the pencil and the point of the compass equals the map distance between the earthquake's epicenter

Table 10.4 Distance to Earthquake's Epicenter from Four Seismograms

Seismograph	*D* Distance (km)
1	
2	
3	
4	

and one of the seismographs.[4] Place the point of the compass on the map at the seismograph location (Fig. 10.22). Draw a circle around the seismograph; the earthquake occurred some place on this circle. Next, set the compass spacing equal to the distance for the second seismograph and draw a circle around that seismograph. Be sure to keep your

....

[4]We will approximate distances to the epicenter as being equal to the distances to the focus. If the focus is very deep and the seismographs are all close to the epicenter, the distances will all be too large. On the other hand, for many earthquakes the focus is quite shallow compared to the distance from the seismograph to the epicenter, which makes this a good approximation.

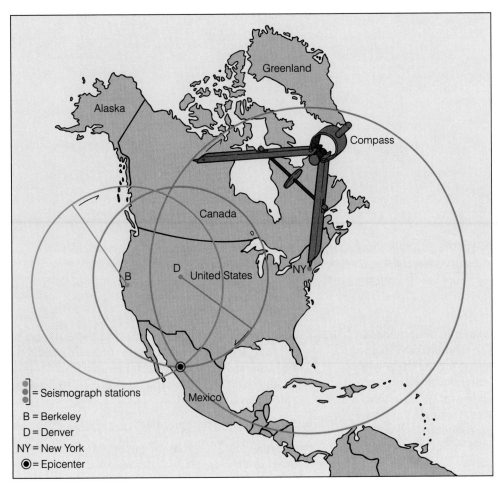

Figure 10.22 Using a compass to pinpoint the location of an earthquake. Once the distance to the epicenter is determined for each seismograph station, the distance is used as the radius of a circle centered at the station. The point of intersection of all three circles is the epicenter.

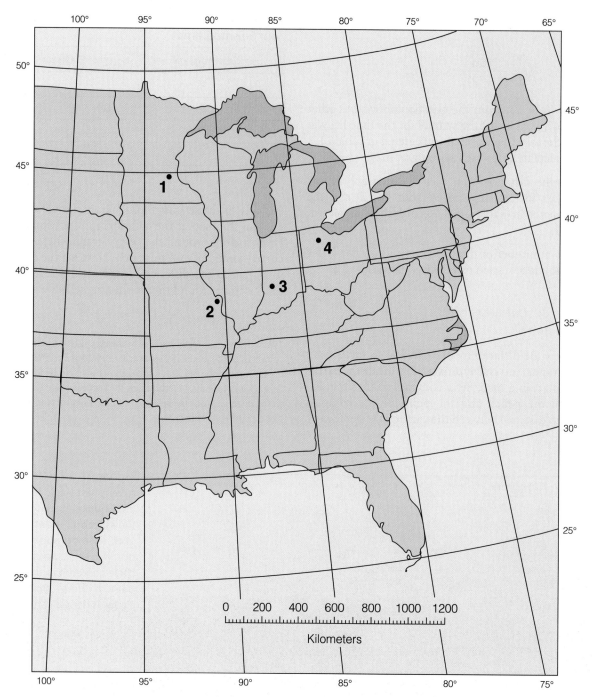

Figure 10.23 Map of the eastern half of United States showing the locations of four seismographs: 1 = Minneapolis, Minnesota; 2 = St. Louis, Missouri; 3 = Bloomington, Indiana; 4 = Bowling Green, Ohio

seismographs straight. You don't want to use the distance for Seismograph 1 on Seismograph 2. The earthquake's epicenter should fall on this circle, too. Repeat for the third seismograph. All three circles should intersect at one point, which is the epicenter as illustrated in Figure 10.22. In practice, the circles do not always intersect at a single point because seismic waves travel at different speeds through different rocks. In such a case, the region where they nearly intersect is likely to contain the epicenter.

22. a. Follow the prior instructions using the distances in Table 10.4 and the map in Figure 10.23. You could do this with any three of the seismographs, but do it for all four to double-check your answer. Clearly label the epicenter on the map.

b. In what state is the epicenter? _____

Notice in Figure 10.24 that the earthquake you just plotted occurred in the New Madrid seismic zone.

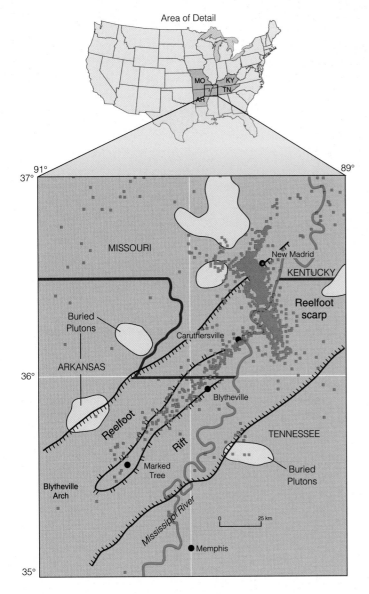

Figure 10.24 Map of parts of Missouri, Kentucky, Arkansas, and Tennessee showing the New Madrid seismic zone. An old rift beneath thousands of feet of sedimentary rocks is a zone of weakness in the North American Plate, where numerous earthquakes occur (red squares). Earthquakes of magnitude greater than 8 have occurred here historically.

Time of the Earthquake

Another question of interest is "When did the earthquake happen?" We have a pretty good idea from the time of arrival of the P waves given in Table 10.5 that the earthquake occurred shortly before 3:11 Central Standard Time, but the P waves had already traveled for some time before they got to the seismographs. In fact, we have four different times recorded because each seismograph was a different distance from the earthquake's focus. When did the P waves leave the focus? We have calculated the distance to the earthquake's focus, and we know the average speed of the P wave. Just use that simple equation again: time = distance/speed. Time in this equation is the travel time of the wave, not the time of the earthquake. Subtract the travel time from the P wave arrival time to obtain the time of the earthquake.

23. For each station, calculate what time the earthquake occurred; enter the answers in Table 10.5. Should all the answers have come out the same or different?

   _____

   Explain. _____

   _____

   _____

   _____

24. Why is the answer for Seismograph 4 an hour different from the other answers?

   _____

   _____

Table 10.5 Time of Initiation of the Earthquake as Determined from Four Seismograms

Seismograph	P Wave Arrival Time as Hours, Minutes, and Seconds (h:min:s)	P Wave Travel Times (s)	Time of Earthquake (h:min:s)
1	3:12:53 Central Standard Time		
2	3:10:50 Central Standard Time		
3	3:11:11 Central Standard Time		
4	4:12:06 Eastern Standard Time		

Topographic Maps

We learned in Lab 1 that maps are among the most important of the visual and conceptual tools Earth scientists use to study the physical surface of the Earth. Topographic maps, which show the shape of the Earth's surface (hills, valleys, etc.) are essential in the study of most areas, whether for Earth science research, land development, recreation, or other land use.

Topography and Contours

The term **topography** refers to the shape of the physical features of the land surface. A topographic contour map, familiar to many people because of its widespread use, is the representation of the actual topography of a given area. Although drawn in two dimensions only, it uses contour lines to depict three-dimensional objects such as hills, valleys, and mountains.

Each **contour line** (see 1 on the map in Fig. 11.1) connects points on a map that have the same elevation on the Earth's surface. A single contour line does not divide or branch, nor does it cross others.[1] Contours must be continuous, stopping only at the edge of a map or where they are briefly interrupted to label the elevation. The elevations are measured above some datum plane, usually mean sea level. An island's 0-ft contour line would be the shoreline. Assuming that the island is a single high hill, a footpath located exactly 20 ft above sea level and continuing entirely around the island at this same elevation would coincide with the 20-ft contour

....
[1]Contours may appear to join and divide in a vertical cliff, but these must be contours of different elevations. Contours that cross occur only where there is an overhanging cliff and the lower contours will be dashed in this case.

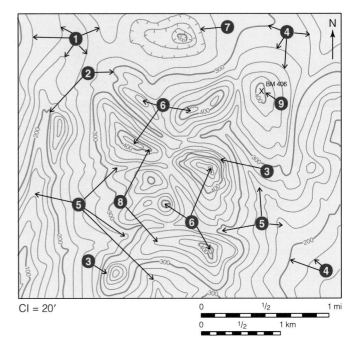

Figure 11.1 A topographic contour map of a hypothetical area with hills, valleys, and other topographic features. 1 = contour lines; 2 = index contours; 3 = steep slopes; 4 = gentle slopes; 5 = valleys; 6 = hills; 7 = depression; 8 = ridges; 9 = bench mark.

CI = 20'

on a map. In this case, the **contour interval,** or difference in elevation between two adjacent contour lines, would also be 20 ft; the next higher contour would connect all points that were 40 ft above sea level. If every fourth or fifth contour is drawn with a heavier line, usually labeled with a number, it is known as an **index contour** (see 2 in Fig. 11.1). Wherever the actual terrain is steep, map contour lines are closely spaced (3). Conversely, gentle slopes are portrayed by contours spaced widely apart (4). Contour lines run horizontally parallel to valley walls and hill sides and curve up a valley, cross a stream or gully at right angles (5), then double back down the valley, often making a V shape on the map that points upstream. Hills (6) are represented by a series of more or less concentric contour lines forming continuous loops. A hollow topographic depression or basin is depicted by means of one or more lines that encircle the basin and have hachure marks

(7); the hachure marks point in toward the center of the depression. A ridge (8) is a long narrow area of higher elevation often found as an extension of a hill or mountain, or as a connection between mountains. It has contours that double back on themselves, with the higher elevations toward the center and lower outside (8). In various places on topographic maps are **x**'s that show the locations of bench marks (9). A **bench mark** is a permanent marker on the ground that has precisely surveyed location and elevation measurements. The **x** on the map is labeled BM and its elevation is written next to it (9). The difference in elevation between the highest and lowest point in an area is known as **relief.** An area with high relief will have either closely spaced contours or a large contour interval or both; an area with low relief will have a combination of widely spaced contours and/or a small contour interval.

 ## Topographic Maps Computer Exercise

■ Start **TASA Topographic Maps**.
■ Click stop (■) to get to the main menu, then click **7. Principles of Contour Lines** and click (▶) to advance through this section answering the following questions:

1. Complete the following sentence:

   A _____ line is a line drawn on a topographic map (usually in _____ on USGS maps) that _____ _____.

2. If you walked along a contour line, would you be going up, going down, or staying level? _____

3. What reference plane (or datum) is usually used? _____

4. Complete the rules of contour lines given by the program:

   Rule 1: Every point on a contour line is the same _____.

   Rule 2: Contour lines never _____ or _____.

   Rule 3: Widely spread contour lines mean _____.

   Rule 4: Closely spaced contour lines mean _____.

   Rule 5: Hills are shown with _____, _____.

   Rule 6: A contour line forms a "V" pointing _____ when crossing a river or stream.

Rule 7: _____ on the land are shown with closed contours with hachures (short lines) that point _____.

5. The change in elevation between any two different, adjacent contour lines is the _____ (_____).

6. If the *contour interval* is 40 ft, which elevation below would be on a contour line?

   (a) 20 ft   (b) 30 ft   (c) 200 ft   (d) 300 ft

7. What is meant by *index contour*? _____ _____

8. Answer the three questions in this section:
   a. What is the highest elevation of the land shown on the area represented by this topographic map? Select the most appropriate answer. _____
   b. What is the lowest elevation of the land shown on the area represented by this topographic map? Select the most appropriate answer. _____
   c. Toward what direction is Salt Creek flowing? _____

■ To stop, click the stop (■) button twice, then click **Quit.**

■

Making a Topographic Map of a Volcano

Your instructor will supply a plastic model for the following exercises. The purpose of these exercises is to demonstrate the concept of contour lines and therefore improve your understanding of topographic maps. It is important to remember that contour lines are composed of points of equal elevation, and they (1) never intersect or divide, (2) are continuous (never end except at the edge of a map, but often form enclosed shapes), and (3) are at right angles to the direction of slope.

Procedure for Map Construction

9. With the volcano inside the box, pour water in the box up to the first elevation, which represents 100 ft elevation on the ground.

10. Place the top on the box and a clean acrylic sheet over it. Tape the sheet to the box lid with a minimum of tape. Looking from above, trace the water line onto the acrylic sheet. *Label* the contour line with the elevation by leaving a short gap in the contour and inserting "100" there.

11. Repeat this procedure labeling *each* of the contours in increments of 100, until the water reaches "700 ft" eleva-

tion. At elevations 600 ft and 700 ft, check to see whether you need to draw two contours. If so, label both with the correct elevation. What symbol will you use on the contours of the crater? _____

12. Trace the contour map on tracing paper for the next exercise. Include the contour interval at the bottom of the map. Don't forget to label the contours on your copy.

13. Clean up as directed by your lab instructor!

Volcano Map Questions

14. Describe the general shape of the contour lines at 400 to 700 ft.

_____

_____

15. What general statement can be made about the closeness of the contour lines and the steepness of slope?

_____

_____

Topographic Profiles

A profile shows how a landform would appear if you could see it in silhouette. It adds an upward dimension, giving you an accurate picture of the topography along the line from which it is drawn, therefore making it easier to imagine the area as it actually appears. Profiles that are drawn with the vertical scale identical to the horizontal scale (which is given on the map) are *normal profiles*. If the relief is low in the area or if topographic features need to be accentuated, the vertical scale is increased in relation to the horizontal scale; this is an **exaggerated profile.**

In this exercise, you will draw a topographic profile from the map that you constructed of the volcano model. Follow the procedure described next and watch carefully as your instructor demonstrates the method.

Procedure for Profile Construction

16. Draw a straight line 6 in. long across the short dimension through the center of the volcano on the map and clearly mark the end points of the line. Label one end point A and the other A′. This is your reference (or base-) line for your topographic profile.

17. Take a piece of scratch paper, which will become a *profile gauge,* at least as long as the base line drawn in 16 and lay one edge along the line as shown in Figure 11.2a. Mark the placement of the end points and label them A and A′ on the scratch paper profile gauge. Where contours intersect the edge of the profile gauge paper, make a small tick mark on the profile gauge perpendicular to the paper's edge. Label each tick mark with the elevation of the contour. For your volcano, these elevations will be 100 ft, 200 ft, and so forth. It is usually helpful to label each high point and each

low point along the profile gauge as shown in Figures 11.2 and 11.3. You should be able to recognize when you have reached a high or low point because the same contour elevation crosses the profile line twice in succession and that makes the same elevation repeat along the gauge.

18. Use the lines in Figure 11.4 to construct your topographic profile. At the left side of the lines, draw a straight vertical line (perpendicular to the lines). At the top, label this A (see Fig. 11.3). Lay the marked edge of the profile gauge horizontally along a line of your profile; match the tick mark labeled A on the gauge with the line labeled A on your profile. Now mark the location of A′ at the other end of the profile, draw a vertical line at this location, and label it A′. You should now have two vertical lines 6 in. apart on your profile in Figure 11.4. These two lines are the edges of your profile. Starting on the left of the lowest line, moving upward label every horizontal line with the elevations 100 ft, 200 ft, 300 ft, and so forth. You are now ready to draw your topographic profile.

19. Lay your profile gauge horizontally along the line for 700 ft elevation in Figure 11.4. Match your A and A′ end points (see Fig. 11.3). Mark a tick on your profile for each 700 ft tick on your profile gauge. In Figure 11.3, this step is illustrated for the map in Figure 11.2 for a 380-ft contour and then for a 360- and 260-ft contour (Fig. 11.3a, b, and c, respectively; *Note:* You will *not* have a 380-, 360-, or 260-ft contour on your profile). Next move the profile gauge down to the 600 ft line and mark the 600 ft tick marks on Figure 11.4. Continue this process until you have a mark on your profile for every tick mark on your profile gauge.

20. Draw a continuous smooth curve through all the ticks in Figure 11.4, as shown in Figure 11.2b. Your curve should pass through the intersection of the horizontal lines with each tick mark. Connect the curve at hilltops *above* the horizontal line and at valley bottoms *below* the line. You can, in this case, check your profile against the volcano model. In later labs, you will construct profiles to visualize topography of an area that you have never actually seen.

21. How does your profile compare to the volcano model?

_____

_____

Using and Constructing Topographic Maps

Raised relief maps are topographic maps that are drafted onto a plastic sheet that is molded to give the viewer an idea of the topography in the map area. Many beginning students find it difficult to imagine what an area looks like by reading the contours on a two dimensional map surface; the raised relief map is a sort of mental bridge that aids in this visualization. Compare the topographic map provided by your instructor with a raised relief map of the same area. Practice reading the map so that it will become possible for you (after

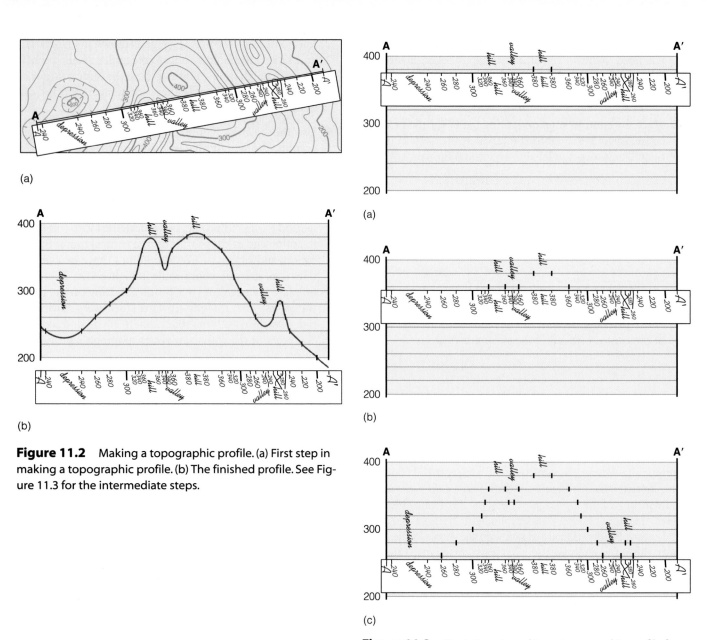

(a)

(b)

Figure 11.2 Making a topographic profile. (a) First step in making a topographic profile. (b) The finished profile. See Figure 11.3 for the intermediate steps.

(a)

(b)

(c)

Figure 11.3 Next steps in making a topographic profile for map in Figure 11.2. (a) Second step. (b) Third step. (c) Eighth step.

Topographic Profile of Volcano Model

Figure 11.4 Draw a topographic profile of your volcano map on the lines here following the instructions in Exercises 17–20.

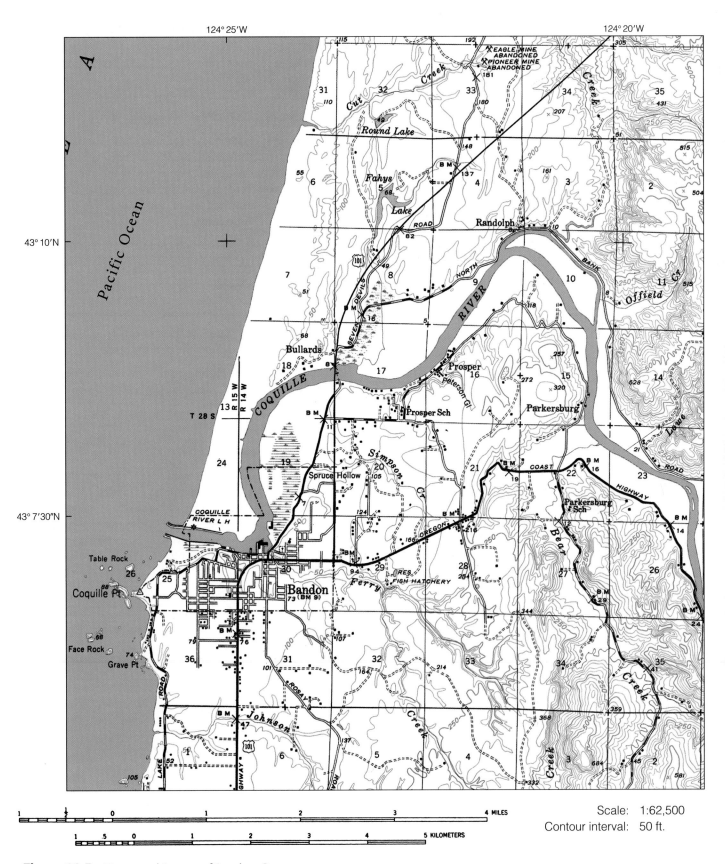

Figure 11.5 Topographic map of Bandon, Oregon

finding a few elevations as points of reference) to imagine the shape of the land that is implied by the contours. Next, apply your knowledge of contours to answer questions about a flat topographic map.

22. Study the map of Bandon, Oregon in Figure 11.5.

   a. Describe the topography of the area between Prosper School and Bandon, south of the Coquille River. (*Hint:* Is it steep and rugged, gently rolling, nearly flat, or what? Look at the contour arrangement and spacing.)

   _____

   b. Describe the topography along the middle part of Lowe Creek (E part of map). Contrast this area with the area you described in the last question.

   _____

   c. How are the flat surfaces of the marsh in Section 19 expressed by contour lines? (*Hint:* How many contours do you see within the marsh?)

   _____

   d. Compare the slopes of the sides and tops of the hills in Sections 11, 2, and 35 northeast of the Coquille River.

   _____

   _____

   e. Note the placement of the roads. What topographic considerations (or conditions of terrain) must be accounted for when choosing the path of a road?

   _____

   _____

   f. Note the scale, the contour interval, and the relief; use the appropriate units.

   Scale = _____

   CI = _____

   Relief = _____

23. In each of the boxes in Figure 11.6, sketch a hypothetical contour map of the feature indicated. Use a 20-ft contour interval and label every contour. Not all of these features are found on the Bandon map. Numbers correspond to numbers on Figure 11.1, but try to draw these without looking.

Elevations of Contours on a Topographic Map

Remember that concentric closed loops without hachures on a contour map signify a hill; therefore, the elevations will increase toward the center in steps equal to the contour interval. Also, as you move across contours in one direction, the elevations must continue up or continue down unless you have crossed the same contour twice.

24. Label the elevations of each contour on Antelope Hill on the map in Figure 11.7.

25. What is the possible range of elevations for the top of Antelope Hill? _____

26. Assuming each valley in the map in Figure 11.7 has a stream, draw in the locations of the streams.

27. Locate the steepest slope on the map and label the spot with the letter *S*. Locate the gentlest slope on the map and label it with the letter *G*.

Constructing a Topographic Map

When constructing a topographic map, cartographers draw contours using elevations of points provided by surveying techniques, GPS data, and aerial photography. The following exercises should help you to construct contour maps from a map with elevations given at various points. Do the exercises in order, as the difficulty increases. *Carefully follow the instructions.*

28. On Figure 11.8 connect the **x**'s of equal elevation to make a contour map with a contour interval of 10 ft. Leave a small gap in each contour, and label the elevation there.

Hill (6)	Basin (7)	Valley (5)	Ridge (8)

Figure 11.6 Use a 20-ft contour interval to draw each of these topographic features (Exercise 23).

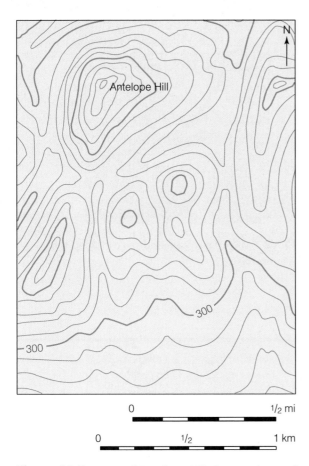

Figure 11.7 Map of Antelope Hill. Contour interval = 20 ft.

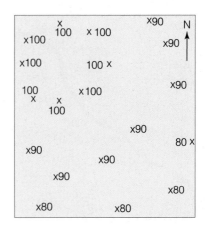

Figure 11.8 Map of a hill. Read the instructions in Exercise 28.

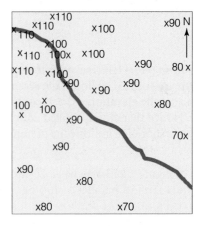

Figure 11.9 Map of a stream valley. Read the instructions in Exercise 29.

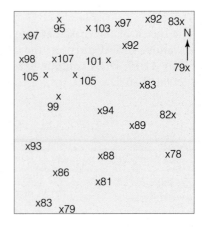

Figure 11.10 Map of a hill. Read the instructions in Exercise 30.

29. Contours are parallel to valley walls and make a V shape when they cross a stream. The V points upstream. For the map in Figure 11.9 the contour interval is 10 ft. Connect the x's for the 110-, 100-, and 90-ft contours first; notice how these contours V upstream. Next draw in the 80- and 70-ft contours with an appropriate V shape. Label the elevation of each contour in a small gap in the contour line. Check your answer with your lab instructor.

30. Cartographers use all elevation data available, not just elevations that exactly match the contour interval, so let's try the same. On Figure 11.10, draw in the contours at a 20-ft contour interval. Remember that each contour must be a multiple of 20. Notice that this time you will not be connecting the x's. Each line must pass between the various x's since no elevation is exactly at the contour interval. To draw your lines correctly, have all of the numbers higher than your contour elevation on one side of the line and all lower on the other side. The contours will probably pass nearer the elevations that are close in value to the contour elevation and far from the x's that are quite different in elevation. Label the elevation of each contour. Some students find it helpful to draw in some x's at the contour elevation by interpolating between elevations on the map and then connecting those x's.

31. Figure 11.11 has a stream valley and a contour interval of 20 ft. When you draw the contours on this map, don't forget to make the contours V upstream. Label the elevation of each contour.

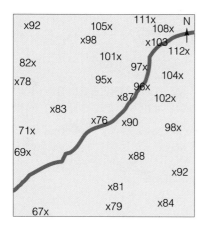

Figure 11.11 Map of a stream valley. Read the instructions in Exercise 31.

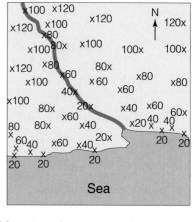

Figure 11.12 Map of a stream and sea cliff. Read the instructions in Exercise 32.

32. The map in Figure 11.12 has a 20-ft contour interval. Draw in the contours. The line at the seashore is the 0-ft contour. Label the elevation of each contour except the shoreline. Notice that closely spaced contours near and parallel to the shoreline indicate the presence of steep sea cliffs.

33. The map in Figure 11.13 also has a 20-ft contour interval. Contours tend to parallel each other, and they run parallel to cliffs. Since the seashore is 0 ft and the cliff is 80 ft high, you must put your 20-, 40-, and 60-ft contours parallel to the shore along the cliff between the sea and the 80-ft elevation points. Draw in the contours, and don't forget that the line at the seashore is the 0-ft contour. Label the elevation of each contour.

34. Examine the sketch of the coastal area in Figure 11.14 and the elevations in feet given on the corresponding map in Figure 11.15. Draw contour lines on the map with an interval of 50 ft.

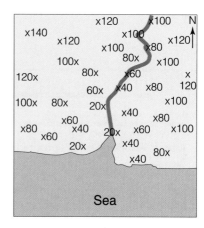

Figure 11.13 Map of a stream and sea cliff. Read the instructions in Exercise 33.

Figure 11.14 Perspective view of a coastal area

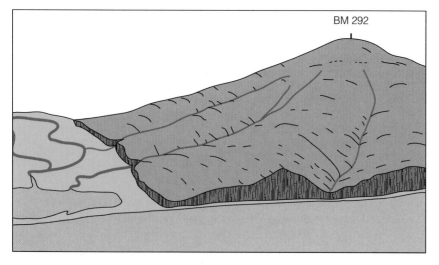

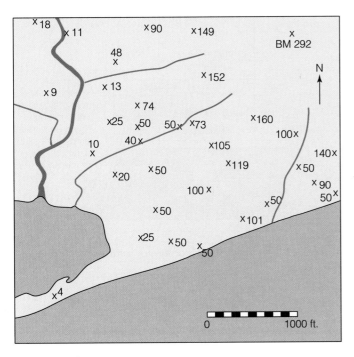

Figure 11.15 Map of the same area shown in Figure 11.14

Computer Check

- Start **TASA Topographic Maps.**
- Click stop (■) then **8. Using the Contour Lines.**
- Click advance (▶) 3 times

Check your answer to Exercise 34. The computer map has a contour interval of 20 ft, but otherwise the right half is the same as the map in Figure 11.15. Check whether you drew contours at the streams correctly. A number of contours come close to the shoreline at the sea cliff.

- To stop, click the stop (■) button.

Air Photos Give a View of the Third Dimension

Aerial photos, photographed in pairs and viewed simultaneously, give a three-dimensional view. These pairs are a major tool in the making of modern topographic maps and are also useful for viewing topography directly. The photos are shot from an airplane with a camera mounted pointing straight down. As the plane flies over an area, the photographer takes the first picture (Perspective 1 in Fig. 11.16); a short distance

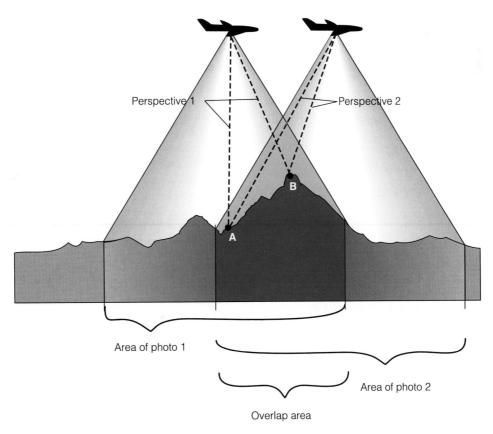

Figure 11.16 Making a stereo pair. A photographer in the airplane takes a pair of photographs, one from Perspective 1 and the other from Perspective 2. The angle between a low point, A, and a high point B is very different in the two perspectives. As a result, in photo 1 points A and B are farther apart than in photo 2.

later he or she snaps the second image (Perspective 2). The two photos must overlap, as in Figure 11.16. When two images of the same area are produced from different points of view (as if they were viewed from the right and left eye), they may be viewed simultaneously with the aid of a **stereoscope** to see a single image in three dimensions (called *stereo*). Your eyes do this naturally, allowing you to see in 3D. One eye sees things from one perspective; the other eye, from a slightly different perspective. Your brain processes the information in such a way that you interpret it as 3D. You may notice using a stereoscope that the view looks like a miniaturized landscape when viewed in stereo. This is because your experience of items that are clearly and obviously three-dimensional is limited to small objects.

Using a Stereoscope with Pairs of Images on the Same Page

Using a stereoscope with a pair of images that are part of a single page, as are those in this lab book, is quite simple. Place one lens of the stereoscope over one photo and the other lens over the other photo so that two similar objects are directly below the lenses (Fig. 11.17). Next look through the stereoscope with your eyes relaxed (as if you are going to sleep). You should be able to see a three-dimensional image.[2]

35. View each of the sketches in Figure 11.18 with a stereoscope for the following questions.

    a. For image (a), do the rectangles in the middle rise up above the outer ones or sink down below them?

    _____

    Is the slope inside the center rectangle steeper or gentler than the outer slopes? _____

    b. For image (b), do the center ellipses rise above or sink below the outer ones? _____

    c. For the contour maps in image (c), are you looking at mountains or depressions? _____ How many? _____ Add hachure marks to the maps if appropriate.

36. Now let's look back at the first topographic map where you were introduced to hills and depressions. It has been reproduced in Figure 11.19 so you can view it in stereo (3D). Do the hills appear as hills and the valleys as valleys, or has the landscape been reversed?

_____

....

[2]A few people will be unable to do this. If you have no depth perception normally, because of a weak eye, blindness in one eye, or perhaps an early eye problem that prevented your brain from learning depth perception, you will still not have depth perception with a stereoscope.

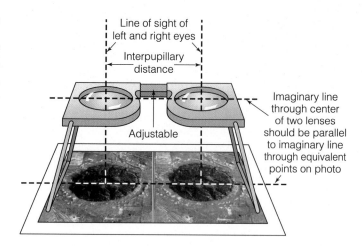

Figure 11.17   Proper positioning for a stereoscope over a stereo pair

37. Next, view the pair of air photos in Figure 11.20.

    a. Describe what you see. _____

    _____

    _____

    b. The topography in this area was formed by glacial (ice) erosion. Is the glacier still there? _____

    c. Read the figure caption. What type of topographic feature do you think an *arête* is? _____

    d. What do you think a *tarn* is? _____

Meteor Crater, Arizona Meteor Crater (at approximately 35°N and 111°W) is a large crater formed by a meteorite. This extraordinary depression near Winslow, Arizona, was formed by the collision of a large mass from outer space with the Earth's surface at some time in the geologic past. The impact produced the large depression shown and elevated the crater rim above the surrounding plain. Many impact sites have small hills in the center of the crater; these are additional evidence of meteorite impact and result from a kind of splashing-back action of the material excavated from the crater.

38. Study the aerial stereo pair of photographs of Meteor Crater, Arizona, in Figure 11.21 using a stereoscope. Then, roughly sketch a topographic profile of the crater from left to right in Figure 11.22. Assume the rim of the crater is 600 ft and the floor is 0 ft.

39. Is the floor of the crater steep or gently sloping?

    _____

    How about the sides of the crater? _____

40. Use your profile to sketch in contour lines (for the crater only) every 100 ft onto the map in Figure 11.23. Take the center mark as 0 and the rim as 600 ft, and label contours correctly. (*Hints:* Use your observations from the

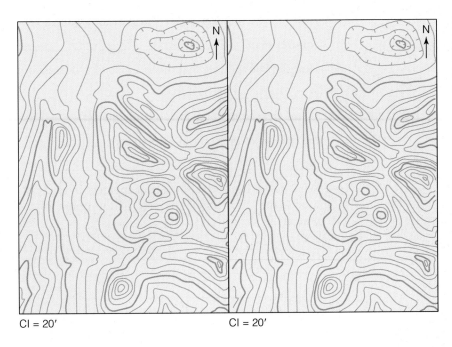

Figure 11.18 Three-dimensional objects for viewing with a stereoscope

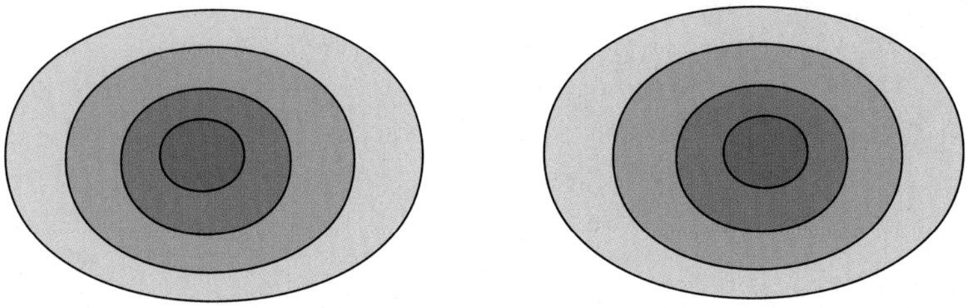

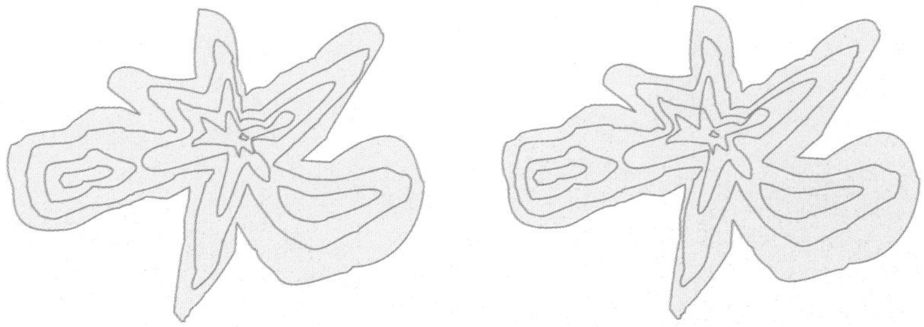

(a)

(b)

(c)

Figure 11.19 The hills and valleys of the first map (Fig. 11.1) have been rendered in three dimensions. Are they right side up or inverted? Use a stereoscope.

N

N

CI = 20′

CI = 20′

Figure 11.20 A pair of aerial photographs of Mount Bonneville, Wyoming. *Arêtes* partially encircle *tarns* in this area.

Figure 11.21 A pair of aerial photographs of Meteor Crater, Arizona

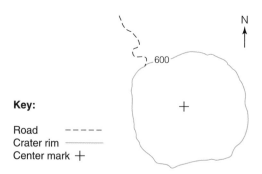

Figure 11.22 Sketch a profile of the Meteor Crater assuming the rim is 600 ft above the bottom.

Figure 11.23 Sketch a topographic map of Meteor Crater, Arizona. The rim contour has been drawn for you, although a special symbol might need to be added.

previous questions. How far apart should the contours be near the crater rim? Near the center?)

41. How did you indicate the steepness of the slope on your contour map? _____

42. How did you indicate that this is a map of a depression? _____

43. If you were to map the area outside the crater, describe the spacing of the contour lines there. _____ _____

44. What details can you see in 3D not apparent in your contour map? What about the area in the center of the crater? List the features.

_____

_____

_____

45. Circle any of the details listed in the previous answer that are evidence of meteoric impact.

Using the Stereoscope with Pairs of Separate Images

Air photos are commonly available at university libraries and commercially, but the pairs are separate photos that must be positioned properly for viewing.

- Examining the stereo pairs (usually two aerial photographs with overlapping areas), find a point that is common to both.
- Place your left index finger on the point you just found in the left image and your right index finger on the point on the right image. Line up the images under the stereoscope so that the common points (and your index fingers) are level (on a horizontal line); while looking through the scope, move your fingers along with the images horizontally either together or apart until your two fingers appear to merge into one image. Relax your eyes (pretend to fall asleep, then open your eyes) and look through the viewer from directly above. Figure 11.17 shows the correct configuration.
- Now carefully remove your fingers, and, keeping your eyes relaxed (close and open them again if necessary), look at the image. The overlapping area should now appear as one image. If not, adjust the stereopairs slightly until the common points appear to merge; be careful that like objects in the image remain horizontally aligned, and that movement is horizontal. There will be two additional images, from the nonoverlapping areas, that are not three-dimensional—one to the right and one to the left; ignore these.
- When using the *geoscope* (the mirrored stereoscope), follow the prior directions starting with your fingers about 8.5 in. apart with the magnifying lenses, and about 7 in. without. The lenses provide greater magnification, resulting in more detail and greater vertical exaggeration, which is important when viewing stereo photographs with gentle slopes.

46. Your instructor will provide a pair of aerial photographs. Use the procedure for pairs of separate images to view them. Describe what you see in the stereo pair and answer any questions provided by your instructor.

_____

_____

_____

_____

_____

_____

Shorelines and Oceans

L A B **12**

OBJECTIVES

1. To learn the features of passive and active continental margins: continental shelf, continental slope, continental rise, abyssal plain, and oceanic trench
2. To understand coastal sediment transport processes
3. To become familiar with the surface ocean currents and be able to sketch their locations
4. To learn what El Niño means and be able to distinguish between ocean surface temperature patterns in the eastern equatorial Pacific during El Niño, La Niña, and normal conditions.

When studying plate tectonics, we learned how rock types and their densities affect ocean crust formation. Because the oceanic crust is denser and therefore lower in elevation than continental crust it becomes the basins that hold the oceans. The oceans cover about 71% of Earth's surface (Fig. 9.4). In fact, the ocean water overfills the ocean basins somewhat and spills over onto the edge of the continental crust along the continental shelves. The edges of the oceans have a distinctive pattern of elevation change as a result of this. The bathymetry of the oceans also has patterns that are influenced by plate tectonics. Mid-ocean ridges occur at divergent plate boundaries and oceanic trenches at convergent boundaries (Fig. 1.2).

Solar energy, mostly sunlight, heats the oceans and atmosphere, producing weather. Wind creates waves in the ocean. Where the oceans impinge on the land surface, at the shoreline, the energy from waves powers erosion, sediment transport, and sometimes destruction of coastal properties. Wind also creates predictable patterns of ocean currents. Due to dynamic and changing aspects of the climate that we do not fully understand, the normal patterns of wind and ocean currents sometimes fluctuate, causing climate variations such as the El Niño. El Niño and La Niña have been topics of the news media, but how much do we understand about how they affect us through climate variations? The ocean not only is a large part of our Earth's surface, but touches the lives of those who never come in contact with it; in this lab we will investigate these effects.

The Edge of the Oceans

The edges of the oceans, where ocean and continent meet, known as *continental margins*, are important from human and geological standpoints. This is where humans come in contact with resources and hazards from the sea. Shoreline processes and humans interact here. It is also where vast quantities of sediment (16.5 billion tons per year by rivers alone) from the continents are transported to the sea onto the continental margins. Much oil and natural gas production occurs from these sediments. Eventually, over geologic time, the sediments accumulate, lithify, and become incorporated as part of the continents, often by plate tectonic activity. A significant proportion of rocks on continents originated at the edges of the oceans. Both the Grand Canyon and the top of Mount Everest have lithified marine sediment from continental margins.

Continental margins may be of two basic types. A **passive continental margin** is one without a plate boundary, such as on either side of the Atlantic Ocean and the western part of the Indian Ocean. An **active continental margin** occurs where the continent's edge coincides with a convergent plate boundary. The Pacific Ocean is surrounded by active margins. At a passive margin, the bathymetry (topography under water) progresses from a continental shelf to a continental slope to a continental rise to the abyssal plain, but at an active margin the continental slope leads to an oceanic trench and then to the abyssal plain. Figure 12.1a shows a hypothetical ocean basin with an active margin on the left and a passive margin on the right. The following features of the sea floor are also shown in Figure 12.1:

- The **continental shelf** is a gently sloping area of shallow water adjacent to a continent (Fig. 12.1a), and it holds the largest volume of sediment in the world. Where the continental and oceanic crust meet, the solid surface of the Earth slopes quite steeply downward away from the land, and the edge of the ocean water partially covers the edge of the continental crust (Fig. 12.1b). This creates a continental shelf.
- Toward the ocean from the continental shelf is the **continental slope,** marked by a dramatic increase in slope as one progresses seaward (Fig. 12.1a). The continental slope is near the boundary between the continental and oceanic crust.

229

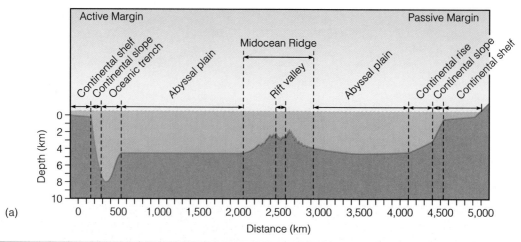

(a)

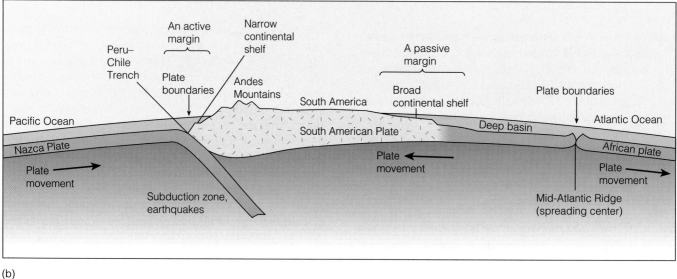

(b)

Figure 12.1 (a) Bathymetric profile of an ocean with an active margin on the left and a passive margin on the right. Vertically exaggerated. (b) Plate tectonic cross section of active and passive margins across the South American continent.

- Along a passive continental margin, farther toward the deep sea from the continental slope, the seafloor slopes more gently again at the **continental rise** (Fig. 12.1a).
- The broad expanse of deep sea that is nearly flat is known as the **abyssal plain.** Over 70% of the ocean floor is abyssal plain (Fig. 12.1a).
- At active margins with convergence, subduction occurs. Here the continental slope dives down into an **oceanic trench,** before rising again to reach the abyssal plain (Fig. 12.1a and b). The deepest trench in the world is deeper than the highest mountain (Fig. 9.4) is tall.
- Farther out into the ocean, and not part of a continental margin, the seafloor may rise up to a **mid-ocean ridge** such as the Mid-Atlantic Ridge (Fig. 12.1a and b) or the East Pacific Rise.

Ocean Margin Profiles

The profiles or cross sections of passive and active continental margins are quite different. The East Coast of the Americas (western side of the Atlantic Ocean) is primarily an example of a passive continental margin. The West Coast of the Americas (eastern side of the Pacific Ocean) is an active mar-

gin. In the following exercises, you will draw profiles of these margins, one near Long Island, New York and the other near El Salvador, Central America.

1. Using Figure 12.2, draw a bathymetric profile in Figure 12.3 from A to A′ on the bathymetric map of the East Coast near Long Island, New York. This is similar to drawing a topographic profile as we did in Lab 11. Remember that the shoreline is the zero contour.

2. Circle which of the following are present here:
   A continental shelf
   A continental slope
   A continental rise
   An abyssal plain
   An oceanic trench
   Label those present on your profile.

   Table 12.1 shows the thickness of sediments occurring along the continental margin off the East Coast of the United States.

3. Use the information in Table 12.1 to draw the location and thickness of sediments on your profile in Figure 12.3. Keep in mind that values in Table 12.1 are in feet

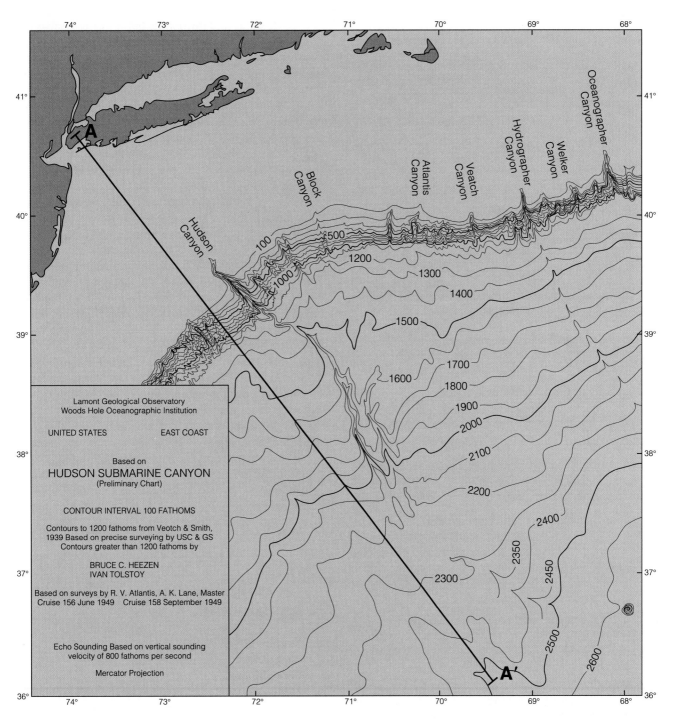

Figure 12.2 Bathymetric map of part of the continental margin of eastern United States, near Long Island, New York. Contour interval = 100 fathoms; 1 fathom = 6 ft.

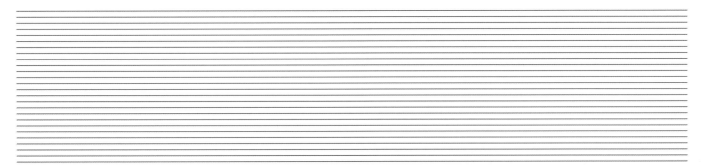

Figure 12.3 Bathymetric profile of the continental margin off the coast of Long Island, New York, corresponding to the line from A to A′ on Figure 12.2.

Table 12.1 Approximate Thickness of Sediments Occurring off the East Coast of the United States

Location	Sediment Thickness (ft*)
Shoreline	300
In the middle of the continental shelf	2400
At the shelf-slope break (the edge of the continental shelf)	15000
At the slope-rise break	4000
In the middle of the continental rise	>9000
Abyssal plain	6000

*The actual values vary from one location to another.

but your profile is in fathoms. Choose a color for sediments, and color them in on your profile.

4. Using new colors, color in the continental crust and oceanic crust where they should occur beneath the sediments. Include a key explaining the colors on Figure 12.3.

5. What do you think would be the source of these sediments? _____

_____

The Central American continental margin, Eastern Pacific Ocean, occurs at an active continental margin. Figure 12.4 is a bathymetric map of part of this region.

6. Draw a bathymetric profile in Figure 12.5 across the Central American Pacific continental margin (Fig. 12.4) from A to A'. You will have to insert the 100-fathom contour between the 0- and 500-fathom lines at the correct position in the profile. Circle which of the following are present here:
   A continental shelf
   A continental slope
   A continental rise
   An abyssal plain
   An oceanic trench
   Label those present on your profile. For any of these that are not present, what takes its place? _____

7. Refer back to the plate map in Figure 9.7 and locate the area shown in Figure 12.4. What type of plate boundary occurs here? _____ What are the plates involved? _____ _____

Shorelines

Longshore Drift

When you go to the seashore, you probably go to a beach with sand. Where did all that sand come from? It may ap-

pear to be stable and immobile, but it moves continually as if it were a slow-moving river of sand. Some of the sand up above the high tide level is moved by wind, but the sand in the surf zone, where the waves come in and go out, is moved by the waves. Waves usually wash up the beach at an angle to the shoreline then run directly down the slope of the beach. Since the slope of the beach is perpendicular to the beach, the water does not wash back exactly where it washed up the beach. This continues repeatedly, causing the water to move along the shore, generating the **longshore current** (Fig. 12.6). Each wave also carries some sediment up and down the beach. Because the water does not generally slosh back the same way it moved up the beach, the sediment tends to move along the shore. The movement of sediment parallel to the shore is called the **longshore drift** or **beach drift**. At the mouths of bays, sediment tends to be deposited across the opening, creating **spits,** which go part way across the bay mouth, or **bay-mouth bars,** which completely cross the mouth of the bay (Fig. 12.7). Where an island close to shore blocks the wave action, deposition occurs, creating a sandy connection between the island and the shore called a **tombolo** (Fig. 12.8a). Attempts to control beach erosion take advantage of longshore drift. A **groin,** (Fig. 12.8b) built perpendicular to the shore, acts like a dam to the longshore drift so the sand builds up on the "up-current" side, but it also tends to erode on the "down-current" side. A **breakwater** (Fig. 12.8b) built parallel to the shore helps sand deposition between it and the shore in a similar manner to a tombolo. Another way to protect a shoreline is to import sand to the beach.

Longshore Drift Computer Exercise

- Start **In-TERRA-Active.**
- Click **SURFACE** then **Shorelines & Coasts.**
- Click **Explore.**
- Set the orange sliders on **Shallow, Sand, Low** and **NW.** Click **CHECK IT.** Then move the wind direction slider to **SW,** and check it.

8. In what way is the shoreline different for these two configurations? Why?

_____

_____

_____

_____

- Click **Quit** and **yes.**

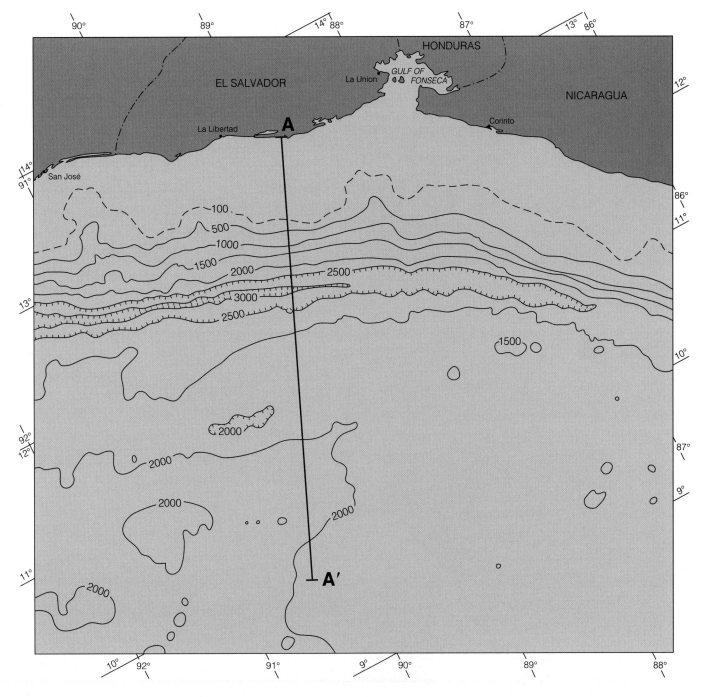

Figure 12.4 Bathymetric map of part of the Middle Americas Trench, Eastern Pacific Ocean. Contour interval = 500 fathoms; 1 fathom = 6 ft.

Figure 12.5 Bathymetric profile of the continental margin across the Middle Americas Trench, from A to A′ on Figure 12.4.

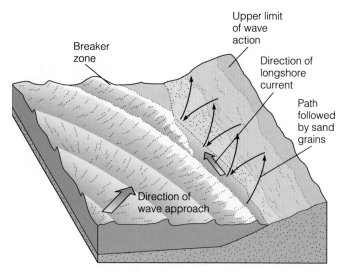

Figure 12.6 Development of longshore current

9. Study Figures 12.7b, 12.8a, and 12.9, looking for spits, tombolos and any other features that might tell you the direction of the longshore current and drift. Draw on each figure the location and direction of the longshore drift and the probable direction of predominant waves. Explain your reasoning and list the various pieces of evidence for each figure.

   a. Figure 12.8a _____

       _____

   b. Figure 12.8b _____

       _____

   c. Figure 12.9. For this figure also answer the question: Why has the beach on the right side of the photo disappeared?

       _____

       _____

       _____

10. Now apply what you know.
   a. Sketch on Figure 12.8b what the distribution of sand will be after a period of longshore drift.
   b. What happens to the beach in front of Joe Smith's

      house?_____

      In front of your house?_____

      In front of Sue Green's house?_____

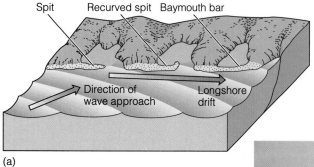

(a)

(b)

Figure 12.7 Development of spits and bay-mouth bars. (a) Spits and bay-mouth bars form where the longshore current carries sand across the mouth of bays or estuaries. A spit crosses the mouth part way and a bay-mouth bar crosses all the way. (b) A spit across the mouth of Klamath River, California.

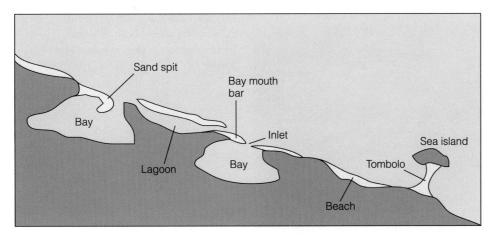

Figure 12.8 (a) Use the depositional features along this imaginary coastline to determine the direction of the longshore drift (Exercise 9a). (b) Beach structures near Sue Green's and Joe Smith's houses. Follow instructions in Exercises 9b and 10.

(a)

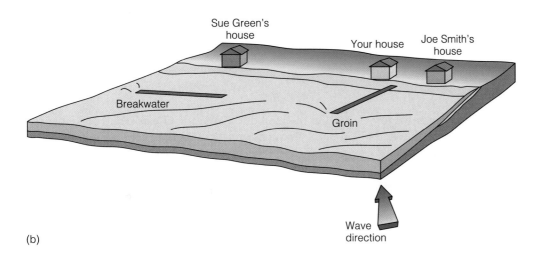

(b)

c. What would you need to do at your house to compensate for the groin built by Joe Smith?

_____

_____

Headland Erosion

Because of the persistence of longshore current and **refraction,** or the bending of waves toward the shoreline, wave energy and erosion are concentrated at **headlands,** or promontories that stick out into the water (Fig. 12.10). The angle of the waves causes the longshore drift, on average, to move from the headlands toward bays. This moves sediment into the bays, gradually filling them up. As the headlands erode and the bays fill with sediment, the shoreline tends to become straighter and straighter.

Stream Table or Sand Tub Experiment In a stream table (or a wide tub of sand), arrange the sand so it makes two headlands with a bay between (Fig. 12.11) as indicated by your instructor. Your lab instructor will also explain how to generate waves in your "sea." Work as a group.

11. Read through the whole experiment before you start. Perform the following steps:
   a. Place the acrylic sheet on the stream table, and mark the stream table and the acrylic with washable markers so you will be able to replace the acrylic in exactly the same position.
   b. Trace the shoreline, or where the water meets the sand, onto the acrylic sheet. Look for the reflection of the light off the water, and then move your head as you sketch so it is directly above the shoreline. Also trace the edge of any cliffs. Remove the acrylic.
   c. Everyone select a waterfront home sight. Sketch the position of your house on the map in Figure 12.11.
   d. Start making waves as instructed.

12. After about 1 min, what happened to your house?

_____

_____

13. Remove the houses, and replace and align the acrylic to its former position on the stream table.
   a. Draw arrows on the acrylic showing the direction the waves came in toward the shore and the movement of the sand.

Figure 12.9 A groin (perpendicular to the shore) built to slow beach erosion at Manasguan, New Jersey

b. Trace the new shoreline on the acrylic. Remember for best accuracy, mark where the water meets the sand and have your head directly over the place you are drawing. Also trace the cliff positions.

c. Mark on the acrylic where erosion (**x**'s) and deposition (dots) occurred.

d. Copy the arrows and erosion and deposition patterns drawn on the acrylic onto Figure 12.11.

e. Write a summary of coastal sediment transport, erosion, and deposition.

_____

_____

_____

_____

_____

_____

Ocean Currents

Ocean currents are driven by winds and, like winds, help to redistribute solar energy from the equator toward the poles (Fig. 15.4). Because global winds have a particular pattern (see Lab 15, Fig. 15.5), ocean currents tend to develop a specific pattern as well. The **trade winds** are easterly winds; that is, they blow from the east. They move masses of air between about 25° latitude and about 5° latitude, and generate ocean currents from east to west just north and south of the equator. The resulting currents, called north and south **equatorial currents,** (Fig. 12.12) are warm due to heating of the surface water from more direct sunlight.

14. On the map of the Atlantic Ocean (Fig. 12.13), use red and blue pens or pencils to make a key to indicate warm (red) and cold (blue) currents. Keep room in the key for 4 more items. Locate the equatorial currents on the map of the Pacific Ocean in Figure 12.12 and use them as a guide to draw in the warm north and south equatorial currents in their correct positions in the Atlantic Ocean on Figure 12.13. Label the currents.

As the equatorial currents move away from a continent on the east side of the ocean, upwelling will occur. **Upwelling** is when deep cool ocean water moves upward. This upwelling helps replace the water pushed away from the east side of the ocean by the trade winds.

When an equatorial current encounters a continent on the other side (western) of the ocean, two things happen: the warm water piles up, and the water is deflected north and south. At the coast, where the north equatorial current turns south and the south equatorial current turns north, water from both equatorial currents impinge on each other, and an equatorial countercurrent develops. This current flows from west to east along the equator where winds are very light. So an **equatorial countercurrent** develops between the equatorial currents flowing in the opposite direction. The equatorial countercurrent also helps to replace the water that is blown away from the east side of the ocean.

15. Add colors to your key to indicate warm water piling up from the equatorial currents running into a continent and where cold water would occur due to upwelling. Color these warm and cold regions in on the maps in Figures 12.12 and 12.13.

16. Locate an equatorial countercurrent in the Pacific map in Figure 12.12 and use it as a guide for drawing the equatorial countercurrent on the map of the Atlantic Ocean (Fig. 12.13). Label this current.

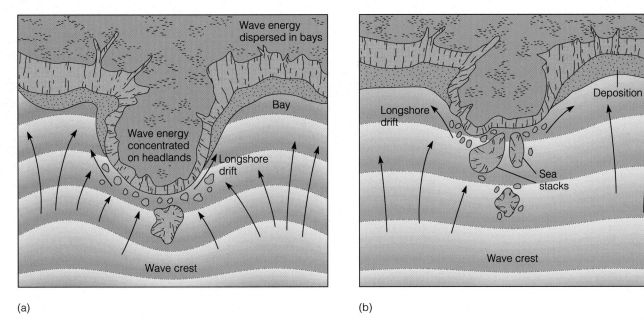

(a)

(b)

Figure 12.10 Headland erosion. (a) Wave refraction causes waves at a headland to bend toward the promontory concentrating their energy there. The angle of waves impinging on both sides of the headland causes longshore drift to carry sediment toward the bays. (b) After many years have passed the headland has been reduced to sea stacks and the bays have filled with sediment making the shoreline straighter.

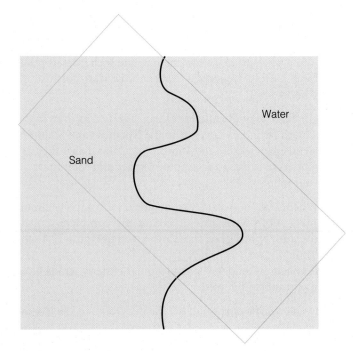

Figure 12.11 Configuration of sand as two headlands and a bay in the stream table or sand tub. If your stream table is narrow, you may need to arrange the sand as shown with the diagonal box as the edges of your stream table.

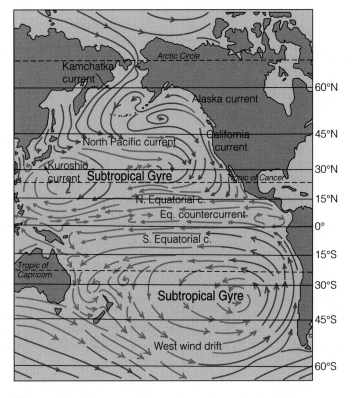

Figure 12.12 Map of ocean currents in the Pacific Ocean. Red arrows indicate warm currents, and blue arrows cold currents.

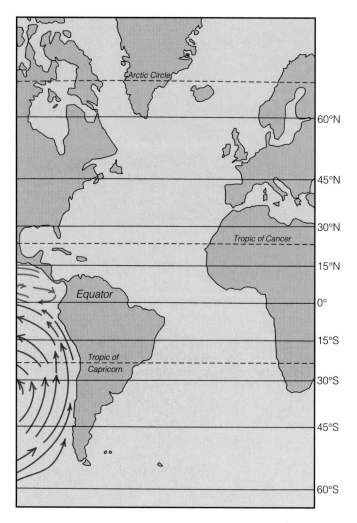

Figure 12.13 Map of the Atlantic Ocean
Key:

Where the two equatorial currents impinge on the continent and diverge away from each other, **western boundary currents** form. In the Northern Hemisphere, the western boundary current flows northward along the west side of the ocean, the east coast of the continent; in the Southern Hemisphere, it flows south along the west side of the ocean. The Kuroshio Current off the coast of Japan is the western boundary current for the North Pacific Ocean (Fig. 12.12). The western boundary currents flow away from the equator and are, therefore, warm currents and tend to warm the coastal areas and produce a warm moist climate. The *Gulf Stream* in the North Atlantic is an excellent example of a western boundary current as it carries warmer water northward along the eastern coast of the United States, making the coastal region more temperate than areas farther inland.

17. On the map of the Atlantic Ocean (Fig. 12.13), draw in the western boundary currents in the Northern and

Southern Hemispheres. Label these and add a color to your key and map to label moist coastal areas.

The westerly winds that blow between 30° and 60° latitude (see Lab 15, Fig. 15.5) produce ocean currents that flow from west to east at about 40° to 50° north and south latitudes. The *West Wind Drift* and *North Pacific Current* are examples (Fig. 12.12) of currents generated by westerly winds. Although these currents have cooled considerably compared to the equatorial currents, they are still coming from the equator and are warmer than the surrounding water. They are still considered warm currents.

18. The West Wind Drift (Fig. 12.12) flows around what continent (not shown). _____

19. On the map of the Atlantic Ocean (Fig. 12.13), draw in the east-flowing currents in the Northern and Southern Hemispheres.

Where the currents driven by the westerly winds encounter a continent on the east side of the ocean, they diverge to the north and south. Such currents that head toward the equator, in either the Northern or Southern Hemisphere, are called **eastern boundary currents,** because they move along the eastern portion of the ocean basin. As these currents head toward the equator, they carry colder water into warmer areas, so these are now considered cold currents. Where cold currents travel along a continental margin, a coastal desert is likely to develop. Coastal deserts such as the Atacama Desert in Chile and the Namib Desert in southwestern Africa result from such cold ocean currents along the coast (Fig. 12.14).

20. Along what coast of a continent would such a desert result from eastern boundary current: east/west/north/south (choose one)?

21. On the map of the Atlantic Ocean (Fig. 12.13), draw in the eastern boundary currents in the Northern and Southern Hemispheres using the appropriate color. Label the currents. Add dry coastal areas to your key and color them on your map.

Where a set of ocean currents makes a complete circuit around a subtropical area of ocean, the complete loop is called a **subtropical gyre;** these are shown in Figure 12.12 for the Pacific Ocean.

Between 60° and 90° latitude, polar easterly winds blow ocean currents to the west.

22. Locate such a current in the Pacific Ocean in the Northern Hemisphere. _____ In the Southern Hemisphere the East Wind Drift flows west around Antarctica south of 60°S.

23. Draw in comparable west-flowing currents in Figure 12.13.

24. Now compare your map (Fig. 12.13) with one showing the actual Atlantic Ocean currents in Figure 12.15, and name the following currents.

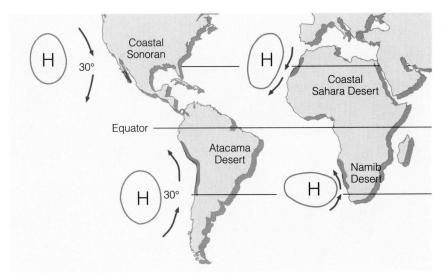

Figure 12.14 Coastal deserts form adjacent to eastern boundary currents. These cold currents chill and dry the air.

a. North Atlantic western boundary current

_____

b. South Atlantic western boundary current

_____

c. North Atlantic eastern boundary current

_____

d. South Atlantic eastern boundary current

_____

Any time a current encounters a continent, it can be deflected either north or south along the coast, or both. The geometry of the coastline may have some influence, as it does in eastern South America.

25. Notice the current in Figure 12.15 that flows from the south equatorial current in the Atlantic, into the North Atlantic along the coast of Brazil. Why do you think the water does this?

_____

_____

_____

26. Now that you are familiar with ocean currents, draw in the ocean currents on the map of the Cambrian world of about 520 Myr ago in Figure 12.16. Use the following list to help you:

- Equatorial currents
- Equatorial countercurrents
- Western boundary currents
- Eastern boundary currents
- East-flowing currents between 30° and 60° latitude
- Subtropical gyres
- West-flowing currents between 60° and 90° latitude

Ocean Salinity

Seawater has a salinity of 35%. This means that for every 1000 g of seawater, there are 35 g of salt (Fig. 12.17). Or, if you evaporated 1000 g of seawater, the solid left behind would have a mass of 35 g. Mix up a cold salt solution with salinity equal to seawater.

27. What is the volume of solution your instructor indicates you should make up? _____ ml. How many grams of salt do you need to make this amount of solution? _____ g. Of water? _____ g. Show your calculations:

28. Check your answer with your instructor before proceeding. Now make some water twice as salty as seawater by doubling the amount of salt you determined earlier. Weigh or measure out the correct amounts of salt and ice water (0°C water), and mix them together. Stir until the salt has dissolved. Place your solution in the ice bath to be ready for use in the next experiment.

29. Where on Earth would there be a good source for icy cold seawater? _____

How could this water become deep water in equatorial regions? See Figure 12.18a and b for hints.

_____

_____

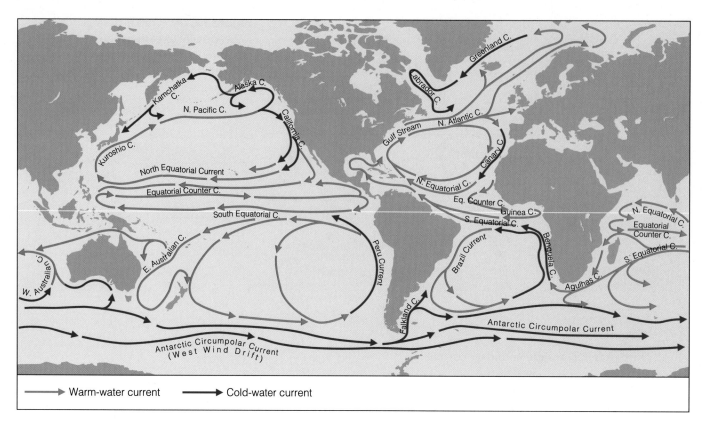

Figure 12.15   Ocean currents of the world

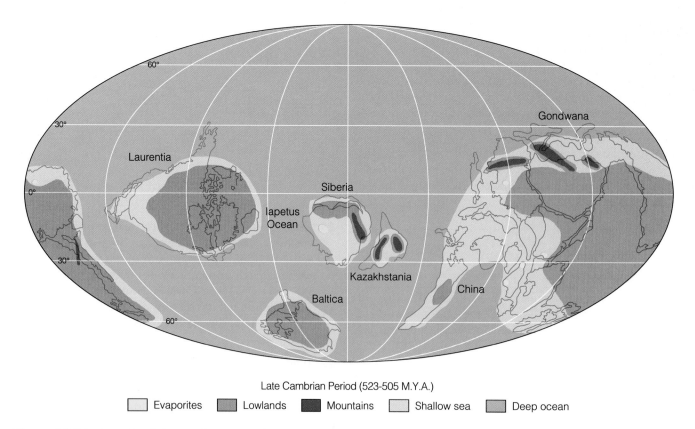

Late Cambrian Period (523-505 M.Y.A.)

☐ Evaporites   ☐ Lowlands   ■ Mountains   ☐ Shallow sea   ☐ Deep ocean

Figure 12.16   Late Cambrian configuration of the continents (523–505 Myr ago). Draw in the ocean currents as they would have existed at that time.

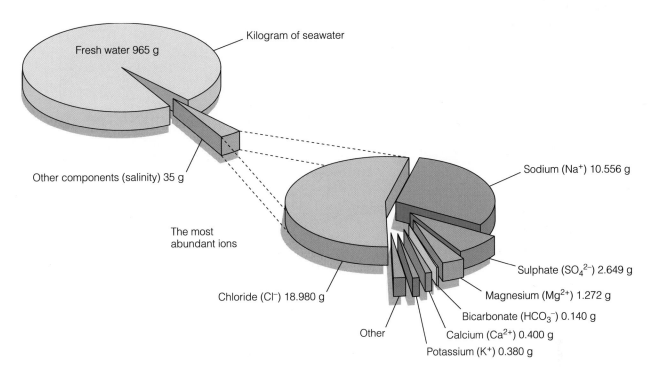

Figure 12.17 Average ocean salinity and chemistry

Nutrients in Seawater

Where light is available, nutrients in the surface waters of the ocean are rapidly used up by plants and phytoplankton (microscopic single-celled organisms that perform photosynthesis). When organisms die, three things may happen: (1) they are eaten by other organisms, (2) their bodies begin to settle but decay and release nutrients into the water before they reach the bottom, or (3) their bodies settle to the bottom of the sea to become sediment. Sunlight does not penetrate into the deeper parts of the ocean; therefore, photosynthesis cannot take place in the deep ocean.

30. At what levels in the ocean would there be abundant nutrients? Why?

_____

_____

_____

_____

El Niño—La Niña

El Niño and La Niña are oceanic phenomena that influence the worldwide climate. Discussions in the news and science media on this subject have become so prevalent that "blame it on El Niño" (or La Niña) is now part of the common vernacular as a comment when unusual weather takes place. El Niño and La Niña have been held responsible for weather disturbances, even though they are neither storms nor droughts. While they are not themselves weather phenomena or climate patterns, they do influence the climate and the weather.

El Niño is the occurrence of unusually warm water in the equatorial eastern Pacific Ocean, off the coast of Peru and Ecuador. **La Niña** is a more extreme version of the normal current pattern of cold water along the same coast but is also disruptive to the weather when it occurs. How, then, do such simple things as warm or cold water have such dramatic influence on the climate? The answer to that question is not completely understood. We will first try to understand what El Niño and La Niña are and what produces them. Then we will look at some of the ideas about how they influence the climate.

To understand an abnormal condition such as El Niño, you first need to understand what is normal (Fig. 12.19a). Normally, the trade winds blow and generate the north and south equatorial currents in the Pacific Ocean (Fig. 12.12) that pile up warm water in the western Pacific near Australia, New Guinea, and Indonesia. As the equatorial currents leave the eastern side of the Pacific along the South American coast, water must flow in to replace the water that has moved away. Some of this water comes from the south—the eastern boundary current known as the Peru or Humbolt Current. But some of the water comes from upwelling. Upwelling water is colder than the surface water, as you indicated on Figure 12.12, and brings nutrients to the surface. The upwelling nutrient-rich waters are great for fish, making fishing excellent. This is why Peru was the largest exporter of fish, until its fishing industry collapsed during the 1982–83 El Niño. The collapse cannot be blamed entirely on El Niño, since the coastal waters of Peru were also extensively overfished that year.

Trade winds, upwelling, and cold water are normal features of eastern Pacific circulation. What about El Niño (Fig. 12.19b)? El Niño occurs when the trade winds slacken.

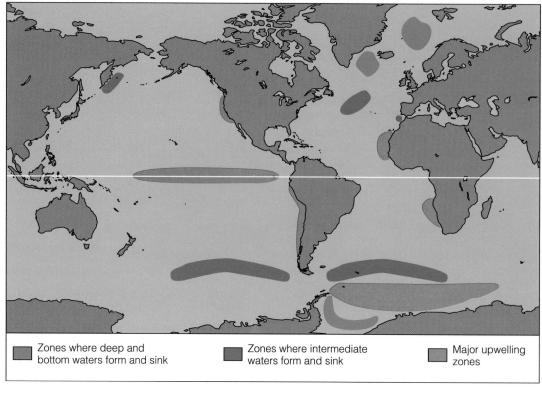

(a)

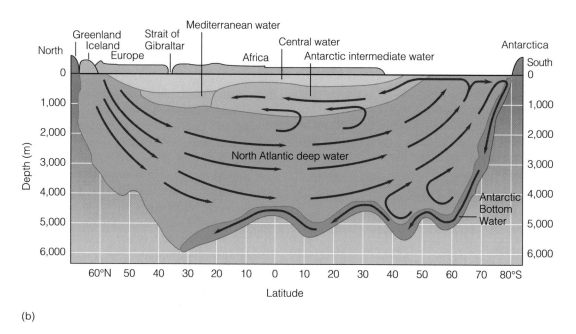

(b)

Figure 12.18 The sources and movement of deep ocean water. (a) Zones of sinking and upwelling of deep and intermediate waters. (b) Movement of deep waters in the Atlantic Ocean shown in a cross section extending from Greenland to Antarctica.

The equatorial currents slow; upwelling diminishes. Some of that warm water that was piled up in the western Pacific now has a chance to slosh back. The equatorial countercurrent strengthens, bringing warm water to the eastern equatorial Pacific. Since the upwelling stops and the water warms up, the fish die or move elsewhere to cooler waters, and the fish catch plummets. Peruvian fishermen originally noticed this phenomenon and named it after the Christ Child—El Niño—because the warming water and declining fish were first noticed in December.

In contrast to El Niño, La Niña is an extreme case of the normal wind and water circulation. When the trade winds blow more strongly than normal, the equatorial currents are stronger, piling more warm water up in the western Pacific, and resulting in stronger upwelling in the eastern Pacific during La Niña.

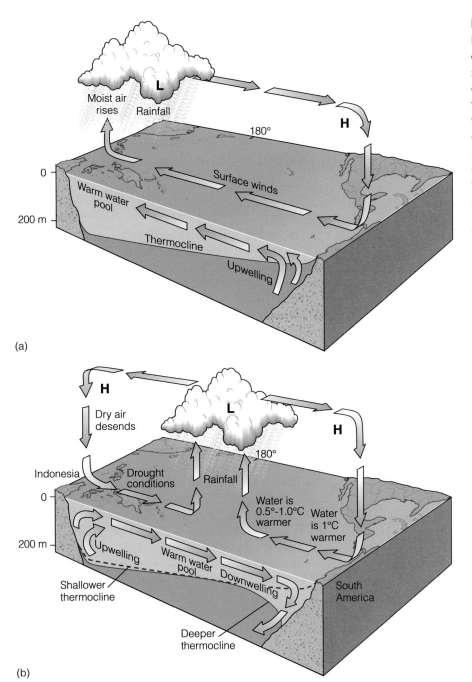

Figure 12.19 Normal and El Niño conditions. (a) In normal conditions the trade winds blow strongly forcing warm water west causing a warm moist climate in the western Pacific. Cooler upwelling waters in the eastern Pacific produce a normally dry climate there. (b) During El Niño conditions the trade winds are weak, the warm pool moves east along with clouds and unusually heavy rains. Upwelling occurs in the west and downwelling in the east.

El Niño/La Niña Experiment

Materials needed:

- Clear, wide, flat water container
- Food coloring: yellow and blue
- "Seawater" in ice bath from Exercise 28
- Hot plate
- Two beakers
- Thermometer
- Straw
- White-background cloth or paper

Find out from your instructor how many drops of food coloring you will need for your two solutions so that the yellow is fairly intense and the blue is distinctly colored but not medium or dark. Add the specified amount of blue food coloring to the saltwater solution in the ice bath. Stir.

Using a hot plate, heat the same quantity of tap water to 40°C. Stir in the specified number of drops of yellow food coloring. Place the water container on the white background. Pour the yellow solution into the container. Prop up one end of the container so the solution drains away from that end. Smoothly pour the blue solution into the container on the propped up end. Smoothly lower the propped up end.

If this is done correctly, you should have two layers of liquid, a blue layer and a yellow layer. If your solutions mixed, either start over or work with a group that successfully kept the colors separate.

31. Which layer is on top, yellow or blue? _____

Measure the temperature of each layer:

yellow: _____   blue: _____

32. Why is one layer on top and the other on the bottom? Why didn't you just get the color green?

_____

_____

Now simulate normal conditions (or La Niña). Without touching the water with the straw, blow across the top of the two liquid layers; this represents the trade winds in the Pacific Ocean.

33. What do you observe? _____

_____

34. While you were blowing, you should have observed a patch of differently colored water; where did it come from? _____ What is the term used for this movement in the ocean? _____

35. From your measurements of water temperature, what is the water temperature probably like in this patch compared to the surrounding water? _____

36. This would be analogous to normal / El Niño / La Niña conditions in the ocean. Circle which.

37. When you stopped blowing, what happened to the patch of water?

_____

_____

38. How would you expect the water temperature would change as the patch goes away and the other water comes back? _____

39. The flowing back of other water would be analogous to normal / El Niño / La Niña conditions in the ocean. Circle which.

Fish like the cool upwelling water because it is high in nutrients and near the surface where plants and phytoplankton can grow to feed the fish. When upwelling stops, the fishing is bad, for both sea birds and Peruvian fishermen.

40. If your experiment were actually the ocean, where and when would you go fishing?

_____

_____

El Niño—La Niña Temperature Maps

41. If you were a National Oceanographic and Atmospheric Administration (NOAA) scientist, what would you measure to detect El Niño? _____

42. Trade winds blow strongly during which conditions?

_____

43. What would you measure to predict that an El Niño was coming? _____ How would you measure it?

_____

NOAA scientists have installed buoys in the equatorial Pacific Ocean to measure these things. Seventy buoys are arranged along lines of longitude spanning 8000 mi of the equatorial Pacific Ocean in a grid or array called the Tropical Atmosphere-Ocean (TAO) Buoy Array. The map in Figure 12.20 shows ocean surface temperatures for normal conditions such as existed January 1993, measured by the buoy array of NOAA's TAO Project in the Pacific Ocean.

44. Enhance the readability of the water temperatures on the map in Figure 12.20 by coloring them with 2°C color intervals as described next. Select six gradational colors from cool colors to warm colors to represent water temperatures for every 2°C from 20° to 32°C. Use light-colored pencil or highlighters coloring these in on the key first. Use the same color scheme for each of the following maps. Start by coloring all numbers within one range but not including the end members of the range; so, for example, color in yellow for numbers from 26.1 to 27.9. Color the area both east-west and north-south between any two adjacent numbers within the same range. Switch to another color and do the same for the new range, until all but the round even numbers (e.g., 26.0) have been colored. The even numbers are the edges (where contour lines will occur), so color the two appropriate colors up to the even number. If time permits or as indicated by your instructor, draw in temperature contours using a 2°C or 1°C contour interval. If you are drawing a 1°C interval, draw in the 2°C interval first then fill in the intermediate temperature contours to make your final map. Finish coloring in each color up to the contours. Keep in mind that since temperatures are continuous, large jumps in temperature will still have a continuous temperature change; so, for instance, at a place where 23° is next to 27°, the temperatures 24°, 25°, and 26° must be in between.

45. What are the most striking features of normal conditions in terms of water temperature? Where is the water warm and where is it cooler? Can you explain why the contours

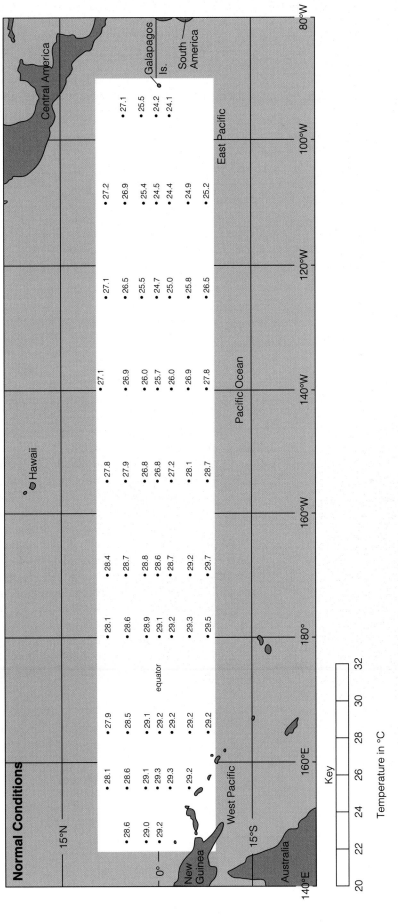

Figure 12.20 Normal Conditions. Sea surface temperature data are from NOAA TAO Array for January 1993.

from 160°W to 90°W are so much different in pattern than those from 150°E to 160°W?

_____

_____

_____

_____

_____

_____

46. What temperature is the surface water near the Galapagos Islands? _____ °C. To convert from degrees Centigrade to degrees Fahrenheit, multiply the °C by 9/5 and add 32°F. What is the temperature in Fahrenheit? _____ °F.

47. The next map (Fig. 12.21) shows ocean surface temperatures for El Niño conditions during January 1998. As before, color (and contour) the temperatures using the same colors as in the previous map.

48. What are the most striking features of El Niño conditions? How do they differ from normal conditions?

_____

_____

_____

_____

_____

49. What temperature is the surface water near the Galapagos Islands? _____ °C. What is the temperature in Fahrenheit? _____ °F

50. The map in Figure 12.22 shows ocean surface temperatures for La Niña conditions for December 1998. Color (and contour) the temperatures using the same colors as in the previous maps.

51. What are the most striking features of La Niña conditions? How do they differ from normal conditions?

_____

_____

_____

_____

_____

52. What temperature is the surface water near the Galapagos Islands? _____ °C. What is the temperature in Fahrenheit? _____ °F

53. The map in Figure 12.23 shows ocean surface temperatures for September 1996. Color (and contour) the temperatures using the same colors as in the previous maps.

54. What temperature is the surface water near the Galapagos Islands? _____ °C. What is the temperature in Fahrenheit? _____ °F

55. What conditions were prevalent at that time?

(a) Normal conditions   (b) El Niño conditions
(c) La Niña conditions

Climate, El Niño, and La Niña

Beside the changes in the seasons, El Niño and La Niña produce the largest short-term worldwide climate change (Fig. 12.24). The energy contained within the warm water of El Niño is equivalent to 500,000 20-megaton hydrogen bombs, and that is enough energy to change the climate. Where the warm water normally occurs, in the western Pacific near Indonesia and Australia, the climate is normally wet. During 1997, warm water that would result in the next El Niño traveled 150 miles each day toward the west coast of South America. Warm water like this creates a mass of warm moist air above it, which rises and condenses into clouds. The rain usually associated with the warm water migrates eastward with the warm water. Australia and Indonesia experience a drop in seawater temperatures, drought, and an increase in forest fires. Rain falls in the eastern Pacific and western South America where normally the climate is dry. The warm moist air rises 6–10 mi into the atmosphere, displacing the jet streams that have to flow around it. The deflected jet streams change worldwide climate.

La Niña is even more poorly understood than El Niño, but some of its effects are more severe than those of El Niño. For example, it tends to be associated with an increase in strength and frequency of hurricanes in the Atlantic in the summer.

56. Summarize the El Niño/La Niña phenomenon: fill in Table 12.2 with the correct water temperatures, atmospheric pressures, wind, and rainfall patterns for normal conditions, El Niño, and La Niña. Note that drought signifies unusually dry not just normally dry conditions. Words in parentheses in each column indicate choices.

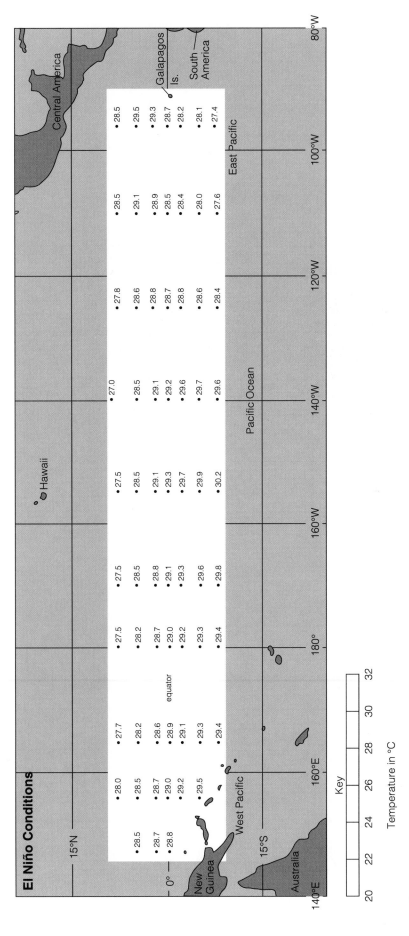

Figure 12.21 El Niño Conditions. Sea surface temperature data are from NOAA TAO Array for January 1998.

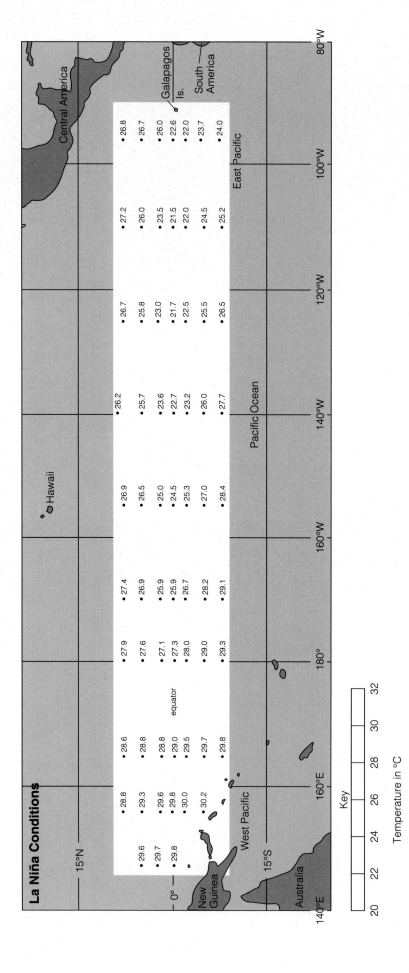

Figure 12.22 La Niña Conditions. Sea surface temperature data are from NOAA TAO Array for December 1998.

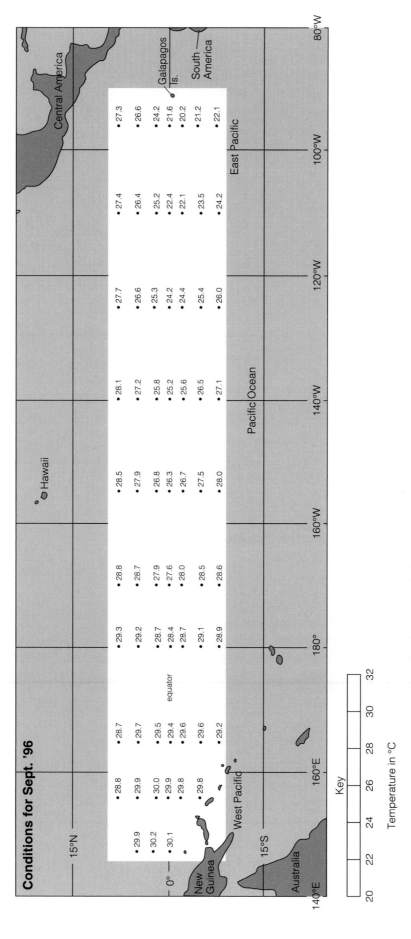

Figure 12.23 Sea surface temperature data are from NOAA TAO Array for September 1996.

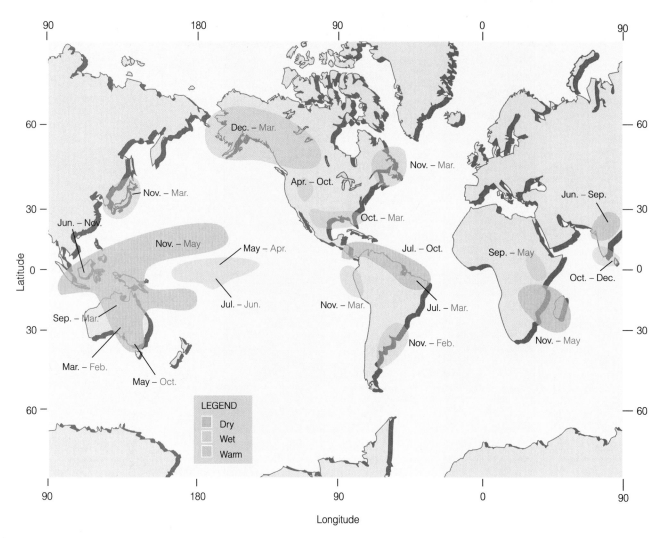

Figure 12.24 Climate abnormalities during El Niño. Months in red are the following years.

Table 12.2 Summary of El Niño and La Niña Conditions

Conditions	WESTERN EQUATORIAL PACIFIC			EASTERN EQUATORIAL PACIFIC		
	Water Temperature (warm, cool, or cold)	Atmospheric Pressure (high or low)	Rainfall (heavy rain, dry, or drought)	Water Temperature (warm, cool, or cold)	Atmospheric Pressure (high or low)	Rainfall (heavy rain, dry, or drought)
Normal						
El Niño						
La Niña						

Groundwater and Karst Topography

OBJECTIVES

1. To learn the difference between porosity and permeability
2. To understand how groundwater flows, its importance as a resource, and the risks of contamination
3. To understand the origin and geologic hazards of karst topography and their relationship to groundwater

Water from rain, snowmelt, or streams that enters the ground is called **groundwater.** Once in the ground, the water continues to move downward through soil, sediment, and porous rock until it reaches the zone where all spaces or **pores** in rock and sediment are filled with water. This zone is known as the **zone of saturation.** The top of the zone of saturation is the **water table.** The term *groundwater* refers to any water in the ground, but especially water in the zone of saturation.

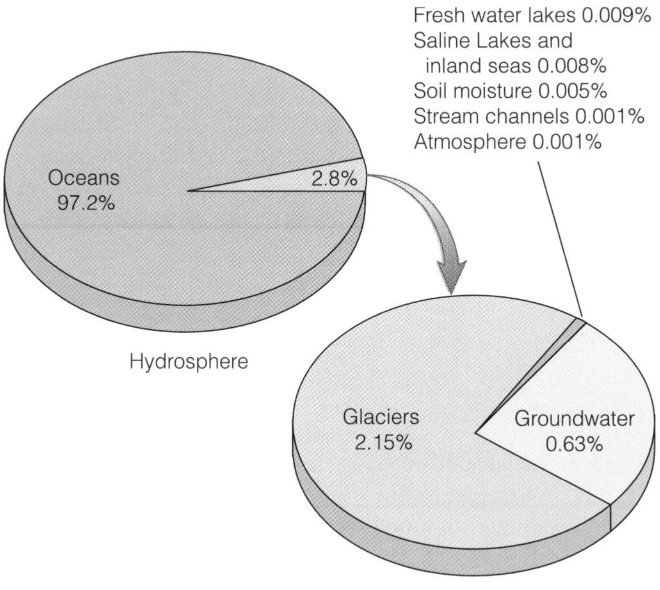

Groundwater is the third largest reservoir of water on Earth and the largest reservoir of fresh liquid water (Fig. 13.1). Only the oceans and ice in icecaps and glaciers contain more water. This makes groundwater an important resource for drinking water, irrigation, and industrial purposes. After it reaches the water table, groundwater continues to flow in response to gravity at a rate that is slow in comparison to streams—generally meters per day or less for groundwater compared to kilometers per hour for streams. Because of its slow flow, groundwater can be a limited resource. It may flow through pores, cracks, and caves (Fig. 13.2). As it flows, groundwater may dissolve material through which it passes, making the openings larger and eventually forming caves. In caves the flow of water can be as fast as in surface streams, but cave flow is not common except in Karst regions (discussed later).

Fresh water lakes 0.009%
Saline Lakes and
  inland seas 0.008%
Soil moisture 0.005%
Stream channels 0.001%
Atmosphere 0.001%

Oceans 97.2% 2.8%

Hydrosphere

Glaciers 2.15%   Groundwater 0.63%

Figure 13.1   Percentages of water in different reservoirs in the hydrosphere (top left), and nonmarine hydrosphere (2.8%, bottom right). At 0.63% groundwater makes up by far the largest quantity of fresh liquid water. Glaciers here include ice caps.

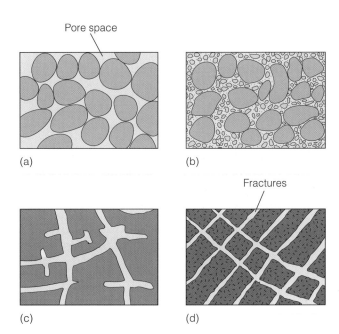

Pore space

(a)   (b)

Fractures

(c)   (d)

Figure 13.2   Different types of porosity. (a) Pores between grains of sand or other sizes of well-sorted clasts. (b) Porosity is considerably reduced in poorly sorted sediment. (c) Soluble rocks such as gypsum or limestone have porosity in solution cavities. (d) If present, porosity in crystalline igneous or metamorphic rocks generally takes the form of fractures.

Table 13.1 Porosity of Two Sediments

SEDIMENT NUMBER	1	2
Brief description of sediment		
Volume of water (ml)		
Volume of sediment (ml)	80	80
Porosity (%): $\dfrac{\text{vol. water}}{\text{vol. sediment}} \times 100$		

Porosity

Porosity is the percentage or proportion of pore spaces in a volume of rock or sediment.

1. Examine Sediment 1 and enter a brief description in Table 13.1. Measure the porosity of Sediment 1 by placing some of it in a 100-ml beaker up to the 80-ml mark. Tap the beaker gently on the table to pack the sediment down. Add or remove sediment if necessary and repack, measuring carefully. Fill a graduated cylinder with exactly 50 ml of water. Slowly pour water from the cylinder down the side of the beaker into the sediment and measure how much water is needed to reach the 80-ml level. As you pour, be sure the water is soaking in, not just running across the surface. Don't pour too quickly, or you will trap air bubbles in the sediment and not get an accurate measure of the porosity. Read the volume remaining in the graduated cylinder and enter it in the following equation to determine how much water you added to the sediment:

   50 ml − _____ ml = _____ ml.

   Enter this number in Table 13.1.

   Divide this volume of water by volume of sediment and multiply by 100 to get a percentage. Fill the information in Table 13.1.

2. Examine Sediment 2 and enter a brief description in Table 13.1. Measure the porosity of Sediment 2 in the same way as for Sediment 1. Fill this information in Table 13.1.

   When you are finished with the sediment, empty the beakers into the wet sediment container for the correct sediment, and clean the beaker for the next person. Do not mix wet and dry sediments or Sediments 1 and 2.

3. How do the porosities of the two sediments compare?

   Make a numerical comparison. _____

   _____

Permeability and Flow Rate

How fast water travels through rock or sediment depends on (1) the porosity, (2) the size of the pores, (3) how well they are connected, and (4) the pressure of the water. **Permeability,** related to the first three, is a measure of the ability of the rock or sediment to transmit fluids. Generally more porous materials are also more permeable, but the pores must be connected and large enough for the water to flow through rather than cling to the pore surface. Larger pores allow water to flow more readily. The higher the water pressure, or hydraulic head, the faster water will flow through permeable material. It is common among sedimentary rocks for groundwater to flow faster parallel to bedding than across the layers.

4. Examine the coquina limestone and the pumice samples. The solid parts of these rocks have similar densities, yet there is a dramatic difference in the density of the whole rock.
   a. Which of these rocks has the lower density?

   _____

   b. What gives this rock a lower density than the other

   has? _____

   _____

5. Now compare the relative permeability of pumice and coquina. With a plastic squeeze bottle, squirt water on the samples as you hold the rocks over a plastic tray or over the sink. Do not allow any water to flow around the edge of either rock.

   a. Which is more permeable? _____

   b. What makes this rock more permeable than the

   other? _____

   _____

   c. Next check whether the coquina is more permeable parallel or perpendicular to the bedding. Gravity is pulling water downward. Use this fact in testing the sample. Hold the sample so the bedding is vertical to test how fast the water flows parallel to bedding and horizontal for perpendicular. Which flows faster?

   _____

Experimental Determination of Flow Rate in Sediment

When we use groundwater, an important consideration is how fast it flows to replenish the supply. The flow rate is restricted by the permeability, but if no pressure is forcing the water through the permeable material, there will be no flow. **Hydraulic head** is the height of a column of water above the discharge point or spring. Gravity—the weight of the water—creates the pressure that provides the force for the water to flow. The next experiments examine the effects of permeability and hydraulic head on flow rate. These experiments use two columns containing the same sediments for which you measured porosity for Exercises 1 to 3. In one, the sediment is finer grained than in the other.

6. Read all the instructions before you start the experiment of flow rate in sediment. In some cases your lab setup will have soft plastic tubing. It is important not to squeeze the tube if you are impatient for the water level to drop so you can start your experiment. By squeezing the tube you would be purposely trying to alter the very thing you are setting out to measure. What characteristics of the sediment would squeezing change? How?

_____

_____

_____

7. Your lab instructor will assign you two or three measurements to make from the same column. Place a beaker under the appropriate column. Saturate the sediment by adding water until it is flowing (or dripping) out of the column and into the beaker. If other students have already made some measurements, the sediment may already be saturated. Once the sediment is saturated and dripping, fill the column with water slightly above the level assigned. When the level of the water reaches the assigned level, immediately place a graduated cylinder so it will catch every drop of water flowing out of the column during a 30- or 60-s period. At the end of the assigned time, replace the previous beaker under the column. Enter the volume and time in the appropriate places in Table 13.2.

8. Now allow one more drop to fall into the graduated cylinder. Read the new volume and subtract the first volume measurement to get an idea of how accurately you are able to measure the volume. If you see no difference, what are the smallest increments you can measure with your graduated cylinder? Whichever is greater of these would be the minimum error in your experiment:

_____ To really establish the error in your experiments, you would have to repeat the experiment a number of times to see the variation in the results. (However, this is a time-consuming process.)

9. Calculate the flow rate by dividing the volume by the time. Enter these figures in the table on the blackboard or overhead or computer spreadsheet to share with the rest of the class. (Think of this as the communication stage of the scientific method.) If more than one group measures the same column and hydraulic head, average all their results.

10. Copy the data collected by other class members from the blackboard/overhead or spreadsheet into Table 13.2.

11. Use the graph paper in Figure 13.3 to graph your results from Table 13.2. If your instructor prefers, use a spreadsheet's graphing feature; he or she will supply the method. It is conventional to place the independent variable (hydraulic head in this case) horizontally and the dependent variable (flow rate) vertically. After choosing your scale for each axis, double-check that the spacing between 0 and the first number is equivalent to the spacing for the next two numbers (this is a very common graphing error). Label 0 on each axis of your graph. Label your graph.[1]

12. Draw a straight line through your data for each sediment. Does (should) the line pass through the origin?

_____ Determine the equation for each line as follows,

or have the spreadsheet graphing feature add a best fit line to the graph and give you the equation:

What is the flow rate on the line at 0 hydraulic head?

Sediment 1 _____ Sediment 2 _____

What is the flow rate on the line at 60-cm hydraulic head?

Sediment 1 _____ Sediment 2 _____

The equation for a line is then given below. The flow rate and hydraulic head are your two variables; the other items are numbers you should substitute into the

··········

[1]Every graph needs four or five things: (1) a title or caption, (2) a verbal description for each axis, (3) numbers for each axis, (4) units for each axis (unless the numbers are unitless), and (5) the points on the graph. Some graphs need additional information such as a key or labels for different parts of the data.

Table 13.2 Rate of Flow of Water through Sediment

SEDIMENT NUMBER	Unit of Measure	HYDRAULIC HEAD				
		10 cm	20 cm	30 cm	40 cm	50 cm
1	Volume (ml)					
	Time (s)					
	Flow rate (ml/s)					
2	Volume (ml)					
	Time (s)					
	Flow rate (ml/s)					

Figure 13.3 Graph paper

equation. The final equations should have the form $y = mx + b$, where y is flow rate, x is hydraulic head, and m (slope) and b (y intercept) are numbers.

$$y = \frac{(\text{flow at } 60 - \text{flow at } 0)}{60} \times x + \text{flow at } 0$$

a. What is your equation for Sediment 1?

_____

b. What is your equation for Sediment 2?

_____

13. Think of your equations as hypotheses for each of the sediments. You are hypothesizing that there is a linear relationship between hydraulic head and flow rate as expressed by your equations. Now use the equations to predict the flow rate at 70 cm, if your columns go that high, or 45 cm, if they don't. What flow rate do you predict?

Sediment 1 _____

Sediment 2 _____

14. Test your hypothesis for either Sediment 1 or Sediment 2. Pour water into the sediment column to set the hydraulic head to the appropriate value and see whether you get the flow rate you predicted. How close was your answer? Subtract your answer from the predicted value.

_____ Is it within the error of the experi-

ment? _____ If not, the next step would be to

reevaluate the error in the experiment or reevaluate the hypothesis.

15. Look back at the results in Table 13.1 and at your graph. How do the flow rates at 50-cm hydraulic head compare for the two sediments?

_____

Is the difference in the porosity of the two sediments sufficient to explain the difference in their flow rates? Yes / No

What else could account for the difference in the flow

rates? _____

_____

_____

16. Summarize your results:

a. How does porosity influence flow rate?

_____

_____

b. How and why does sediment size influence flow rate?

_____

_____

c. How and why does hydraulic head influence flow rate? _____

_____

d. Is there a mathematical relationship? If so, what is it?

_____

_____

Water Table, Groundwater Flow, and Wells

The *water table* is the level of the top of the saturated zone (Fig. 13.4). Where the water table intersects a natural hole or depression in the land surface a lake exists. In stream valleys of humid regions, groundwater may flow out of the ground (discharge) into the stream. Because stream flow is faster than groundwater flow, the stream may draw down the level of the water table to the stream level (Fig. 13.5). For this reason, it is common in humid regions for the water table to mimic the surface topography, but with lesser slopes.

An important source of water in many regions is well water. A **well** is a hole drilled into the zone of saturation, below the water table (Fig. 13.6a). The lining of the well is permeable so water flows into the well from the surrounding rock or sediment. The level of water in the well will correspond to the water table, unless the water is under pressure in a confined aquifer. When water is pumped out of the well, the water level drops; thus, the water table around the well also

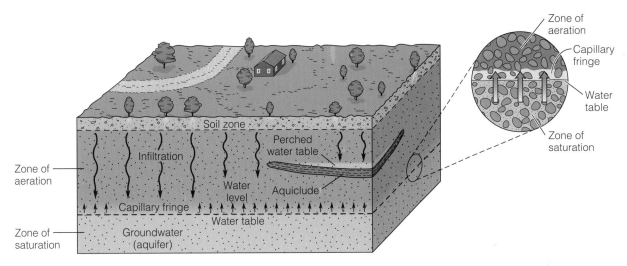

Figure 13.4 Groundwater in the subsurface is found in a number of forms: as soil moisture; as dampness in the zone of aeration where both water and air occupy the pore spaces; as water in the pores in the zone of saturation, which may be perched above an impermeable layer in some places; and as water in the capillary fringe that is drawn up by the surface tension of water. The largest quantities are in the zone of saturation.

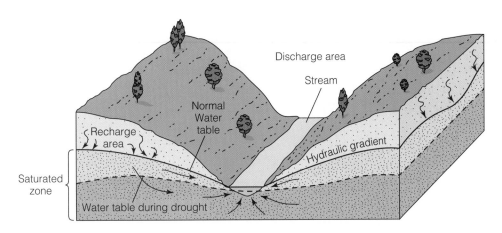

Figure 13.5 In humid areas the recharge of groundwater is generally sufficient so the water table intersects the ground in valleys and the water flows out of the ground, or discharges, forming streams. The rate of flow in the streams is faster than in the ground. As a result, the stream draws down the water table, creating a difference in the height of the water table. Areas of higher water exert a greater pressure, and the difference in pressure causes the groundwater to flow toward the stream.

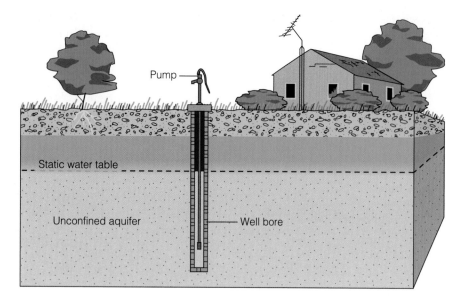

(a)

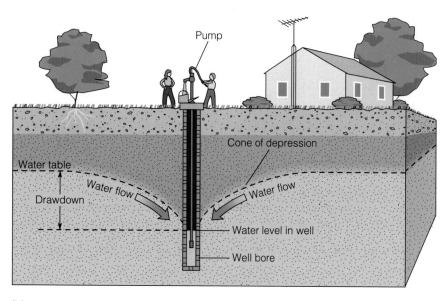

(b)

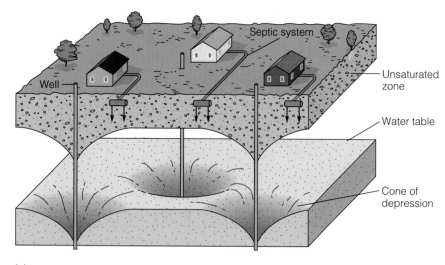

(c)

Figure 13.6 Development of cones of depression and resulting lowering of the water table. (a) The water table around an unpumped well. (b) A cone of depression forms around a well while it is being pumped. The drawdown creates a hydraulic head causing water to flow toward the well. (c) Cones of depression can overlap, lowering the general water table.

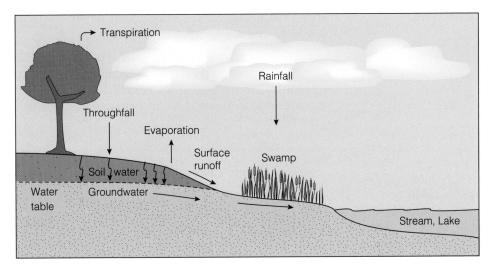

Figure 13.7 Water on the land such as swamps, streams, and lakes often indicates that the water table is intersecting the ground surface at that location. This information can be useful in determining the level of the water table.

drops, forming a **cone of depression** (Fig. 13.6b and c). The water level in lakes, swamps, streams, and wells in a region may all indicate the level of the water table (Fig. 13.7). Knowing the level helps people judge how deep to drill for water when they put in a well. Since the price of drilling a well is usually by the foot, this could be very useful information. An **aquifer** is a permeable body of rock or sediment below the water table that can sustain a productive water well.

Groundwater flows from high pressure to low pressure. The experiments with the sediment columns show that the pressure is higher and groundwater flows faster when the water level (or water table) is higher. A sloping water table produces a difference in pressure and causes groundwater to flow from areas where the water table is high to low water-table areas.

Interaction between the Water Table and Surface Water

In the next experiment we will observe the interaction of surface and groundwater in a stream table. Reduce the water on the outflow end of the stream table to its lowest level. Move all of the sand into one third of the stream table on the inflow side. Make a level area of sand at the inflow end except for a depression for a lake right at the inflow. Make a steep cliff in the sand on the other side. Near one side of the stream table, as shown in Figure 13.8, make a stream canyon starting a few inches from the inflow lake and ending in the outflow lake. Dig a series of small to medium holes in the sand as shown by the circles in Figure 13.8. Briefly turn on the water flow of your stream table until the water in the lake at the inflow end is as high as possible without spilling over. Turn on the flow repeatedly, or set the flow to a slow trickle to keep the inflow lake at this level.

17. Observe or measure the water level as indicated by your instructor in all the lakes and the stream, including the inflow lake and the outflow lake. If measuring, write the values on the map in Figure 13.8. Try to keep a nearly constant level in the inflow lake while you do this.

18. What feature in the sand is indicated by the levels of the lakes and the stream? _____

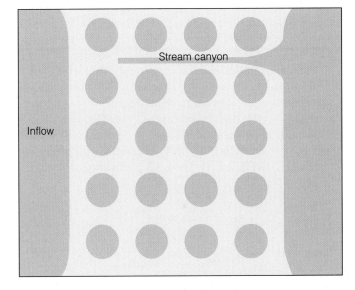

Figure 13.8 Map of stream table configuration. Circles represent holes in the sand. You can use a soup spoon to dig out the holes.

If you measured the lake levels, draw water-level contours on the map in Figure 13.8 using a contour interval that gives you at least four contours. What surface have you contoured? _____

19. Is the groundwater flowing? How do you know?

_____

_____

_____

20. Is the water table sloping? How do you know?

_____

_____

_____

21. How does the stream influence the water table?

_____

_____

Water Table and Potentiometric Surface in a Groundwater Model

The column experiment demonstrated that some sediments allow water to flow through them faster than others. Highly permeable materials make good aquifers because they allow faster flow. If the material is less permeable or hardly permeable at all, then it is called an **aquitard** or an **aquiclude,** respectively. An aquifer may be **confined** or sandwiched between aquicludes, which may allow the aquifer to become pressurized. Such an aquifer is called an **artesian aquifer** (Fig. 13.9). The pressure in an artesian aquifer allows the water to rise above the aquifer anywhere the aquifer is intersected by a natural break or a well. The height the water rises is known as the **potentiometric surface** (Fig. 13.9). It is analogous to the water table for an unconfined aquifer.

Recall that groundwater flows in response to a difference in pressure caused by a sloping water table. Similarly, water in an artesian aquifer flows from where the potentiometric surface is high—from higher pressure to lower pressure.

The groundwater model shown in Figure 13.10 should have blue dye in Wells B, PW1, F, AW, and G and green dye in Wells A, C, D, PW2, E, H, and I. All drains should be closed

at first. (The following exercises can only be done with an appropriate model.)

22. Examine the model and the samples of sediment used in it. How many different types of sediment are present? List them in order of their permeability. How did you decide this? Which would be the best aquifer, and which would be the best aquiclude?

_____

_____

_____

_____

_____

_____

23. Examine the water level in all the wells, tank (the circle), and lakes or streams. Where do you think the water table is? What is the shape and position of the water table at present? _____

Is the water table sloping? Is the groundwater flowing? Yes / No. How do you know? _____

_____

_____

_____

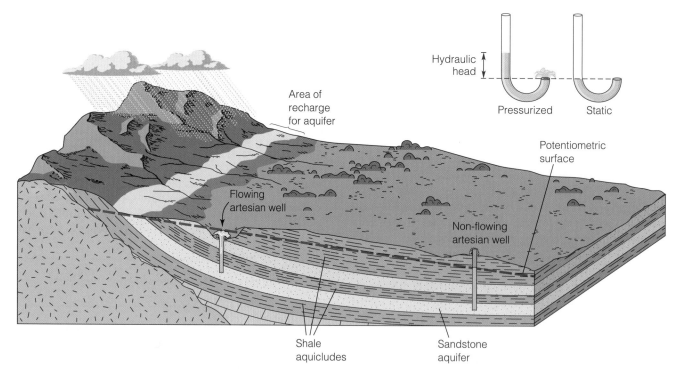

Figure 13.9 Artesian aquifer. Where an aquifer is confined between _aquicludes_ and where the recharge area is at relatively high elevation, the water in the aquifer is under pressure. The _potentiometric surface_ is the level to which the water would rise as a result of this pressure if the aquifer were pierced by a well. If the ground surface is lower than the potentiometric surface, the water will fountain out of the ground, forming a _flowing artesian well_. If the ground surface is higher than the potentiometric surface, the well will be a _nonflowing artesian well_ with the water level equal to the potentiometric surface.

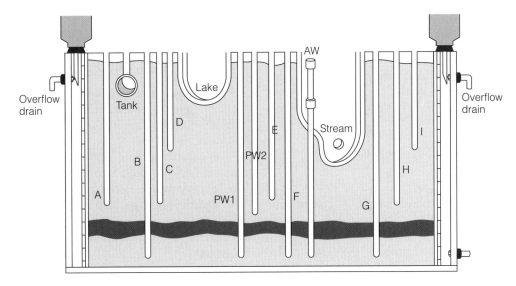

Figure 13.10 Groundwater model

24. What is different about the wells with blue dye from the wells with green dye?

_____

_____

25. Now open the valve in the stream (near Well AW). Watch the model for a minute or two until the water levels stop changing. Periodically check that the stream is draining properly; sometimes air bubbles in the tubing impede the flow. Jiggle the tube to get the flow going again.

a. Sketch on the diagram of the model (Fig. 13.10) the level of water in the lake, tank, and stream, each well with green dye, and each end of the model. Draw in the approximate water table level within the sand between these features.

b. Where is the water table the highest (or highest hydraulic head)? _____

c. Is the water table higher or lower near the stream compared to the area surrounding the stream?

_____

How does the stream influence the water table and why?

_____

_____

26. In a different color, sketch in the level of the water in each well with blue dye.

a. How does the level in these wells compare with the wells with green dye? _____

b. Which sediment layer(s) has(ve) blue water?

_____

c. What do we call this(ese) layer(s) of sediment?

_____

d. Use as straight a line as possible to connect the levels for the blue wells. What surface have you just sketched?

(See Fig. 13.9 if needed.) _____

Groundwater Flow

27. Now study the streaks of green dye in the model. Color them in on your diagram (Fig. 13.10). What direction is the groundwater flowing?_____

Summarize the relationship between the direction of groundwater flow and the slope of the water table.

_____

_____

Is there any evidence that groundwater can flow upward? If so, what evidence? _____

_____

_____

28. Color in the streaks of blue dye on your diagram (Fig. 13.10). What direction is the groundwater flowing here?

_____ Summarize the relationship between the direction of groundwater flow and the slope of the potentiometric surface. (Look carefully to detect the slope in the potentiometric surface.) _____

_____

_____

_____

Wells Leave the stream outlet on the groundwater model open for the remainder of the experiments.

29. Use a syringe to pump water out of Well PW2. (Please be *careful*; the syringe will break if you do not insert it and hold it vertically while you do this.)

   a. What happened to the water table in Well E when you did this? _____

   b. What happened to the water table in Well PW1? _____

   c. What direction is the groundwater flowing when you extract water from the well? _____

   d. Identify the cone-shaped drop in the water table around a well. _____

   e. Why did the water in Well PW1 behave differently from the water in Well E? _____
   _____

30. Use a syringe to pump water out of Well PW1.
   a. What happened to the water level in Wells F, AW, and B when you did this? _____

   b. What happened in the other nearby wells? _____

   c. What direction is the groundwater flowing when you extract water from the well? _____

31. Based on what you have learned, what do you think will happen if two families have wells near each other that obtain water from the same aquifer, and one family starts extracting much more water from their well? _____
   _____
   _____

Groundwater Contamination

Although surface water is much more easily contaminated than groundwater because it is more accessible, groundwater may be contaminated from a number of sources:

- Leachate from landfills
- Salt and oil from roads
- Infiltration of herbicide
- Pesticide and fertilizer runoff from agricultural fields
- Infiltration from polluted rainwater
- Leakage from sewer lines
- Normal drainage from septic tanks
- Industrial chemicals from waste ponds or basins
- Leakage from buried liquid storage tanks
- Acid mine drainage

- Metal contamination from a mining site
- Leakage from faulty casings of hazardous waste injection wells or where the hazardous waste may leak through fractures or wells piercing the confining aquicludes

These sources of contamination are illustrated in Figure 13.11. In general, contaminants will tend to flow with the groundwater, although some float on top and some sink below it.

Let's discuss one of these, leachate from landfills, a bit more. If rainwater gets into a landfill, it can leach various toxic chemicals out of the landfill material into the water (Fig. 13.12), forming a solution known as **leachate**, which you could think of as "garbage juice." If the leachate leaks out of the landfill, it may enter the aquifer and flow along with the water.

Landfill Location Computer Exercise

- Start **In-TERRA-Active.**
- Click **SURFACE** then **Groundwater.**
- Click **Apply.**
- Locate the best site for a landfill by following the instructions in the program. As you do this, answer the following questions:

32. Which rock is the least permeable? _____

33. Which rock forms the aquifer? _____

34. Which of the nine subdivisions is the best site for the landfill: NW, N-center, NE, W-center, center, E-center, SW, S-center, SE? Circle one.

35. Summarize the factors that make this the best site and make other locations poor sites.
   _____
   _____
   _____
   _____
   _____
   _____

- Click **Quit** and **yes.**

■

Using Water Table Contours

Where enough information is available about the water table from elevations of lakes, streams, swamps, and wells, it may be possible to draw a contour map of the water table as in Figures 13.13 and 13.14 and as you may have done for the stream table. The depth of wells and direction of groundwater flow can be determined from such a map.

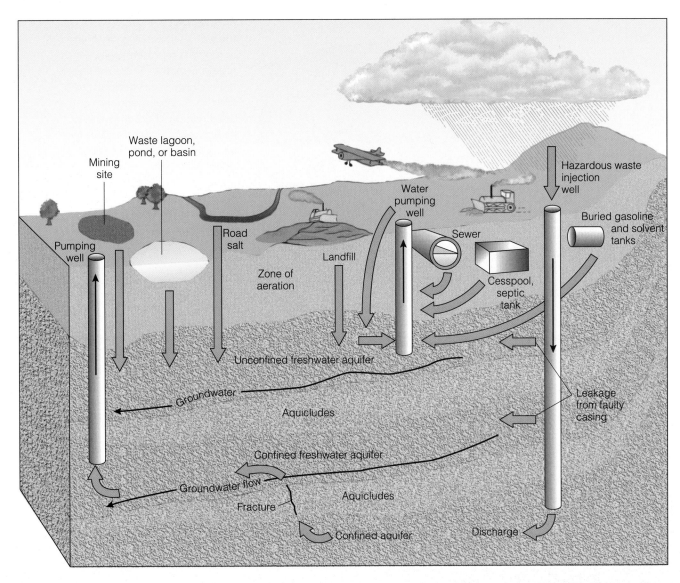

Figure 13.11 Some sources of contamination of groundwater. The two pumping wells extract groundwater that may be contaminated from a number of sources such as those described in the text.

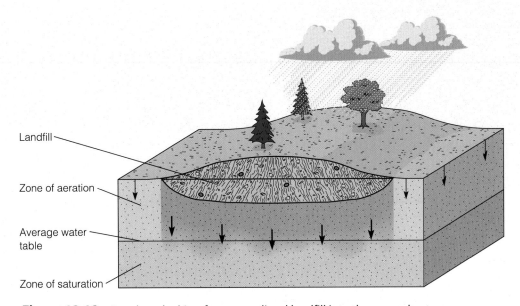

Figure 13.12 Leachate leaking from an unlined landfill into the groundwater

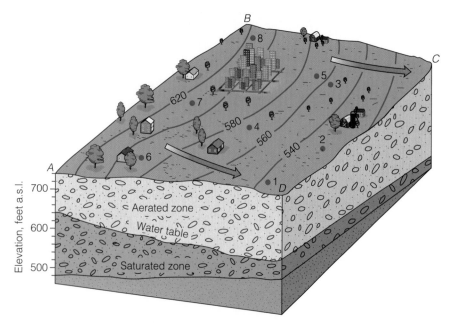

Figure 13.13 Contour map of the water table near a city. Numbered dots represent wells, which give information about the water table, allowing a hydrologist to draw the contours and determine the direction of groundwater flow. Arrows show the groundwater flow direction, which is perpendicular to the contours. Information about the subsurface geology may also be determined at the time the wells are drilled. Elevations are in feet above sea level.

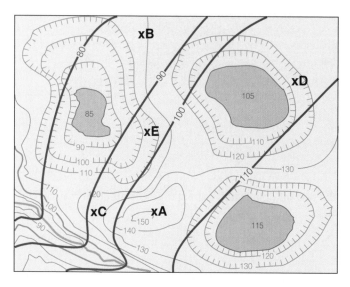

Figure 13.14 Topographic map of an area of karst topography with sinkhole lakes. Contours of the water table surface are shown as heavy dark blue lines. Contour interval = 10 ft.

The depth of the water table at any given location is less than the depth needed to drill a well, because a well must penetrate the zone of saturation. To calculate the depth of the water table, first determine the elevation of the land surface at the location. Next, determine the elevation of the water table, and then subtract. For example, the location marked with an **x** at A in Figure 13.14 has a land elevation of 150 ft, and the water table is 104 ft. This makes the depth to the water table 150 ft − 104 ft = 46 ft.

36. What is the depth to the water table at each of the other lettered locations on the map?

B: _____ − _____ = _____

C: _____ − _____ = _____

D: _____ − _____ = _____

E: _____ − _____ = _____

The direction of flow of groundwater is important for a number of reasons. It helps to determine where water in a well comes from and what happens to contaminants and pollution that get into the groundwater (Fig. 13.11). Because of gravity and the resulting pressure, water in the ground flows in the direction of the steepest downward slope of the water table carrying contamination, if there is any, with it. Flow of groundwater, then, is *perpendicular* to the water table contours. The arrows in Figure 13.15 show the groundwater flow direction.

37. Study Figure 13.15. If toxic waste were dumped into the lake labeled 115, draw or shade in on the map the path that would be taken by the toxic waste as it traveled through the ground in the groundwater.

  a. Which, if any, of the Lakes 85 or 105 or locations labeled A, B, C, D, or E would become contaminated?

    _____

  b. Would this toxic waste contaminate the part of the stream shown on the map? Yes / No

  c. If toxic waste were dumped into Lake 105, which, if any, of the locations labeled A, B, C, D, or E would become contaminated? _____

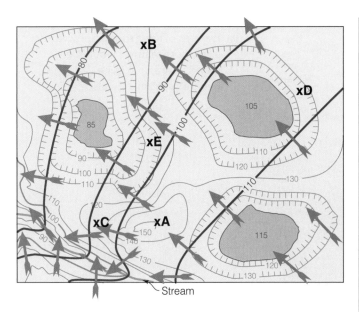

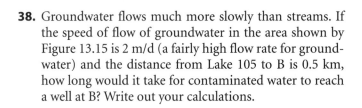

Figure 13.15 Topographic and water table contour map of an area with sinkhole lakes. Arrows show the direction of flow of the groundwater.

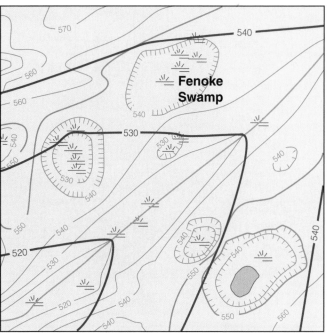

Figure 13.16 Topographic and water table contour map of an area with swamps and sinkholes. Contour interval = 10 ft.

38. Groundwater flows much more slowly than streams. If the speed of flow of groundwater in the area shown by Figure 13.15 is 2 m/d (a fairly high flow rate for groundwater) and the distance from Lake 105 to B is 0.5 km, how long would it take for contaminated water to reach a well at B? Write out your calculations.

39. Draw arrows on the map in Figure 13.16 showing the direction of flow of groundwater. Remember that flow direction is perpendicular to water table contours in the downslope direction.

40. If a nearby town decided to make Fenoke Swamp into a landfill, what might happen to the groundwater in this area?

_____

_____

41. Shade the area on the map that might eventually have contaminated groundwater. Would the stream become contaminated? Yes / No. If so, show the contaminated part of the stream on the map.

Groundwater Causes Erosion by Solution

As it moves, groundwater may dissolve a soluble rock, creating cavities and caves in limestone, dolomite, marble, rock salt, or gypsum, and resulting in topography called **karst**. Karst to-

pography is different from stream-dominated topography because much of the high-volume drainage is underground and because solution produces certain unique landforms (Fig. 13.17). Streams often disappear into closed depressions that lead to networks of caverns (Fig. 13.18). A closed depression in karst regions is a **sinkhole** (Fig. 13.17), which may form by solution or when the roof of a cave collapses. Karst topography is most common in regions underlain by limestone. The sudden formation of sinkholes is one of the geologic hazards in some karst areas (Fig. 13.19). Because limestone is susceptible to dissolution, it may be highly permeable and if the porosity is on the centimeter scale rather than cavernous, the limestone may make an excellent aquifer. However, in many karst regions the limestone has cavernous porosity which makes finding groundwater a hit-or-miss proposition.

42. Examine the map of the area near Lake Wales, Florida (Fig. 13.20), a region of karst topography.
    a. What are depressions in this area?

    _____

    b. Why do some have water in them while others don't?

    _____

    c. What is the approximate level of the water table in this area? _____

43. What name is used for the features forming the lakes in the area near Interlachen, Florida, in Figure 13.21?

    _____ What geologic hazard is likely in this area as a result of this type of topography?

    _____

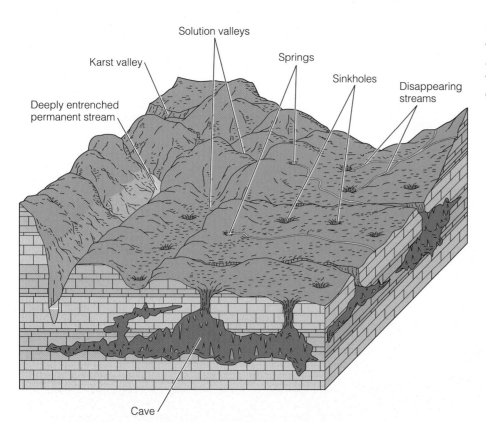

Cave

Figure 13.17 Some common features of karst topography. Not all of these features would be found together in quite this geometry in any one place.

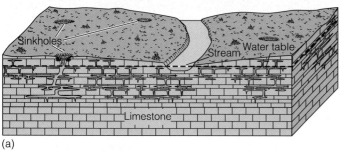

(a)

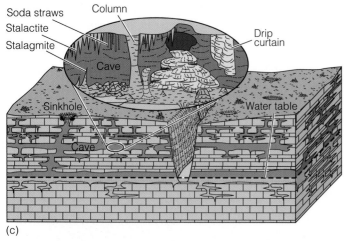

(b)

(c)

Figure 13.18 Karst topography and the formation of caves. (a) Groundwater dissolves the limestone, forming a system of openings and passageways. (b) Groundwater near the water table flows toward surface streams, dissolving more rock and enlarging the openings and extending the passageways horizontally. (c) With uplift and downward erosion by surface streams, the water table drops, leaving interconnected caves above the water table. Once air is present in the caves, features such as stalactites and stalagmites form by the action of dripping water.

Figure 13.19 This sinkhole formed on May 8 and 9, 1981, in Winter Park, Florida, and had a diameter of 100 m and a depth of 35 m. The white rectangle on the left side of the sinkhole about a third of the way down is a truck trailer. The vertical line across the sinkhole is a sewer pipe. Water in the bottom shows the level of the water table.

44. Imagine you own land at the **X** between Junior and Church Lakes near Interlachen. You want to drill a well on your land. Your sister also wants to drill a well on her land at the **x** marked 98 southeast of Interlachen. To aid in determining the depth of the wells and, therefore, the cost of drilling, use the contours of the water table on the map of Interlachen in Figure 13.21 and answer the following questions:

  a. What is the elevation of the land at your well site?

  _____ At your sister's? _____

  b. What is the elevation of the water table at your well site? _____ At your sister's?

  _____

  c. How deep must your well be to reach the water table?

  _____ Your sister's? _____

45. Draw arrows on the map of Interlachen in Figure 13.21 showing the direction of flow of the groundwater.

  a. If the people of Interlachen dumped toxic chemicals into Jewel Lake, would either you or your sister expect to have chemical contamination in your well after sufficient time had passed for contaminated water to flow that distance? _____

  b. If you go fishing in Gum Creek, can you expect your fish to be free of contamination for a long time to come or eventually to be killed off by the toxic chemicals?

  _____

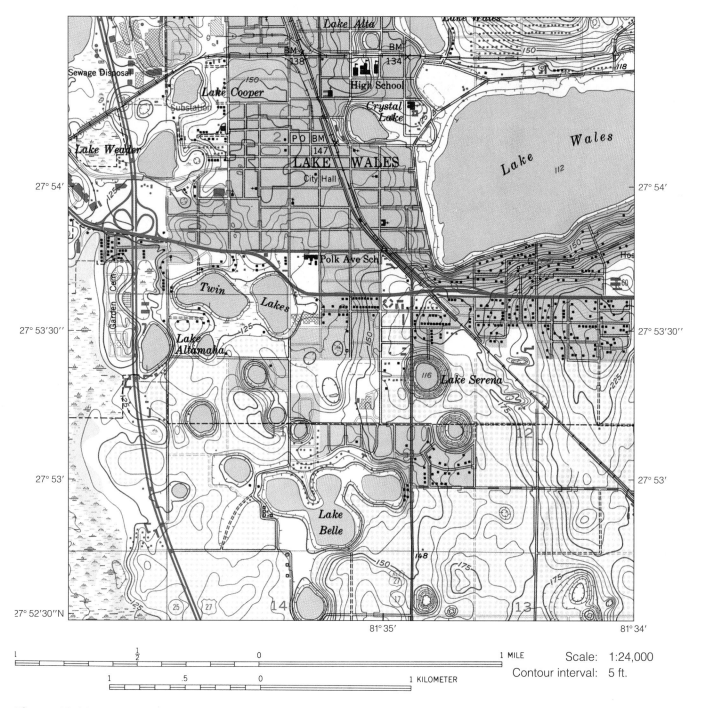

Figure 13.20 Lake Wales, Florida, USGS Topographic Map. Scale 1:24,000; contour interval = 5 ft.

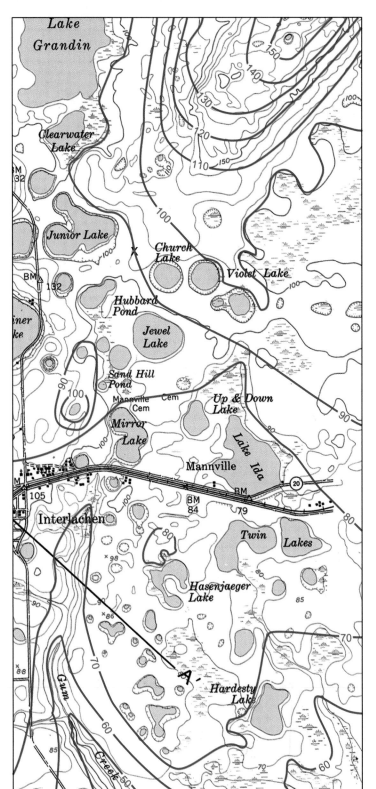

Figure 13.21 Interlachen, Florida, USGS 15-min Topographic Map. Scale 1:48,000; topographic and water-table contour interval = 10 ft. Blue lines are water-table contours.

Streams and Rivers

A **stream** is any body of water that flows under the force of gravity in a relatively narrow channel. A **river** is simply a large stream. All streams behave according to a few simple rules that govern the resulting topography and landforms. Streams may vary in size and appearance, but the principles that control them are the same. The word **fluvial** is often used in discussions about stream. It is an adjective meaning having to do with streams or rivers. For example, fluvial erosion means the same as erosion by a stream.

1. Gravity causes water in streams to flow down slope. The motion of the water is a kind of energy (kinetic energy) that allows the stream to erode rock, transport and deposit sediment, and modify the shape of its valley.
2. The amount of energy, and therefore the amount of work it does, depends on the amount of water and its velocity (speed).
3. The stream's kinetic energy can transport sediment by rolling, sliding, or hopping particles along the bottom (*bed load*), and by suspension in the water (*suspended load*). The chemical properties of water also permit it to carry sediment in solution (*dissolved load*).
4. When a stream's velocity or volume increases, its ability to carry sediment increases, and it will erode material from its bottom or sides.
5. When a stream slows or loses water by infiltration or evaporation, its capacity to carry sediment diminishes, and fluvial deposition begins.

The area occupied by the water in a stream is the stream's **channel.** A stream's **valley,** on the other hand, is the region directly and indirectly eroded by the stream, and it extends well beyond the stream channel. Mass wasting or gravita-

tional movement of material down the slope, such as by landslides or soil creep, erodes the valley until the debris reaches the channel where the stream carries it away. The area drained by a stream is the **drainage basin.** The boundary separating one drainage basin from another is known as a **drainage divide** and occurs at higher elevation than the stream channel.

Stream Gradient

Because gravity is the driving force for stream flow, the steepness of the slope of a stream, known as the **stream gradient** (Fig. 14.1) given in ft/mi or m/km, is important in determining certain characteristics of the stream (Fig. 14.2). In the exercises that follow, observe each stream, its valley, and its gradient to see whether there is a pattern.

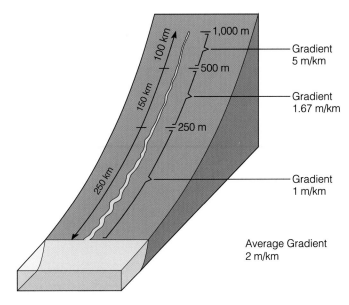

Figure 14.1 Stream gradients are calculated as the drop in elevation divided by the horizontal distance the stream travels. The gradient can be determined for the whole stream—the average gradient—or for a segment of a stream. Streams commonly have decreasing gradients along their course, as shown here.

269

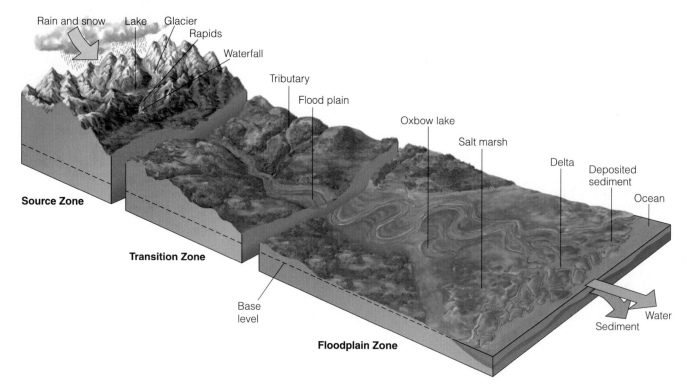

Figure 14.2 The changes in this stream along its course are primarily due to the change in gradient. The gradient in the source zone, near the headwaters of the stream, is steep, and diminishes gradually to very low values in the floodplain zone and delta. Notice the changes in valley and channel shape and width with gradient. Base level, the lowest level to which a stream can erode, is sea level for this stream.

1. Examine the stream in Johnson Canyon near the center of the Cortez, Colorado map (Fig. 14.3) and compare it to the Saco River on the Crawford Notch, New Hampshire map (Fig. 14.4). In your own words, how are these streams and their valleys different?

_____

_____

_____

_____

2. Now, compare the Patoka River on the East Mount Carmel map (Fig. 14.5) to the first two streams. What is different about this stream? About its valley?

_____

_____

_____

_____

3. Of the three, which stream has the steepest gradient (1m = 3.281 ft)? _____ Which has the widest valley? _____ Which has the steepest valley walls? _____ Which has the most winding course? _____

Sinuosity is a measure of how winding the course of a stream is. It is determined by measuring the length of the channel along the stream's path and dividing by the straight-line distance down the valley. Figure 14.6 shows three streams and their sinuosities. A **meandering stream,** discussed later, has a sinuosity of greater than 1.5. Next we will look at some measurements of gradient and sinuosity of a few streams.

4. Before you examine the measurements, think about gravity and slope and how they would affect a stream. What hypothesis can you formulate that would relate the sinuosity of a stream to its gradient? If a stream is flowing on a gently sloping surface, what would you expect its sinuosity to be compared to a stream traveling down a steep slope?

_____

_____

_____

_____

_____

Test Your Hypothesis

5. Examine the appropriate maps for each of the following streams, and describe the shape of the stream. Is it relatively straight, fairly winding, or very winding? Also determine the properties requested for each stream.

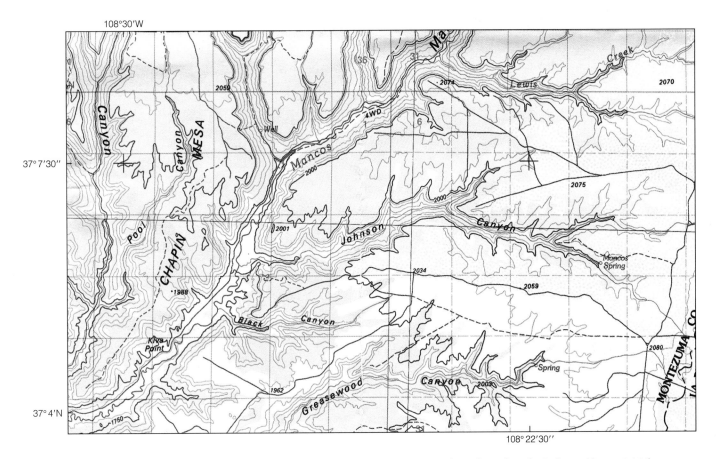

Figure 14.3 USGS topographic map of part of the Cortez 30 x 60 Minute Quadrangle, Colorado. Refer to Figure 1.11 for section numbering. 1 mi = 0.63 in. Scale 1:100,000. CI = 50m.

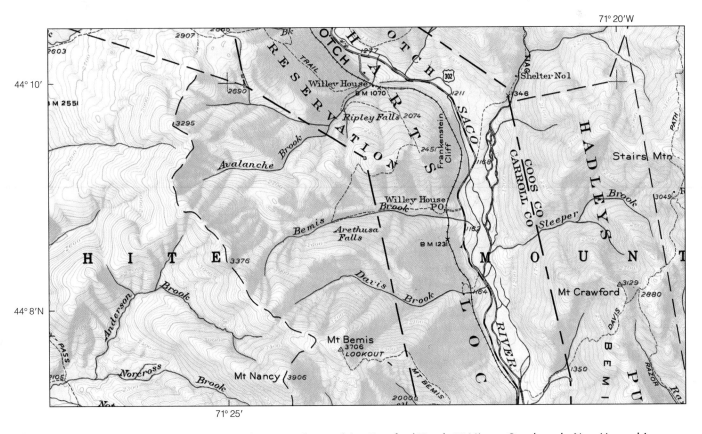

Figure 14.4 USGS shaded relief topographic map of part of the Crawford Notch 15 Minute Quadrangle, New Hampshire. 1 mi = 1.01 in. Scale 1:62,500. CI = 20 ft.

Figure 14.5 USGS topographic map of part of the East Mount Carmel 7.5 Minute Quandrangle, Indiana-Illinois. Inset shows the Patoka River between Section 31 and the town of Patoka. Scale 1:24,000. CI = 10 ft. See Figure 14.7 for scale bars.

Scale 1:100,000
CI = 20 m

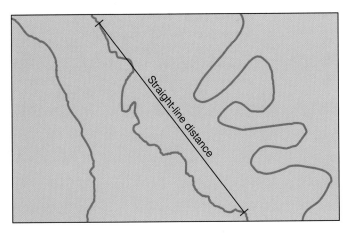

Figure 14.6 Three streams of different sinuosity. The sinuosities from left to right are 1.11, 1.37, and 3.10, respectively.

Streams:

a. South Platte River in Figure 14.7, starting at the 5090 contour and ending at the 5060 contour.

Stream shape: _____

Elevation change: _____

b. Henrys Fork of the Snake River in Figure 14.8, starting at the plus sign west of the word *Henrys* and ending at the plus labeled Mile 5. Note that each + is one mile away from the last + along the length of the winding channel. Stream shape: _____

Channel length along the river's course between the pluses mentioned: _____

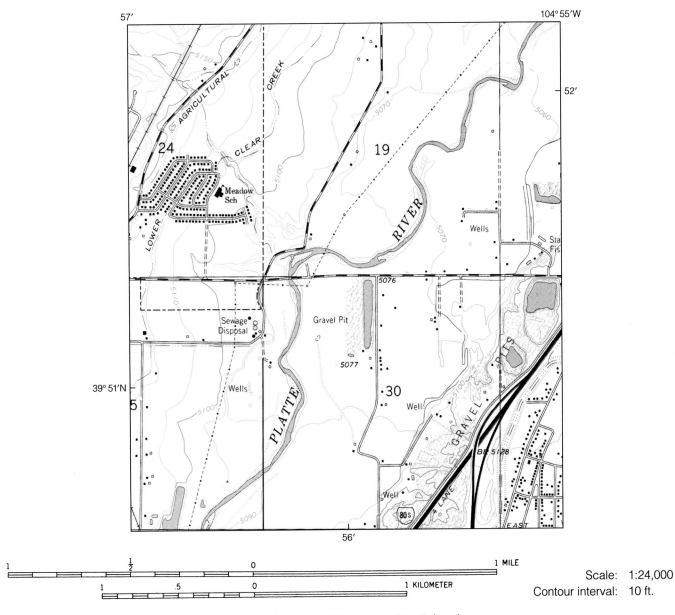

Figure 14.7 South Platte River in USGS topographic map of Commerce City, Colorado

Scale: 1:24,000
Contour interval: 10 ft.

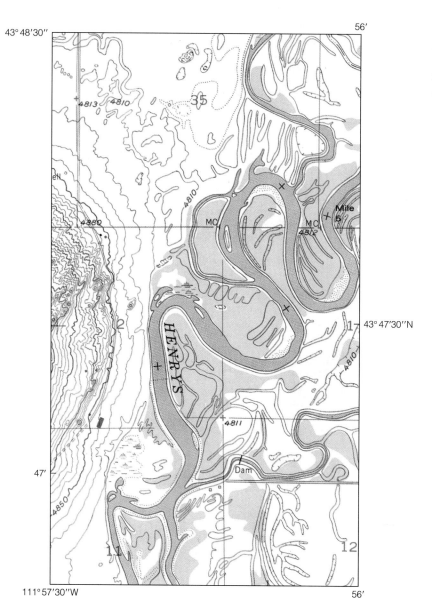

4813 4810

35

4810

4810

MC
4812

Mile
5

MC

HENRY'S

43° 47'30"N

4810

4811

Dam

11

12

Figure 14.8 Henrys Fork of the Snake River in USGS topographic map of Menan Buttes, Idaho. Scale 1:24,000. CI = 10 ft. See Figure 14.7 for scale bars.

c. The North Fork of the Swannanoa River in the Mount Mitchell, North Carolina map in Figure 14.9, starting at the 3000 contour and ending at the 2300 contour. Stream shape: _____

Elevation change: _____

d. Patoka River in Figure 14.5 inset, from the town of Patoka (390 ft) to Sec. 31 (380 ft).

Stream shape: _____

Elevation change: _____

e. Saco River in Figure 14.4 from the 1260 contour (near the number *1277*) to the 1000 contour (near the second R in *RIVER*). Stream shape: _____

Elevation change: _____

f. Mancos River in Figure 14.3, from Sec. 10 (5720 ft) to Sec. 28 (5560 ft; see section numbering in Fig. 1.11).

Stream shape: _____

Elevation change: _____

6. Enter the numerical data you just determined in the appropriate places in Table 14.1. Also fill in the other blanks in Table 14.1 using the calculations indicated.

7. If you had to measure the channel length along the winding path of the stream on a map, how would you go about it?

_____

_____

_____

_____

_____

8. Using the data in Table 14.1, graph stream gradient (horizontal axis) versus sinuosity (vertical axis) using the

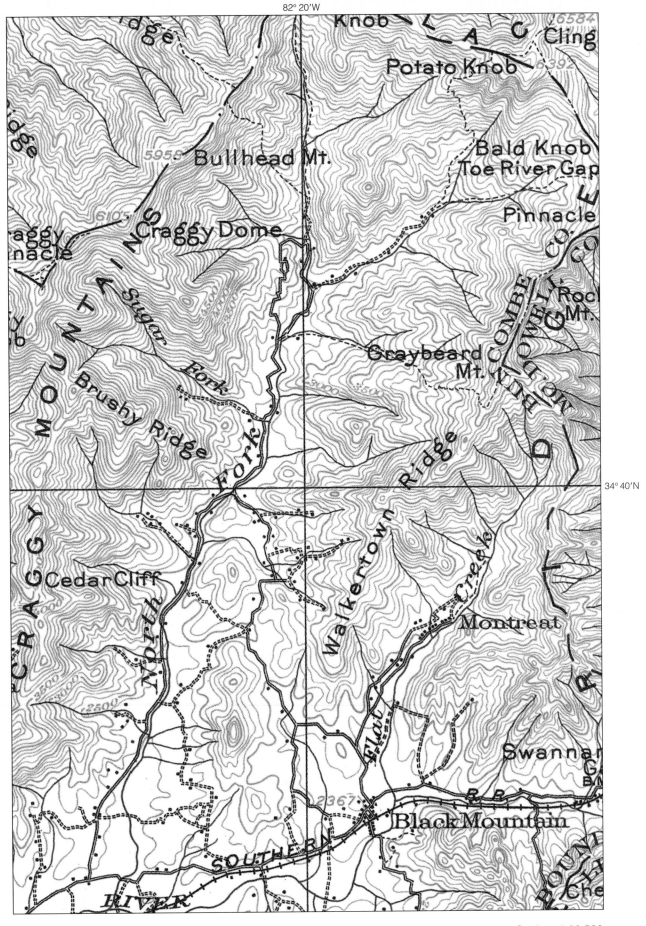

Figure 14.9 North Fork of the Swannanoa River, south of Craggy Dome, in USGS topographic map of Mount Mitchell, North Carolina. Scale enlarged to 1:62,500.

Table 14.1 Stream Characteristics

Stream	Elevation Change (ft)[1]	Channel Length (mi)[2]	Straight-line Distance (mi)[3]	Gradient (ft/ mi)[4]	Sinuosity[5]	Channel Width (ft)[6]	Floodplain Width (ft)[7]
South Platte River (Fig. 14.7)		3.4	2.57			90	~4300
Henrys Fork of the Snake River (Fig. 14.8)	3.75		0.91			230	~12,000
North Fork of the Swannanoa River (Fig. 14.9)		6	5.5			<16	~700
Patoka River (Fig. 14.5)		8.5	4.87			85	~14,000
Schoharie Creek, New York		6.5	6.1			< 40	~1300
Saco River (Fig. 14.4)		4.16	3.52			30–130	~575–2300
Mancos River (Fig. 14.3)		6.02	3.91			<16	~1000

[1]To determine elevation change, subtract the elevations of the two contours (in feet) indicated in Exercise 5 for each stream.
[2]Channel length is measured along the winding path of the stream.
[3]The straight-line distances were measured between the two points of known elevation indicated in Exercise 5 where contours cross each stream. This is the *same* part of the stream measured for channel length.
[4]To determine the gradient, divide the elevation change by the channel length.
[5]Sinuosity is measured as channel length divided by the straight-line distance.
[6]Channel width was measured from the width of the water at an average spot along the stream segment.
[7]Floodplain width was measured at an elevation of 20 ft above the stream.

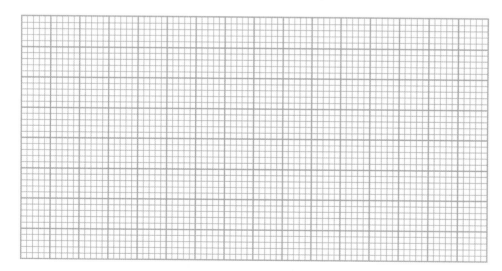

Figure 14.10 Graph paper

graph paper in Figure 14.10. Completely label your graph. Should 0 be included on the vertical axis? _____ Why or why not? (*Hint:* Think about how sinuosity is calculated and what the possible values might be.)

_____

_____

9. Was your hypothesis in Exercise 4 correct?

_____

10. Modify your hypothesis if necessary by answering the following question: What is the general relationship between stream gradient and sinuosity as seen in your graph?

_____

_____

11. Examine the data in Table 14.1 and make some additional generalizations:

a. What is the general relationship between stream gradient and channel width?

_____

_____

b. What is the general relationship between stream gradient and floodplain width?

_____

_____

The next section explores the relationships of stream characteristics as observed in the prior exercises.

Stream Erosion: Downward or Sideways?

Streams are very effective at shaping the landscape and cause more erosion than any other process. A stream cannot erode downward below the level of its mouth. This level is known as **base level** (Fig. 14.2), and it is sea level for those streams that flow into the ocean. Lakes and reservoirs create temporary base levels for streams, but are eventually filled with sediment. Once this happens, the stream can begin to erode down through the lake deposits, and base level shifts downstream to the next lake or reservoir, or, if none, then to the ocean. If base level remains the same, then as the stream erodes downward, its gradient must decrease with time. When the gradient gets low, downward erosion slows and sideways erosion (lateral erosion) becomes more important.

12. Based on this information and your new hypotheses in the previous questions, how would you expect a stream's valley to change with time in terms of the channel width, the valley width, and the sinuosity?

_____

_____

_____

_____

Base Level Computer Exercise

- Start **In-TERRA-Active.**
- Click **SURFACE** then **Running Water.**
- Click **Explore.**
- Click **Adjust Sea Level.**
- Examine the diagram and read the information as you do the exercise so you can answer the following questions:

13. What happens when sea level is lowered?

_____

_____

_____

14. What happens when sea level is raised? _____

_____

_____

- Click **Quit** and **yes.**

Consider a large block of land that is uplifted high above its surroundings to form a plateau. At first the area is high above its base level and can have a steep stream gradient.

15. While the stream gradient is steep, which type of erosion is more likely, downward or lateral? _____

16. Later, the gradient decreases. Which type of erosion is more likely then? _____

Downcutting

A stream can only remove material or erode where the water is in contact with the material. If a stream were the only eroding force, the stream would develop a slot canyon with vertical cliffs as shown in Figure 14.11. However, vertical cliffs are not common; mass wasting causes the slopes to fall, slide, slump, flow, or creep, bringing sediment to the stream to be carried away and making the slope gentler. These other processes contribute to the erosion of the surface, producing a valley that looks quite different than that in Figure 14.11. Let's simulate this with a simple experiment.

17. Tip and jiggle the tray of dry sand until the sand is smoothly sloping from one end to the other. Imagine a straight stream running down the middle of the sloping sand. Use your finger to simulate the stream's erosive force and move or scrape away sand along the stream's

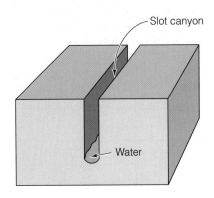

Figure 14.11 Cross section of a downward cutting stream with no mass wasting of the valley walls

imaginary path. Do this repeatedly until a valley forms. Does your finger touch every area where sand moves?

_____ The same is true of the water in a stream.

What other force moves the sand? _____ This

same force acts to make stream valleys.

18. Sketch in, on Figure 14.11, the shape of the stream valley you formed in the sand. In cross section or profile, what could you call this shape valley? Similarly shaped valleys result from fluvial downcutting.

_____

Imagine an area uplifted to a plateau. Stream erosion in such an area undergoes changes with time, which are divided into three main stages of stream erosion: early, middle, and late stage. The main or most active processes for each stage are listed at appropriate points within this lab. These stages and their processes may overlap and are continuous.

Early Stage

Processes:

- *Downcutting* is the fluvial erosion and abrasion that occurs in the bed of the stream. It deepens the channel and occurs when the stream has a steep gradient and carries a maximum size debris.
- *Mass wasting* of the valley walls in the early stages creates a steep V-shaped valley.
- *Headward erosion* lengthens the stream valley at its upper end (head or headwaters of the stream).

Resultant stream characteristics and landforms (see Fig. 14.12; see also Table 14.2):

- Low sinuosity stream
- Steep gradient stream
- Rapids and waterfalls
- Narrow V-shaped valleys
- Plateaus—wide flat divides
- Incomplete drainage—few streams per unit area

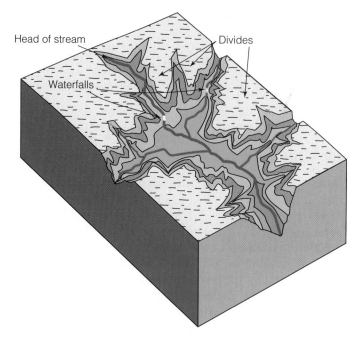

Figure 14.12   Early stage of fluvial erosion

Table 14.2   Features of Different Stage Streams

Stream Stage	Valley Shape	Stream Shape (sinuosity)	Shape of Divides	Proportion of Sloping Area
Early				Small
Middle	or			Largest
Late				Smallest
Rejuvenation				Small to medium

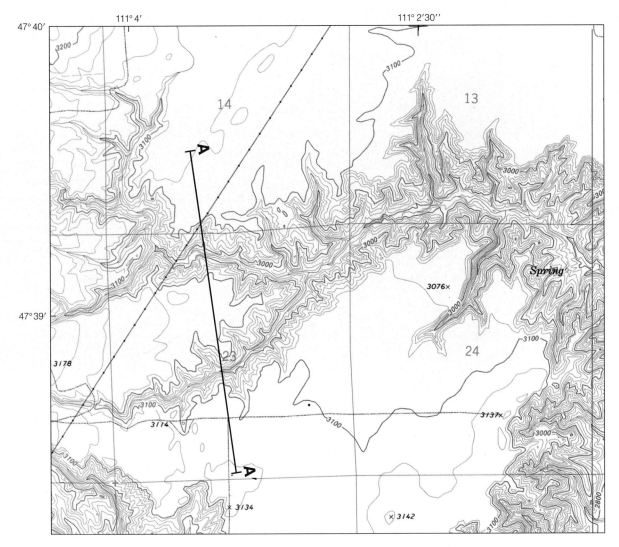

Figure 14.13 Topographic map of an early-stage stream near Missouri River near Portage, Montana. From USGS Floweree 7.5 Minute Quadrangle. See Figure 14.7 for scale bars. Scale 1:24,000. CI = 20 ft.

Figure 14.14 Topographic profile of the early-stage streams in Figure 14.13 (5x vertical exaggeration).

19. Examine the topographic map of an early-stage stream in Figure 14.13. Describe a typical early-stage stream by circling the correct words to complete the following sentences: "In topographic maps of early-stage streams, contours are closer together near divides / streams and farther apart near streams / divides. The stream is fairly sinuous / straight."

20. Draw a topographic profile in Figure 14.14 for an early-stage stream along the line from **A** to **A′** in Figure 14.13.

21. Suppose you are an emergency rescue worker that needs to organize a rescue of a hiker with a broken leg at the point marked 3076X in Figure 14.13. There is an access road to 3137X, but your crew must hike from the road with gear and back with the patient, as no helicopter is available. Using your knowledge of stream valleys and topographic maps, plot the quickest and easiest path possible to accomplish this, and mark it clearly on the map.

Middle Stage

Processes:

- *Downcutting* continues at a lesser rate.
- *Slope retreat* widens the stream valley.
- *Mass wasting* or movement of the land surface, such as by soil creep and landslides, is a dominant process.
- *Reduction of divides*
- *Widening of valleys.* These last two replace headward erosion as a dominant process.

Resultant stream characteristics and landforms (*see Fig. 14.15; see also Table 14.2*):

- Low to moderate sinuosity stream
- Floodplains beginning to develop
- Moderate gradient stream

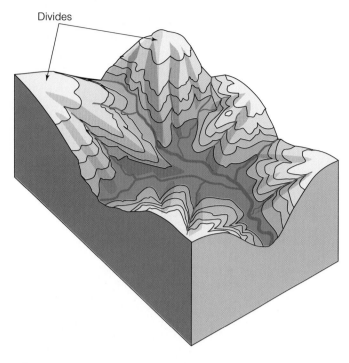

Figure 14.15 Middle stage of fluvial erosion

- Valleys still retain a V-shape, but with the V becoming wider and perhaps developing a slightly rounded or flat floor.
- Rounded or sharp divides—high land areas are mountainous or hilly, not plateaus.
- Maximum relief—little flat area—neither valley floors nor ridges are flat, for the most part.
- More complete drainage—an increase in streams per unit area results in an increase in drainage.

22. Describe a typical map view of a middle-stage stream (Fig. 14.16) by circling the correct words in the following sentences. "In maps of middle-stage streams, the spacing between contours varies greatly / little across the map. Slopes near the stream are steeper than / the same as / gentler than slopes on the divides. Flat areas are rare / common."

23. **a.** Draw a topographic profile in Figure 14.17 for a middle-stage stream along the line from **A** to **A′** in Figure 14.16, using only the index contours. Determine and plot the elevations at each valley and ridge that crosses the profile line before connecting the points on your profile.
    b. Highlight differences between profiles of early- and middle-stage streams on Figures 14.14 and 14.17.

Lateral Erosion and Meandering Streams

Streams that cause lateral erosion also deposit sediment. This occurs when a mature stream with a low gradient near base level has a sinuous course known as *meandering*. **Meandering streams** migrate across their main flow direction, causing

lateral erosion where the flow is fastest and deposition of sediment in the areas of slowest flow. Late-stage streams are meandering streams.

Late Stage

Processes:

- *Sedimentation.* As the stream loses its energy because of a general decrease in gradient toward base level, it begins to drop its load of sediment. Fluvial deposition occurs in point bars (inside bend of meanders) and floodplains.
- *Lateral erosion* and *meandering*—wandering or migration of the stream across its floodplain
- *Flooding* and deposition of fine grained floodplain sediments and natural levees

Resultant stream characteristics and landforms (see Fig. 14.18, see also Table 14.2):

- High sinuosity stream
- Low gradient stream. Very low slope allows for meandering. When currents lose their velocity upon entering a large body of water (a gulf, lake, or ocean), they drop their sediment load, forming a *delta* at the mouth of the stream.
- Meanders and *point bars*. Point bars are new deposits without vegetation on the inside of meander bends. Stream may be *braided* with mid-channel sandbars
- *Oxbow lakes*—cutoff meanders (Fig. 14.19)
- Floodplain well developed, drainage decreasing as land flattens, swamps in floodplain
- *Monadnocks*—isolated hills or erosional remnants of more resistant rock
- Divides nearly flat, close to stream elevation, low relief

24. Imagine you are at the location shown on the map in Figure 14.20. Describe the landscape and stream features.

_____

_____

_____

_____

25. Roughly sketch a topographic profile across the stream in Figure 14.20 in the space below.

Meandering Stream in the Stream Table

Materials needed:

- Stream table with sand and flow-regulated water supply
- Sand bucket
- Scoop or spoon
- Clear acrylic sheet
- Washable marking pens

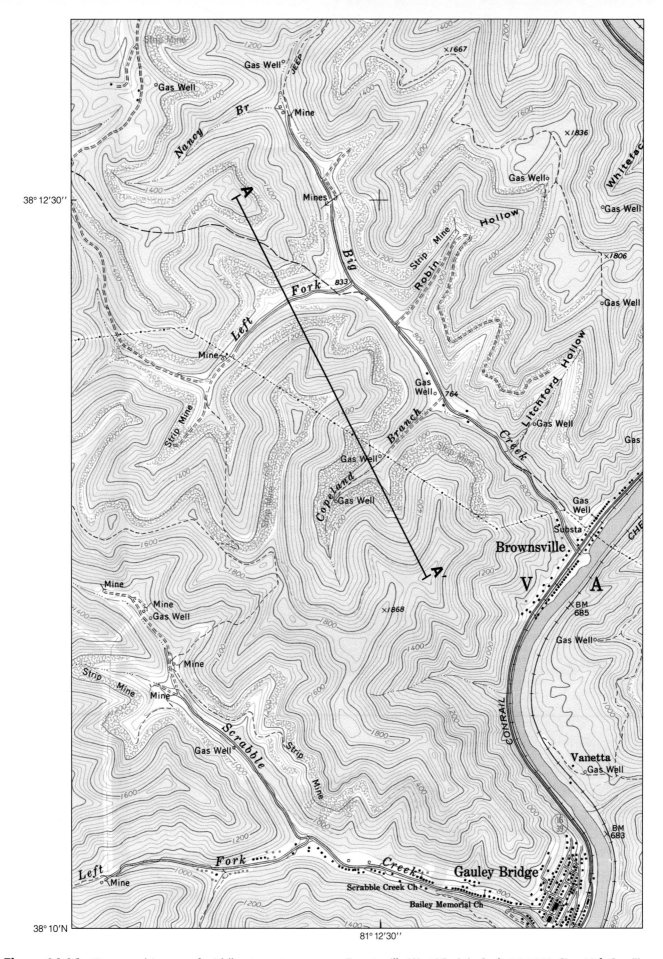

Figure 14.16 Topographic map of middle-stage streams near Fayetteville, West Virginia. Scale 1:24,000. CI = 50 ft. See Figure 14.7 for scale bars.

Read through the whole experiment and the questions before you start. When you start the water flowing, be ready to make the observations needed to answer the questions. Set up the stream table as follows: With the water off, create a smooth, gentle slope from the inflow to the outflow lake. Dig out a small lake at the inflow end. Carefully scoop out a meandering path for the stream from the inflow to the outflow lake, discarding the sand into the sand bucket and leaving the surface smooth. Be sure that the depth of channel along the length is constant at about 1 in. Lay the acrylic sheet over the stream table, and, with washable marking pens, trace the stream channel on the acrylic before you start the flow of

Figure 14.18   Late stage of fluvial erosion

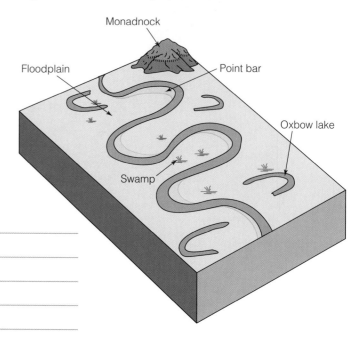

Figure 14.17   Topographic profile of the middle-stage streams in Figure 14.16 (2x vertical exaggeration)

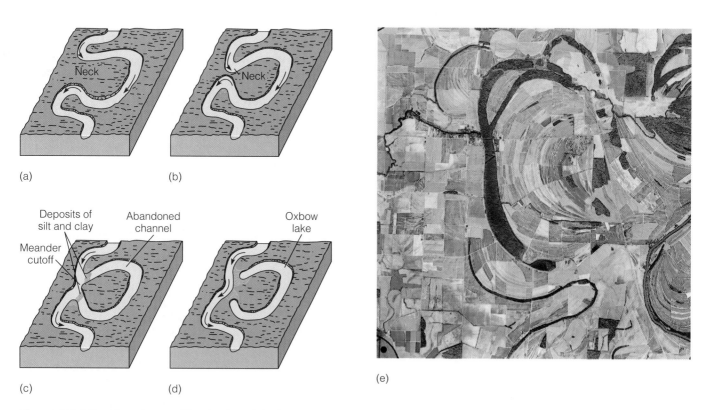

Figure 14.19   Meander cutoff and oxbow lake formation. (a) As a stream meanders, erosion on the outside of the bends where the flow is the fastest causes the meanders to grow. (b) Eventually the meanders become so large that the part of a bend upstream impinges on one downstream. During the subsequent flood events, the water overflows the channel and carves a cutoff between the two bends. (c) Deposition along the sides of the new cutoff channel blocks flow into the old channel, causing it to be abandoned. (d) The abandoned channel develops into an oxbow lake. (e) Oxbow lake left by the Tallahatchie River, Mississippi. Inside the curve of the oxbow lake, old, still visible point bars show where the river gradually migrated before the cutoff.

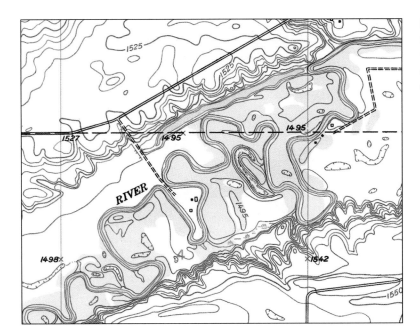

Figure 14.20 Topographic map of the Souris River, North Dakota. USGS topographic map of part of the Voltaire 7.5 Minute Quadrangle, North Dakota. Scale 1:124,000. CI = 5 ft. See Figure 14.7 for scale bars.

water. Set the flow regulator as directed by your instructor. Start the flow of the water. Along the length of the stream, trace the line of fastest flow of the water.

26. Observe the behavior of the stream in the stream table. If you took a canoe on this stream, what path would you follow so your canoe wouldn't get stuck on sandbars or rocks?

_____

_____

27. Where is fluvial erosion occurring? _____

28. Where is fluvial deposition occurring? _____

After about 1 min of flow (or when advised by your instructor), stop the water. Retrace the banks of the stream on the acrylic with a different color pen. You may need to set up the stream table a second time to make all of your observations.

29. Draw a sketch of the original position of the stream.

different colors, and create a key indicating the significance of each color.

31. Based on your observations, formulate a hypothesis about how this stream will migrate in the future.

_____

_____

_____

32. Place houses at various places along the stream bank. Select one of these as your house. Where is it? Choose one: outside a meander, inside a meander, downstream side of a straight reach, or upstream side of a straight reach. If the location of your house is between these points, circle the two it is between.

Turn the water back on for another minute or two.

33. What happened to your house? _____

_____

34. How good was your prediction? Was your hypothesis correct, or do you need to modify it? Write the modifications if needed.

_____

_____

_____

Long-Term Behavior of Meandering Streams There is insufficient time in a single lab to observe the long-term behavior of a meandering stream in the stream table. Instead, to demonstrate some of the patterns of meander migration over many years, we will perform an experiment in which meanders migrate very rapidly.

30. On top of the original position, draw the new position of the stream to show clearly how the stream changed. Use

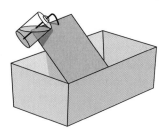

Figure 14.21 Setup for squeeze bottle experiment

Materials needed:

- A stiff flat sheet of metal or plastic
- A water catching basin or sink
- A plastic squeeze bottle containing plain tap water

35. Place the sheet tilted at an angle in the basin, as shown in Figure 14.21. Place the tip of the squeeze bottle against the upper part of the sheet, and squeeze to produce a steady stream of water. Carefully observe what happens.

36. Observe the gradual migration of meanders (curves) as time passes. They migrate (circle two)

   (a) outward   (b) inward
   (c) upstream   (d) downstream

37. What happens to cause a meander to straighten out (known as a *meander cutoff*, Fig. 14.19)? When does this happen?
Hint: Carefully observe the volume of water at the time of the meander cutoff.

_____

_____

_____

_____

Natural meandering streams migrate and behave in a similar manner. One example is the Mississippi River. The Arkansas-Mississippi state boundary was established at the Mississippi River in 1836 when Arkansas was admitted to the Union (Fig. 14.22). When a river changes course gradually over time, a state or county boundary on the river moves with the river, but with a sudden change in the course of the river, the boundary stays where it was.

38. Examine the map in Figure 14.22 of the Arkansas-Mississippi state boundary and the Mississippi River.
  a. Has the Mississippi River changed course since 1836?

    _____

  b. At how many locations on this map does the river deviate from the state boundary? _____

  c. What fluvial event or process would cause such deviations? _____

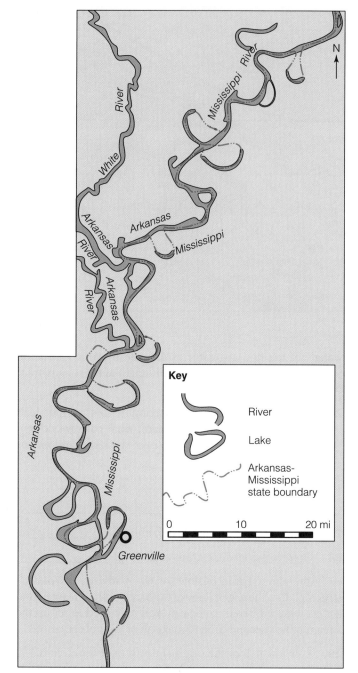

Figure 14.22 Map of the Mississippi River at the Arkansas-Mississippi border showing past meander migration

  d. Where the river and state boundary coincide, has the river been fixed in position since 1836? _____

  e. If not, how has the river's course changed along these coinciding reaches and what fluvial processes have caused these course changes?

    _____

    _____

  f. Have any meander cutoffs occurred along these coinciding reaches of the river since 1836? _____

Mark Twain, in his book *Life on the Mississippi* (1883), wrote about the meandering Mississippi:

> The Mississippi between Cairo and New Orleans was 1215 miles long 176 years ago. It was 1180 after the cutoff of 1722. It was 1040 after the American Bend cutoff: It has lost 67 miles since. Consequently, its length is only 973 miles at present. . . .
>
> In the space of 176 years the lower Mississippi has shortened itself 242 miles. That is an average of a trifle over one mile and a third per year. Therefore, any calm person, who is not blind or idiotic, can see that in the Old Oolitic Silurian Period, just over a million years ago[1] next November; the Lower Mississippi River was upwards of 1,300,000 miles long, and stuck out over the Gulf of Mexico like a fishing rod. And by the same token any person can see that 742 years from now the lower Mississippi will be only a mile and three-quarters long, and Cairo and New Orleans will have joined their streets together, and be plodding comfortably along under a single mayor and a mutual board of aldermen. There is something fascinating about science. One gets such wholesale returns of conjecture out of such a trifling investment of fact.

39. Explain Twain's reasoning. Any problems?

_____

_____

_____

_____

40. The Mississippi has not actually changed its length much. How is this possible?

_____

_____

_____

_____

_____

Rejuvenation: Back to Downcutting Again, But with a Twist

A fourth stage of stream erosion only occurs under special circumstances that revive the erosive downward cutting power of the stream. This stage is called *rejuvenation* (Fig. 14.23) and has similarities to both the early and late stages. Rejuvenation only occurs with a relative drop in base level that can be caused by a drop in sea level or by uplift resulting from plate motions.

Rejuvenation Stage

Processes:

- *Rejuvenation*—uplift of a mature (late stage) meandering stream

....
[1]We now know the Silurian was more than 408 million years ago.

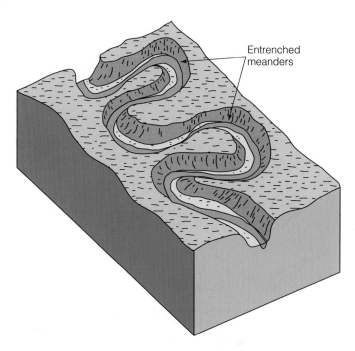

Figure 14.23   Rejuvenation stage of fluvial erosion

- *Downcutting* and *entrenchment* of meanders is the primary active process and results from a relative drop in base level. Lateral erosion is abandoned.
- Initial stages of *mass wasting* create steep V-shaped valleys.

Resultant landforms and stream characteristics *(see Fig. 14.23; see also Table 14.2):*

- High sinuosity—the high sinuosity was inherited from the stream when it was in the late stage.
- Steep gradient—high sinuosity and steep gradient is an unusual combination for a stream.
- *Entrenched* or *incised* meanders: the stream follows the path of the mature stream but cuts down to entrench the stream and form terraces or plateaus along the steep bank. The meandering shape of the path remains as a clue to the stream's past.
- Steep V-shaped valleys or canyons—the stream's meanders no longer migrate laterally.
- Plateaus—wide flat divides

41. Describe a typical map view of a rejuvenated stream (Fig. 14.24) by circling the correct words to complete the following sentences: "In topographic maps of rejuvenated streams, contours are closer together near divides / streams and farther apart near streams / divides. The stream is fairly sinuous / straight."

42. A topographic profile of a rejuvenated stream would be most similar to (a) an early-stage stream, (b) a middle-stage stream, or (c) a late-stage stream.

The Colorado plateau encompasses some of the least-known and most unexplored territory in the continental United States. Much of the area is accessible only by helicopter. Canyon walls are too steep in much of the area for access except by air or water.

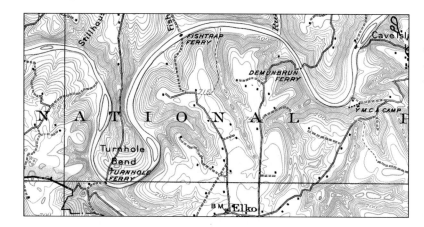

Figure 14.24 Topographic map of a rejuvenated stream, from USGS Mamoth Cave, Kentucky 15 Minute Quadrangle. Scale 1:62,500. CI = 20 ft.

Figure 14.25 Aerial photo pair of the Colorado River. Scale 1:41,900.

43. Examine the aerial photo pair of the Colorado River, Utah, in Figure 14.25. What is the stage of stream erosion? Stage: _____

List evidence: _____

_____

_____

Rejuvenation Computer Exercise

■ Start **In-TERRA-Active.**
■ Click **SURFACE,** then click **Running Water.**
■ Click **APPLY.**

44. Do Problem 1 to help you understand how the landscape along the Colorado River, in Exercise 43, evolved.

First step: _____

_____

_____

_____

_____

Last step: _____

Practice Recognizing Stream Stages

45. What is the stage of stream erosion for:
  a. The Crawford Notch map (Fig. 14.4)?

  _____

  b. Henrys Fork of the Snake River (Fig. 14.8)?

  _____

  c. North Fork of the Swannanoa River (Fig. 14.9)?

  _____

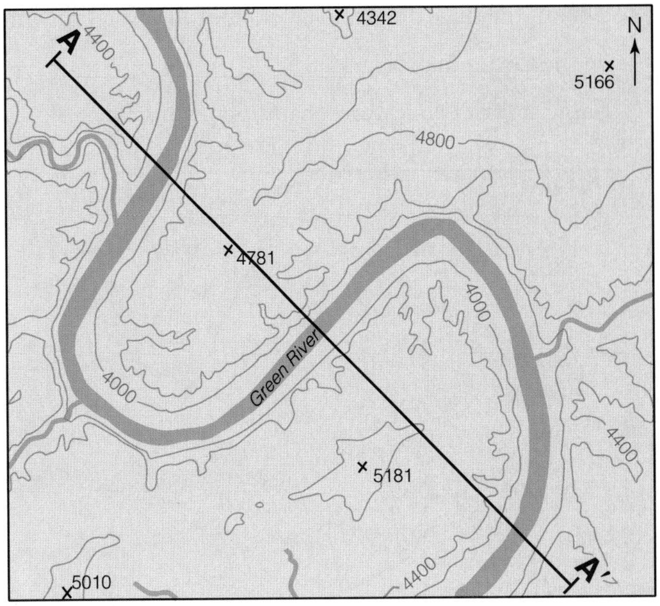

Figure 14.26 Topographic map of meanders in Green River, Utah. Scale 1:45,000.

In the next exercise you will construct a topographic profile across a stream to help you visualize the shape of its valley and interpret its stage.

46. Examine the map of Green River, Utah, in Figure 14.26.
  a. What is the contour interval for the map?

  _____ ft

  b. Label the elevation of each contour on the map.
  c. Use Figure 14.27 to construct a topographic profile from **A** to **A′** along the reference line on Figure 14.26.
  d. Would a person (using binoculars, if necessary) standing at **A** be able to see another person at point **A′**?

  _____

  e. Determine the stage of stream erosion by looking at both your profile and the map (Table 14.2 may help).

  _____

  f. Name at least two features that support your conclusion: _____ _____

Figure 14.27 Topographic profile of the late-stage streams in Figure 14.24. Vertical exaggeration = 1.9x.

3600

47. What is the stage of stream erosion for Johnson Canyon on the Cortez map (Fig. 14.3)? _____

Name at least two features that support this conclusion:

_____   _____

48. What is the stage of stream erosion for area shown on the East Mount Carmel map (Fig. 14.5)? _____

What features associated with this stage are seen on this map?

_____

_____

_____

Wind and the Atmosphere

OBJECTIVES

1. To understand the Coriolis effect
2. To understand Earth's pattern of surface winds
3. To understand relative versus absolute humidity and what they have to do with the weather
4. To be able to read a weather map
5. To learn the chemical composition of the atmosphere and understand the greenhouse effect
6. To understand the origin of the ozone layer and its destruction

The layer of the atmosphere in contact with the Earth's surface and containing water vapor and clouds, is the troposphere. It is the part of the atmosphere that makes our weather. The **troposphere** (Fig. 15.1) is defined as the part of the lower atmosphere where temperature generally decreases with altitude. Heating of this part of the atmosphere is primarily caused by the *greenhouse effect*. Above the troposphere is the **stratosphere,** where temperature increases with altitude (Fig. 15.1). Here ultraviolet radiation from the sun causes this increase in temperature and creates the *ozone layer.*

Within the troposphere, winds develop from differences in air pressure resulting from differential heating of the air. The source of energy is sunlight and, to a lesser extent, solar heat radiation. Wind blows from regions of high atmospheric pressure with cool sinking air to regions of low pressure with warm rising air. But the wind cannot flow straight from high to low pressure because the Earth turns beneath it, causing an apparent deflection called the **Coriolis effect,** which influences any object moving across the Earth's surface. These processes produce a global pattern of winds. As air rises or sinks, it changes temperature and *relative humidity,* which produces condensation or evaporation and is another important factor in weather.

Coriolis Effect

The Coriolis effect results from Earth's rotation. It causes moving objects to veer or turn rather than maintain a straight path relative to the Earth's surface, just as a ball thrown on a turning merry-go-round appears to curve from the perspective of those on the merry-go-round (Fig. 15.2).

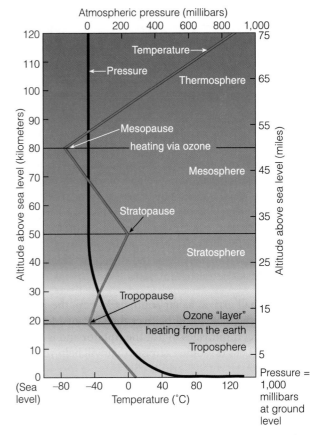

Figure 15.1 Earth's atmosphere consists of layers defined on the basis of temperature changes within the layer. Temperature decreases with altitude in the troposphere and mesosphere and increases with altitude in the stratosphere and thermosphere. Sunlight heats the troposphere indirectly via the greenhouse effect. The stratosphere is heated by ultraviolet radiation from the sun. The ozone layer is contained within the stratosphere where ultraviolet radiation is absorbed. The exact altitudes of the various layers depend on the latitude and the time of year. Atmospheric pressure decreases with altitude.

It influences winds and ocean currents. Actually, the Earth turns underneath the object as the object moves; thus, although it might start heading in one direction, before it gets very far, the Earth has turned beneath it and the direction has changed. Imagine a cannon on an airless Earth that shoots a cannonball from New York City toward Miami. By the time

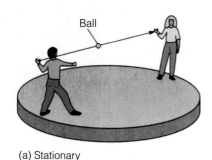

(a) Stationary

Apparent path as seen by
observer on rotating platform

Actual path

(b) Rotation

Figure 15.2 Illustration of the Coriolis effect on a merry-go-round. (a) A ball thrown on a stationary merry-go-round follows a straight path when viewed from above (the actual path of the ball is influenced by gravity so is parabolic vertically—not shown.) (b) If the merry-go-round is moving, the ball appears, from the point of view of people on the merry-go-round, to curve to the side. The platform rotates while the ball is in flight so the ball misses the receiver. An outside observer would still see a straight path.

the cannonball arrives as far south as Miami, the Earth has turned and the ball would be over the Gulf of Mexico. The Coriolis effect diminishes to 0 at the equator and is strongest at the poles. Let's do some exercises that will help you visualize and understand the Coriolis effect.

1. What direction does the Earth turn if viewed from the North Pole, clockwise or counterclockwise?

   _____

   Hint: Use a globe to help if necessary. How must the Earth turn so the sun rises in the east?

2. Place a plastic U-turn ramp in the center of the Coriolis effect rotating platform. Keeping the platform stationary, place a ball on the ramp and let it go. Describe the path of the ball. _____

   _____

3. Turn the Coriolis effect rotating platform in the same direction as your answer in Exercise 1. Place a ball on the ramp and let it go. Describe the path of the ball.

   _____

   If you face the direction the ball was moving, what type of turn did the ball make, right or left? _____

Lift the plastic sheet to erase the ball's track. Try placing the ball in different locations on the rotating platform while continuing to turn the platform in the same direction. Is the direction the ball turns consistent if you always face in the direction the ball traveled?

_____

This exercise demonstrated the Coriolis effect for the Northern Hemisphere. Next let's determine the Coriolis effect for the Southern Hemisphere.

4. If you looked down at the Earth from a point in space above the South Pole, what direction does the Earth

   turn? _____

5. Turn the Coriolis effect rotating platform in this direction, and repeat your previous experiments. Now what

   direction does the ball turn? _____

6. Explain why the ball travels along a curved path while the platform is turning.

   _____

   _____

   _____

   _____

The Coriolis effect will also influence any object traveling through the atmosphere because the Earth turns underneath it as the object moves. This applies to airplanes too, although wind speed and air resistance are much more important factors. For the sake of argument, though, let's ignore these other factors and concentrate on how the Coriolis effect works.

7. Imagine you are taking a flight from Anchorage, Alaska to Honolulu, Hawaii, and that the flight takes about 6 h. Imagine that the pilot simply points the plane in the direction of Honolulu and takes off (neglect wind resistance and speed, or you can think of this as a cannonball shot through a frictionless atmosphere). Examine a globe; what are the latitude and longitude of Anchorage?

   _____ What are the latitude and longitude of Honolulu? _____ What is the difference in longitude (east-west position) between Anchorage and Honolulu? _____ Position the meridian arc or circle of the globe above Anchorage.[1] Rotate the globe the equivalent of 6 h (90° rotation). At the latitude of Anchorage, measure how far the Earth turned (from the meridian arc/circle or your Anchorage pointer), using a string arranged along the parallel. Mark the distance on the string with a fine-point washable marking pen. At the

....

[1]If your globe has no meridian arc or circle, position its axis of rotation vertically and mark the positions of Anchorage and Honolulu with a ring stand and some sort of pointers provided by your instructor.

point where Honolulu was 6 h ago (use the meridian arc/circle plus the difference in longitude from Anchorage or your Honolulu pointer), measure the same distance with the string along the parallel to the east. This is where the plane would arrive because of the Coriolis effect. Where is this?

_____

How far from Honolulu is it? _____ (Review distances on a globe in Exercise 30 in Lab 1 if necessary.)

8. New Orleans is at 30°N and Chicago at about 42°N. The circumference of the Earth at the 30th parallel is about 34,700 km and at the 42nd parallel is about 29,800 km. In 2 h, how far has a person sitting in New Orleans traveled? (*Hint:* What fraction of a day is 2 h? How much of 34,700 km would that be?) _____

In Chicago? _____ Which seated person is traveling faster? _____

How much farther does the faster person travel in 2 h than the slower person? _____ Now let's take a trip from New Orleans to Chicago. The flight takes about 2 h. If a plane took off from New Orleans and headed straight for Chicago without taking the Coriolis effect into account (neglecting wind and air resistance), where would the plane end up in 2 h? _____

_____

Global Winds

Aided by the greenhouse effect, the sun-warmed land or ocean heats the air directly above it, causing the air to rise. The warm rising air causes an area of low pressure. Warming occurs especially at the equator since sunlight streams in from directly overhead rather than at an angle, causing more intense heating than at the poles (Fig. 15.3). At the poles the colder air sinks, creating high pressure. The wind flows generally from the higher pressure at the poles toward the lower pressure at the equator, helping to redistribute heat on Earth (Fig. 15.4). If the Earth did not rotate, the winds would flow straight to the equator, but the winds veer as a result of the Coriolis effect. In fact, this veering is so pronounced that the winds initially equator bound do not reach the equator but are traveling from the east by the time they arrive at about 60° latitude. Here the temperature is relatively warm compared to the poles, so the air heats and rises. Rising air belts at 60° latitude have low pressure and moist climate, similar to the equatorial belt (Fig. 15.5). High pressure occurs where air sinks at belts near 30° latitude. Like the poles, these regions have high pressure and dry climates (Fig. 15.5).

Winds are named for the direction from which they come because you feel rather than see wind. Between 30°N and 30°S, the winds are easterly and blow from 30° toward the equator and are called the **trade winds** (Fig. 15.5). The **westerlies** blow from the west and from 30° to 60° latitude (Fig. 15.5). Flowing from the east and from 90° to 60° latitude are the **polar easterlies** (Fig. 15.5).

Rising and descending air lead to moist and dry climates, respectively (Fig. 15.5). In low-pressure areas, such as the

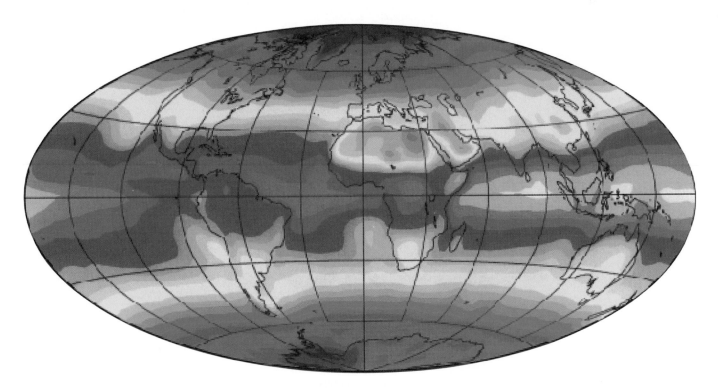

Figure 15.3 Map of the net radiation received over the earth's surface. Net radiation is positive in pinks, reds, and oranges, at low latitude and negative in blues and greens, at high latitudes. Positive net radiation means heating and negative means cooling or heat loss.

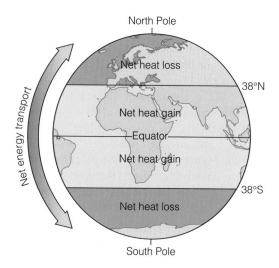

Figure 15.4 More heat is gained on Earth near the equator, where sunlight shines more directly overhead. More heat is lost than gained at the poles. Therefore, heat must be transported from the equator to the poles. Winds and ocean currents accomplish this transport. The necessity for heat transport is what causes the overall global wind and ocean circulation patterns.

equator, the air rises because it is warm and less dense than cooler air. As the air rises and reaches higher altitudes, it cools off. Cold water molecules are less energetic than warm ones and are more likely to condense than evaporate in cold air. As a result cooling air loses moisture by condensation. This causes clouds and precipitation in the form of rain and/or snow—a moist climate. On the other hand, high pressure and warming air come about when cold, denser air sinks. As the air sinks, compression warms it and it reaches lower altitudes where the temperature is warmer. Since the water molecules in warm air are more energetic, evaporation is more likely; you can think of warming air as being starved for moisture/water vapor. This causes evaporation of any water in the vicinity and results in a dry climate.

9. Summarize the climate belts just discussed and shown in Figure 15.5.

   a. Dry climate belts occur at _____ latitudes.

   b. Moist climate belts occur at _____ latitudes.

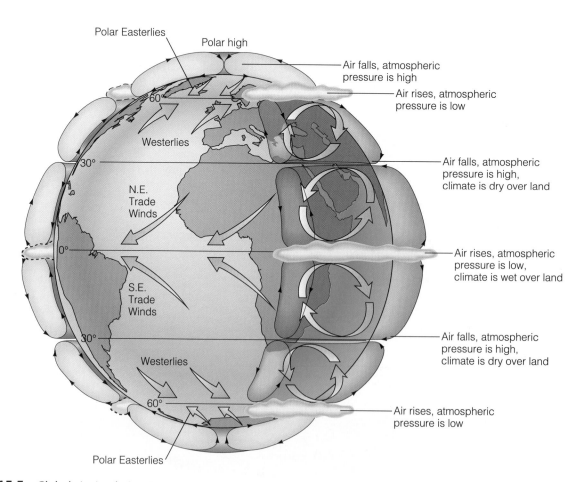

Figure 15.5 Global air circulation. Large convection cells of air circulate in the atmosphere within belts 30° latitude wide (shown on the left and right). The Coriolis effect causes the surface part of these circulation patterns, surface winds, to move east and west as shown in the center of the diagram.

Climate Belts Computer Exercise

- Start **In-Terra-Active.**
- Click **SURFACE** then **Wind & Deserts.**
- Click **Explore.**
- Click the yellow spot at the South Pole.

10. What is the climate like here? _____

How does this compare to your answer in Exercise 9?

_____

_____

- Move the orange slider to 100 and then click the largest yellow spot.

11. Read about this location and answer the following questions.

   a. What type of climate was found here? _____

   b. Zones like this are centered at _____ latitudes.

   c. What heated the air here? _____

   d. What are two locations on Earth today that have similar climates for similar reasons? _____

_____

- Click **Quit** and **yes.**

The map in Figure 15.6 has red arrows showing the direction and strength of the average winds in January for the Earth. The stronger winds have longer arrows, and weaker winds have shorter arrows.

12. In the rectangles on the map (Fig. 15.6), write in the air pressure, either high or low based on the direction of the winds (red arrows).

13. What are the fine lines on the map, and what are the numbers on those lines? _____

Do these lines and numbers correspond to your answers in the previous exercise? _____

14. Notice that the wind does not head directly toward the low-pressure areas or directly away from the high-pressure areas. This is because of the Coriolis effect. What

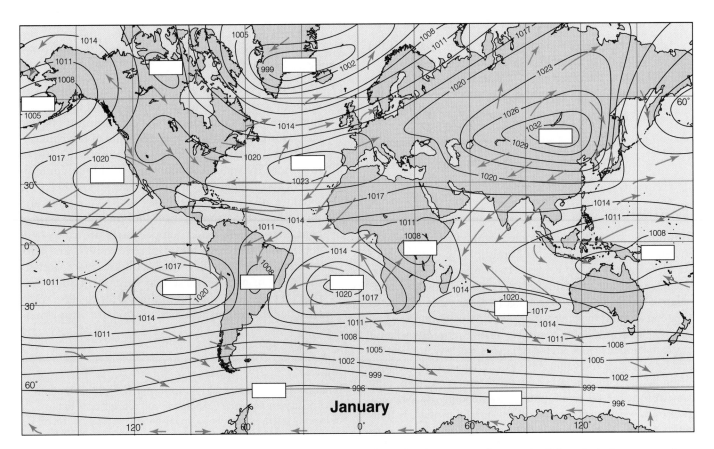

Figure 15.6 Global average January surface winds. Red arrows show the direction and strength of the winds. Average surface barometric pressure is in millibars.

direction, clockwise or counterclockwise, does the wind rotate around high pressure in the Northern Hemisphere? _____

Around high pressure in the Southern Hemisphere? _____ Notice that rotation around low pressure is opposite.

15. On the same map, use different-colored highlighters or pencils to highlight the easterly and westerly winds. Write a key on the map showing what the colors represent. Remember that winds are named for the direction from which they come.
   a. Describe the overall pattern of winds.

   _____
   _____
   _____
   _____

   b. Where do(es) (a) major deviation(s) from the general pattern occur?

   _____
   _____

   c. Approximately what four latitudes make the boundaries between easterly and westerly winds?

   _____ _____ _____ _____

   d. Label each belt according to the correct name of the winds there. Refer to page 291 again if needed.
   e. Look at the direction of the wind just north of the equator and just south of the equator. North of the equator the wind is generally _____,

   and south of the equator the wind is generally _____.

Humidity

Humidity is the water vapor content of the air. It is measured in two different ways. **Absolute humidity** is the total mass of water contained as vapor in a certain volume of air, measured, for example, in grams per cubic meter (g/m^3), whereas **relative humidity** is the percentage of water vapor relative to the total amount of water vapor the particular volume of air can potentially hold. Air has 100% relative humidity when it starts to condense to form fog or clouds. Relative humidity is inversely related to evaporation rate. That is, as the relative humidity goes up, the rate of evaporation goes down. One way to measure relative humidity uses cooling by evaporation on a wet thermometer bulb compared to the dry bulb temperature. The apparatus that does this is called a **psychrometer.** When you feel the humidity or dryness of the air on your skin, you are detecting relative humidity because of its relationship to cooling caused by evaporation.

Warmer air can hold more moisture than colder air. This means that if you start with cold air at 100% humidity and warm it up, the relative humidity drops even though the amount of water vapor in the air (the absolute humidity) stays the same. Conversely, warm air that is cooled down increases its relative humidity and condenses into clouds or fog if it reaches 100% relative humidity.

16. From your own experience, in winter, when the air outside is cold, what is the humidity like inside where the air has warmed? Does it seem moist, or does it dry out your skin? _____

On a warm, very humid day, if you walk outside from an air-conditioned building, your skin may feel clammy; if you wear glasses, they may fog up. Your skin and glasses may be cool enough to cause condensation of some of that humidity. Similarly, at night temperature drops, and as it cools, the relative humidity goes up. At the temperature where the relative humidity reaches 100%, dew or fog forms. This temperature is called the **dew point.** Every batch of air with its particular absolute humidity has a dew point. Simple heating or cooling of air without adding or removing moisture does not change the dew point; with artificially handled air, moisture may be added by a humidifier or a swamp cooler (a cooling unit that uses evaporation of water to cause cooling) or extracted by a dehumidifier or an air conditioner, altering the dew point.

Measuring the Dew Point

The following experiment helps you measure the dew point for the air in your lab.

17. Place some ice water in a glass beaker or metal cup without getting the outside of the container wet. Do the same with some tap water. After a minute or two check for condensation on the outside of the containers. If condensation occurs on one but not the other,[2] mix more ice water and tap water in similar containers to make water of various temperatures. Keep the outsides of the containers dry while you pour and mix. Check for condensation looking for the pair of containers whose water temperatures are closest together and still have two different results, one with condensation and the other without. Place a thermometer in each of these two containers and when the thermometer has equilibrated with the water (reached a constant temperature) record your obser-

....

[2]If moisture condenses on the tap-water container, the dew point is higher than your tap water temperature and you will need to use warmer tap water or heat the water. Do not use water above room temperature, as the dew point is lower than that (unless the room is foggy). On the other hand, if no moisture condenses at 0°C, you will need to make the water colder. Mix crushed ice and salt to get colder temperatures. Only if your air is very dry will you need to do this.

Table 15.1 Dew Point Determination

MEASURING DEW POINT INSIDE		MEASURING DEW POINT OUTSIDE	
Inside Air Temperature (°C) →		Outside Air Temperature (°C) Is It foggy? →	
Water Temperature (°C)	Condensation or Not	Water Temperature (°C)	Condensation or Not
	= dew point		= dew point

vations in Table 15.1. Continue mixing new temperatures of water between the condensing and not condensing container temperatures until you have narrowed the dew point to about 2°C. Whenever you find the closest pairs with opposite results, record the temperatures in Table 15.1.

If the outside temperature is above freezing, measure the dew point outside.

18. Mix up batches of water at about 5°C increments between 0°C and the outside air temperature and put them each in Styrofoam cups. Take some additional, preferably metal, containers and a thermometer with you, taking extra care not to drop or chip the thermometer. Using the batches of water at 5°C intervals, test to see if condensation occurs on the outside of the tin cups. Measure the temperatures and record your results in Table 15.1. When you have the dew point narrowed down to between two temperatures, mix these two water batches to get intermediate temperatures and narrow the dew point measurements further.

19. Examine and summarize your results. Describe in as many ways as you can how the inside air is different from the outside air. In what ways are they similar? Is the dew point different? If so where is it highest and why? How do the relative humidities compare? Can you sense this subjectively? Has moisture been added or extracted from the inside air? How do you know?

_____

_____

_____

_____

_____

_____

Determining the Dew Point from Temperature and Relative Humidity

The equation approximating the dew point from temperature and relative humidity is as follows:

$$D = T + 13.7 \ln\left(\frac{h}{100}\right)$$

where D = dew point in °C
T = temperature in °C
h = relative humidity in percent

The natural logarithm (ln) can be obtained from base 10 logarithms (log) by multiplying by 2.303.

20. Measure the temperature and relative humidity inside and outside using a thermometer and a hygrometer, respectively. Enter the data in Table 15.2. Determine the dew point for both locations by calculation using the equation or reading off the graph in Figure 15.7. If you are reading the graph, you need to determine the temperature difference from the graph then subtract that from the air temperature to obtain the dew point. Enter the dew point in Table 15.2.

21. How do the dew point values in Exercise 20 compare to the values obtained in Exercises 17 and 18? _____

_____

Can you account for any discrepancies? _____

_____

_____

_____

Table 15.2   Dew Point Determination from Relative Humidity

	Temperature (°C)	Relative Humidity (%)	Dew Point (read from graph or calculated)
Inside			
Outside			

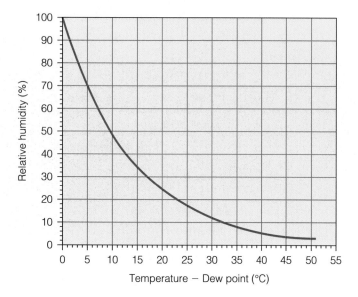

Figure 15.7   Relationship between the difference in temperature and dew point, and relative humidity

Weather

Wind, moisture, humidity, and many other factors go into producing the weather. Two basic contrasting weather systems occur at atmospheric low and high pressure (Fig. 15.8). To make a high-pressure system, air has to sink. When this air moves from high altitude, where air is generally cooler, to low altitude, it warms up.

22. Do you think that the water vapor in this air will condense as it warms up, to make clouds, or will this sinking air mass cause evaporation? _____

Warming air can hold more moisture than in its cooler state, so it tends to dry out its surroundings. Thus, high-pressure systems have dry stable air and tend to have clear skies and no precipitation. These high-pressure systems are also known as **anticyclones.**

A low-pressure system[3] at latitudes higher than about 30° is called an **extratropical cyclone** and has rising and cooling air.

23. What happens to the water vapor in the air as it cools?

_____

Low-pressure systems tend to have clouds and rain or snow and are stormy. For the Northern Hemisphere, winds are clockwise around high pressure and counterclockwise around low pressure (Fig. 15.8). Reverse this for the Southern

....

[3]Tropical cyclones including tropical storms, hurricanes, and typhoons, are low-pressure systems originating at low latitudes, usually between 5° and 20° latitude. These extremely powerful storms obtain their energy from condensation.

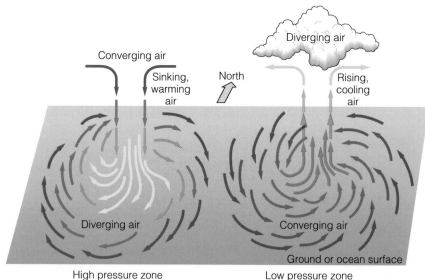

Converging air

Sinking, warming air

North

Diverging air

Rising, cooling air

Diverging air

Converging air

Ground or ocean surface

High pressure zone

Low pressure zone

Figure 15.8   Anticyclone (H = high pressure) and cyclone (L = low pressure) for the Northern Hemisphere showing wind directions and influence of Coriolis effect. High pressure is associated with sinking air that produces clear skies; low pressure is associated with rising air that condenses to form clouds.

Hemisphere. Before weather satellites and television forecasting, people used to predict the weather using a barometer, which is an instrument that measures air pressure.

Making a Barometer

Materials needed:

- A stable platform at least 4 in. $\times$ 12 in.
- 400-ml beaker
- Balloon
- Rubber bands
- Tape
- Paperboard
- Millimeter graph paper
- Straw or wooden skewer

Figure 15.9 shows how to assemble the parts. Cut the narrow blowing end off the balloon. Stretch the remaining piece of balloon over the beaker, and hold it in place with rubber bands tightly wrapped around the lip of the beaker. Cut the end of the straw to make a point, or use a pointed skewer and tape the blunt end to the center of the balloon on the beaker. Tape the beaker to one end of the platform with the straw/skewer pointing toward the other end. Cut a strip of paperboard 1 1/2 in. wide and about twice the height of the beaker. Number every centimeter on a strip of millimeter graph paper equal in length to about half the length of the paperboard strip. Tape the strip of graph paper on one-half of the paperboard. Bend the other end of the paperboard so it will stand vertically when taped to the platform. Attach it to the opposite end of the platform from the beaker. The paperboard acts as a gauge for your barometer. Set a thermometer next to the barometer.

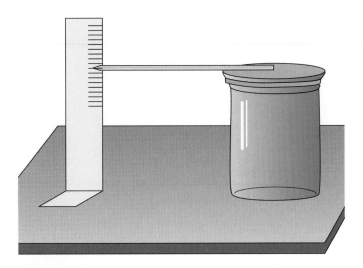

Figure 15.9 Construction of a simple barometer

24. Next, calibrate your barometer.
    a. If possible, find out what the air pressure is today. You may be able to access this information on the Internet or in a newspaper, using a commercial barometer, or your instructor may have the information. Mark on the card the present position of the straw pointer, and label it with the current pressure. Also take down the temperature. _____

    b. Place the barometer and a thermometer where the temperature is different from room temperature: in a refrigerator, a low-temperature oven set to 30°–35°C, or outside. After they have reached equilibrium, record the barometer and thermometer readings and place both instruments back in their original location.

    c. Calculate a linear temperature correction factor for your barometer using the following equation:

    $$CF = \frac{P_1 - P_2}{T_1 - T_2}$$

    where CF is the correction factor

    P_1 is the room pressure from your barometer

    T_1 is the temperature you measured in **a**

    P_2 and T_2 are the other pressure and temperature you measured in **b**

    To calculate a pressure with the correction, use the following equation:

    $$P = P_3 + CF\,(T_1 - T_3)$$

    where P is the corrected pressure

    T_3 and P_3 are the current temperature and pressure readings, respectively

    d. At the end of the lab, see whether the pointer's position has changed. If possible during the next few weeks, mark additional pressures on your barometer's scale, and observe how the readings change with the weather. Be sure to note the temperature and make temperature corrections using the same correction factor and equation in **c**.

25. Examine your barometer.
    a. Explain how it works.

    _____

    _____

    _____

    _____

    b. As pressure increases, will your barometer point higher or lower on the card? _____

c. What would you expect your barometer to do if the weather is getting worse and a storm is approaching?

_____

_____

d. Based on the earlier discussion of air pressure systems, how do you think you could use the barometer to predict the weather?

_____

_____

_____

_____

Fronts and Clouds

In subtropical to polar latitudes, masses of air with warmer or colder temperatures move across Earth's surface. Quite commonly these air masses have fairly sharp or distinct boundaries.

- A **front** is a boundary or transition zone between different air masses of different densities. Usually the density differences are due to a difference in temperature, with cold air being denser than warm air.
- At **warm fronts,** warmer, moist air wedges out cooler air. Condensation and precipitation occur where these air masses are in contact (Fig. 15.10a).

- In a **cold front,** colder air advances underneath warmer air and tends to have a steep, nearly vertical boundary (Fig. 15.10b).
- If air masses move so that neither the warmer nor the cooler air displaces the other, the front is called a **stationary front.** Air flow generally parallels this frontal system.
- If the cold front catches up to the warm front, the warm air is wedged upward, and an **occluded front** results (Fig. 15.11).

On opposite sides of a front, the wind direction, air temperature, and absolute humidity reflected by the dew point are all likely to be different. If sufficient moisture is present in either air mass, clouds are likely at the front.

If a warmer air mass pushes into a region occupied by a cooler air mass, a warm front marks the boundary. Because of the broad, wedge shape of a warm front (Fig. 15.10a), the precipitation associated with it tends to spread out and last longer as steady, sometimes drizzly rain or light snow. Clouds are likely to be fairly flat _stratus_ or _nimbostratus clouds_ (Fig. 15.12g and i) that get lower and lower as the front approaches. Winds change direction as the front passes, from southeast to southwest. A typical warm front is likely to last about 2 1/2 days from its first to its last clouds.

Where a cooler air mass displaces warmer air, a cold front is present. The steep boundary of a cold front (Fig. 15.10b) causes it to have shorter, more dramatic precipitation, with heavy showers and gusty winds. A cold front occurs in a trough of low pressure so that the pressure drops as the front approaches and rises behind it. Wind directions shift from southwest ahead of the front to northwest after it passes. Cold fronts and their associated weather tend to pass quickly, in a matter of a few hours.

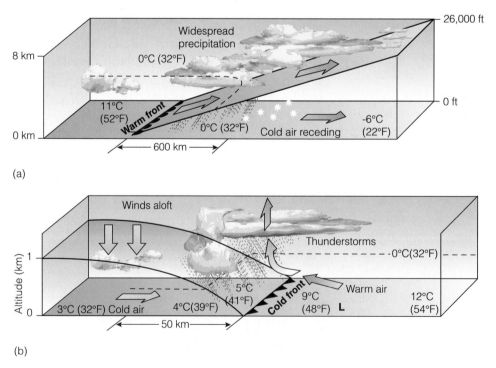

(a)

(b)

Figure 15.10 (a) Warm front, with stratus clouds and light widespread steady precipitation. The warm air mass on the left is overtaking the cold air mass on the right. (b) Cold front, with cumulonimbus clouds, showers, and heavy precipitation. The cold air mass on the left is overtaking the warm air mass on the right.

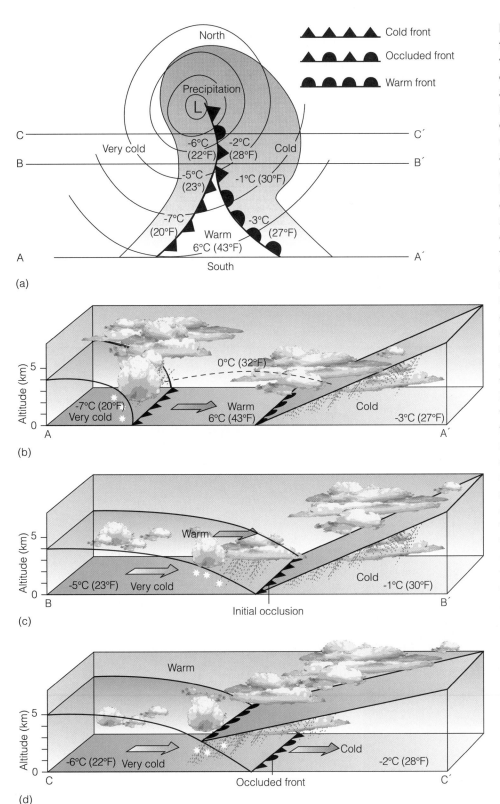

(a)

(b)

(c)

(d)

Figure 15.11 An occluded front produced by converging warm and cold fronts in a cyclonic system. Three air masses are interacting here; the slowest moving air mass is cold air in the east (right) with a warm air mass (middle, above) advancing on it and a colder air mass in the west (left) moving east even more rapidly. (a) A map showing an occluded front near the low pressure (L) of an extratropical cyclone, and separate cold and warm fronts to the south. (b), (c) and (d) can be viewed as a time sequence—first (b), then (c) then (d), or as cross sections through different places on the map in (a), corresponding to line A-A′, B-B′, and C-C′, respectively. (b) At first, a combination develops with a warm front ahead and cold front behind. Air masses listed from west to east (left to right) and from fastest moving to slowest moving are very cold, warm, and cold. (c) As the cold front catches up to the warm front, initial occlusion occurs where the very cold air mass first comes in contact with the cold air mass and the warm air mass is above the two. (d) The very cold air advancing on the cold air mass wedges the warm air upward off the ground producing the occluded front.

■ Warm front clouds progress from *cirrus* well ahead of the front, to *cirrostratus*, to *altocumulus*, to *altostratus*, to *nimbostratus*, to *stratus* and fog right at the front. A few *stratocumulus* clouds (Fig. 15.12) and fog occur behind the front.

■ Cold front clouds are *cumulus, cumulus congestus,* and *cumulonimbus clouds* (Fig. 15.12), and cold fronts have frequent thunderstorms. Ahead of the front *cirrus* then *cirrostratus* clouds (see b and d) give warning of its approach.

At a stationary front, the precipitation and weather disturbance may continue for some time in the same location, but generally they are quite light. A stationary front may gradually develop into a combination warm front ahead and cold front behind in an area of low pressure (Fig. 15.11b). At an occluded front, people on the ground feel two cooler air masses, either cool followed by cold or cold followed by cool, but do not experience the warmer air mass since it has lifted off the ground

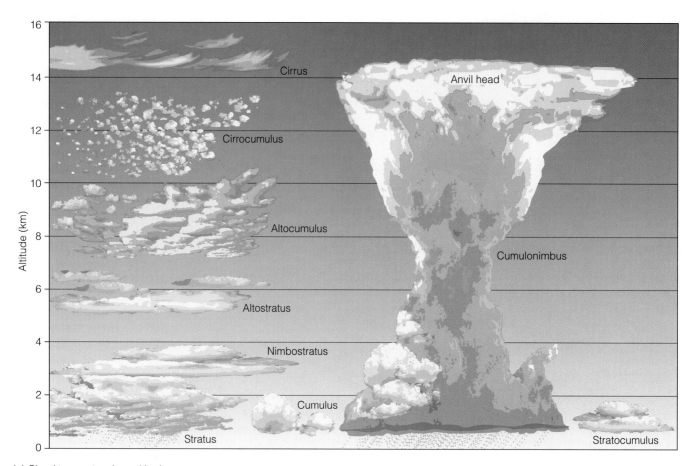

(a) Cloud types at various altitudes

(b) Cirrus clouds.

Figure 15.12　Classification of clouds

(c) Cirrocumulus clouds

(d) Cirrostratus clouds

(e) Altocumulus clouds

Figure 15.12 Classification of clouds *Continued*

(f) Altostratus clouds

(g) Nimbostratus clouds

(h) Stratocumulus clouds

Figure 15.12 Classification of clouds *Continued*

(i) Stratus clouds

(j) Cumulus humilis clouds (fair weather cumulus)

Figure 15.12   Classification of clouds *Continued*

(Fig. 15.11). The weather patterns at an occluded front tend to be similar to a warm front or to a warm front with a cold front at the tail end. Occluded fronts are common in the North Pacific, over the Great lakes, and in the North Atlantic.

26. If it is a cloudy or partly cloudy day, go outside and observe and identify the clouds you see. Refer to Figure 15.12 to classify the clouds. Write down your observations in Table 15.3. Observe and record as many of the other weather variables in the table as you have equipment to measure them. From the clouds and other information you observe, what type of weather system or pattern might be passing through, and where are you within that system or pattern?

_____

_____

Observe the clouds and weather for the next few days to support or disprove this hypothesis. Write down your observations in Table 15.3.

Weather Maps

The main features shown on a weather map are areas of precipitation, rain, snow, sleet, freezing rain, high and low pressure, isobars, temperatures, warm fronts, cold fronts, stationary and occluded fronts, and tropical cyclones (including hurricanes and tropical storms). The weather map will have different symbols for precipitation but will generally give a key showing the meaning of the symbols used. If the map is in color, different types of precipitation may be shown in different colors, as in Figure 15.13. Alternatively, many weather maps use pictorial representations of precipitation such as raindrops for rain and snowflakes for snow. These maps may use color to represent temperature.

27. Which types of precipitation occurred somewhere on the weather map in Figure 15.13? _____

_____

What types of precipitation are in the key but not occurring that day? _____   _____

(k) Cumulus congestus clouds

(l) Cumulonimbus cloud

Figure 15.12 Classification of clouds *Continued*

Table 15.3 Weather Observations

Date and Time	Identified Clouds and Their Probable Altitudes	Temperature (°C)[1]	Humidity or Dew Point	Precipitation	Wind Speed and Direction	Pressure (mb) and Pressure Trend[2]

[1]To convert from Fahrenheit to Centigrade, first subtract 32 then multiple by 5/9: °C = (°F − 32) × 5/9.
[2]Rising, falling, or steady.

The key includes one aspect of weather that is not really precipitation. Which one is it? _____ Did any of this occur anywhere on the map that day? _____

Atmospheric pressure is generally indicated in some way on a weather map, commonly with *H* for a region of high pressure and *L* for a region of low pressure. It may be measured in millibars (mb) or in inches of mercury. Fine lines on the map are commonly **isobars,** lines of equal pressure, somewhat like contours are lines of equal elevation. The pale lines on Figure 15.13 are isobars. Where isobars are close together, there is a rapid change in pressure that leads to high winds. Where they are far apart, the region is calm or has light winds. At the surface, the winds cross the isobars at a slight angle, heading away from highs and toward lows.

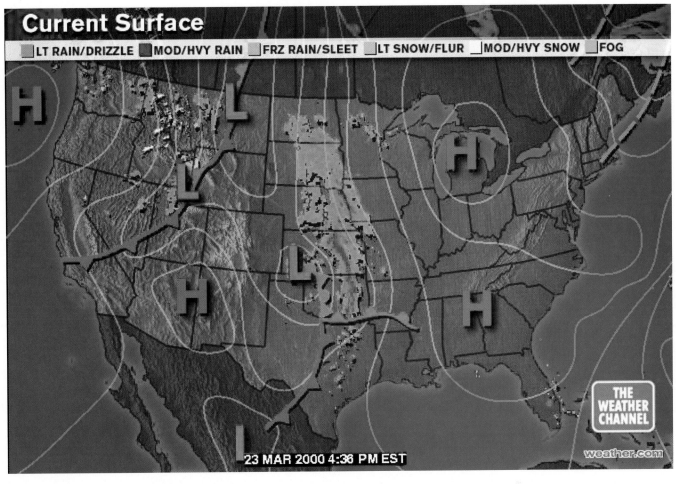

Current Surface

LT RAIN/DRIZZLE MOD/HVY RAIN FRZ RAIN/SLEET LT SNOW/FLUR MOD/HVY SNOW FOG

THE WEATHER CHANNEL

weather.com

23 MAR 2000 4:36 PM EST

Figure 15.13 Weather map for March 23, 2000 with four types of fronts

28. Draw in some wind arrows on the map in Figure 15.13, keeping in mind that winds rotate clockwise around highs and counterclockwise around lows in the Northern Hemisphere (Fig. 15.8).

29. Draw a profile on Figure 15.14 of the air pressure across the United States from the Utah–Arizona state line at Nevada to the Virginia–North Carolina state line at the Atlantic Ocean. The isobar interval is 0.1 inches of mercury (in.Hg),[4] and the isobar nearest the corner of Utah, Arizona, Colorado, and New Mexico is 30.2 in.Hg. Where along your profile was the pressure the lowest on

March 23, 2000? _____

The highest? _____

··········
[4]One millibar = 0.02953 in. of mercury, or 1 in. of mercury = 33.86 mb.

30. What types of fronts are shown on the map in Figure 15.13? Name them and draw the symbol for each.

_____

_____

_____

_____

31. Describe the weather on March 23, 2000, from the weather map in Figure 15.13 in your hometown. (If your hometown is not on the map, describe the weather in

eastern Kansas.) Hometown: _____ Weather:

_____

_____

30.5 ————————————————————
————————————————————
————————————————————
————————————————————
————————————————————
30.0 ————————————————————
————————————————————
————————————————————

Figure 15.14 Pressure profile across the United States from the Utah–Arizona state line to the Virginia–North Carolina state line. Pressure is in inches of mercury (in. Hg).

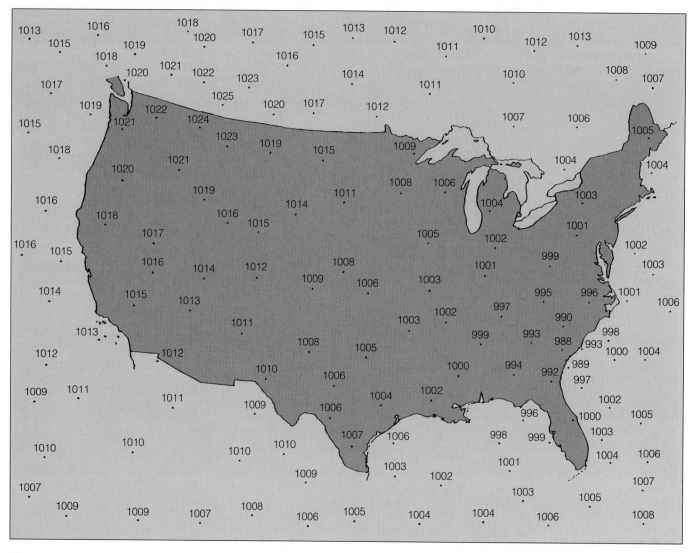

Figure 15.15 Barometric pressure readings across Continental United States. Values are given in millibars adjusted to sea level.

32. Figure 15.15 is a map showing scattered pressure readings in millibars in and around the United States. Draw isobars on the map using a 5-mb isobar interval. Locate the lows and the highs. Draw cold front symbols for any pressure troughs (which would look like valleys if this were a topographic contour map). Where are the winds the fastest? Write "windy" there on the map. Where is it least windy? Write "calm" there on the map. Use a blue pencil to lightly shade in where you think precipitation is falling.

33. Using a weather map for today from the newspaper or from an Internet site, predict what the weather will be like tomorrow, without looking at the weather forecast.

_____

_____

_____

_____

Atmospheric Chemistry

The atmosphere is a mixture of gases, primarily nitrogen (N_2), then oxygen (O_2), and down to argon (Ar) at just under 1%. All other gases amount to substantially less than 1%, except water vapor (H_2O), which varies from 0% to 4%. Carbon dioxide (CO_2), a gas much discussed, would seem to be abundant but is only about 0.037%, although its abundance is rising.

Experimental Measurement of the Amount of Oxygen in the Atmosphere

Read through the whole experiment before you start.

Materials needed:

- Thin candle
- Plastic millimeter ruler
- 1000-ml beaker or jar

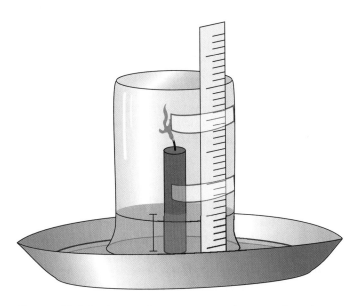

Figure 15.16 Setup of oxygen measuring experiment. Make note of the height of the water inside the beaker at the end of the experiment.

- Large graduated cylinder
- Wide, shallow dish
- Tape
- Water
- Matches or lighter

Securely tape the ruler to the 1000-ml beaker or jar so that one end of the ruler is flush with the top of the beaker. Attach the candle to the inside of the wide, shallow dish with a little melted wax. Pour water in the dish around the candle so the water is an inch or so deep. Light the candle, if you haven't already.

34. Smoothly and without hesitation, turn the beaker upside down over the candle. All parts of the beaker's rim should touch the water at the same time to trap a beaker full of air. Watch what happens as the candle uses up the oxygen in the beaker. When the candle goes out, wait until the water level inside has reached its maximum. (The setup is shown in Figure 15.16.) Then record the level of the water inside the beaker using the ruler's gradations.

_____

35. We need to know the volume of the whole candle and the volume of the bottom part of the candle that was immersed in water after it went out. Measure the length

_____ and diameter _____ of the candle to calculate its volume. Use the formula $\pi LD^2/4 = 3.14LD^2/4$, where L is a length measurement and D is the diameter. Or measure the volume directly by immersing the candle in a graduated cylinder. The volume of the whole candle is _____

36. Hold the candle next to the ruler on the beaker, and mark the water level from Exercise 34 on the candle. Immerse the candle up to this mark to determine the volume of the candle that was underwater at the end of the experiment. Or you can measure the length of this part of the candle and use the formula in Exercise 35 to determine the volume.[5] What is the volume of the candle that was immersed in water at the end of the experiment?

_____

37. Use the graduated cylinder to carefully measure the volume of the beaker to the pour spout _____

and the volume to the ruler value in Exercise 34.

_____

38. Determine the initial gas volume: beaker volume − candle volume _____

39. Determine the volume of gas lost (replaced by water): beaker volume − volume to ruler measurement determined in Exercise 37 − immersed candle volume

_____

If wax burns completely, it uses up three oxygen molecules and gives off two carbon dioxide and two water molecules. The water starts out as vapor, but most of it quickly condenses into liquid water. The result is that the burning process uses up oxygen that is replaced by another gas, CO_2, in the proportion 3:2. The volume loss during the burning process is one gas molecule lost for every three original oxygen molecules. This means that ideally the proportion of oxygen in the atmosphere is three times the proportion of volume lost in the experiment. Of course, in practice the wax does not burn perfectly. Black soot coming from the candle, especially at the end of the experiment, is carbon, which is solid. For every carbon instead of carbon dioxide molecule, the ratio 3:1 goes down. Also, carbon monoxide may be produced, which would reduce the volume loss ratio.

40. Calculate the percentage of oxygen in the atmosphere as 3 × (volume of gas lost)/(initial gas volume) × 100.

_____ All answers should be recorded on the blackboard or overhead. Average all the answers obtained. _____

41. If all the other gases in the atmosphere add up to 1%, what is the percentage of nitrogen in the atmosphere?

_____ Now look up the correct answers in your textbook, or get the values from your instructor.

·············

[5]Unless your ruler starts at 0 right at its end, the measurement in Exercise 34 will not be the length of the candle underwater.

Oxygen % = _____ Nitrogen % = _____

How do these values compare to your results?

_____

What could account for variability of the results and any difference from the correct value?

_____

_____

_____

_____

Greenhouse Effect

The **greenhouse effect** is a process that causes warming of the atmosphere (Fig. 15.17) and is closely tied to atmospheric chemistry and specific atmospheric gases called *greenhouse gases*. The three main constituents of the atmosphere—nitrogen, oxygen, and argon—are not greenhouse gases. This means that less than 1% of the atmosphere is made up of the important, climate-influencing greenhouse gases. The greenhouse effect has to do with Earth's energy budget, so understanding it requires knowledge of the energy that comes to Earth from the sun and the energy that leaves Earth for outer space. For Earth to maintain a constant temperature, the energy that comes in must balance energy going out (Fig. 15.17). Both of these are forms of energy that are part of the electromagnetic spectrum, which includes gamma rays,

X rays, ultraviolet rays, visible light, infrared radiation, microwaves, and radio (and broadcast television) waves. All these different rays and waves are part of a continuum with different wavelengths, where gamma rays have the shortest wavelengths and are the most energetic and dangerous, and radio waves have the longest wavelengths and are the least energetic. Let's start by looking at the different types of energy coming from the sun and Earth.

Incoming and Outgoing Radiation

Radiation from the sun that reaches Earth is mostly visible light (Fig. 15.18a). This radiation adds energy to the Earth when it is absorbed and converted to heat at the Earth's surface. Except for reflected light, the Earth radiates energy toward space within the longer wavelengths of the far infrared (Fig. 15.18a). The longer wavelengths from Earth interact differently with Earth's atmosphere than the shorter wavelengths from the sun. Also the incoming and outgoing radiation must balance if Earth is to maintain a constant temperature (Fig. 15.17).

42. Use Figure 15.18a to answer the following questions:
   a. More than 80% of radiation from the sun is what two

   types of radiation? _____ and

   _____, with wavelengths rang-

   ing from _____ to _____ microns

   b. What type of energy is reradiated (not reflected) from

   the Earth's surface? _____

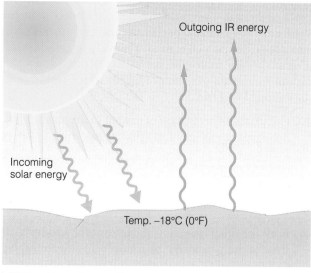

(a) Without greenhouse effect

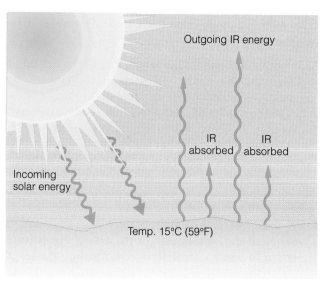

(b) With greenhouse effect

Figure 15.17 Greenhouse effect. (a) In a world without a greenhouse effect, outgoing infrared radiation (IR) would not be trapped, and the balance between incoming short-wavelength visible light energy from the sun and outgoing longer-wavelength IR radiation from the Earth would produce an average surface temperature of −18°C (0°F)—cold enough for the oceans to freeze. (b) With greenhouse gases in the atmosphere, the atmosphere absorbs some of the outgoing IR radiation and reradiates it back to the surface. This produces the present average surface temperature of 15°C (59°F).

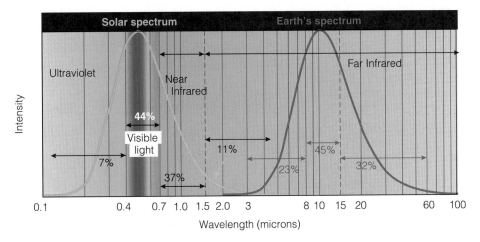

(a)

Figure 15.18 (a) Spectra of the sun and Earth. The sun's maximum radiation is visible light at about 0.5 microns, equivalent to blue-green. Earth's maximum is at about 10 microns, in the far infrared. (b) Atmospheric absorption spectra of greenhouse gases in the atmosphere. For each gas, the vertical axis indicates the proportion of absorption from 0 to 1, where 1 would be 100% absorption. Gray areas show the wavelengths that are absorbed by the particular gas. Colored regions have little or no absorption. The lowest section of the diagram, labeled "atmosphere," shows all the greenhouse gases added together.

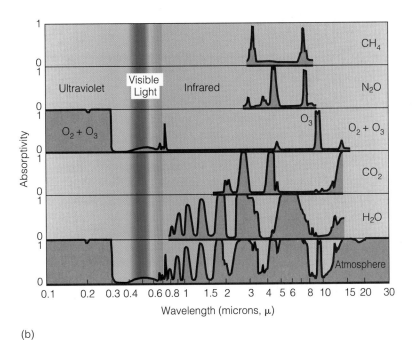

(b)

c. What is Earth's peak radiation wavelength? _____

The greenhouse effect is a trap that keeps some solar energy absorbed by the Earth from returning to space. Without the greenhouse effect, the oceans would have frozen billions of years ago, and life as we know it would not exist on Earth. Any trap needs to allow the thing being trapped to come in but then needs to prevent it from going back out. To be a trap, the greenhouse effect depends on the transformation of the energy from mostly visible light as it comes in from the sun to infrared radiation as it leaves the surface. Particular gases in the atmosphere, called greenhouse gases, have the property that allows shorter-wavelength visible light through but absorbs the longer-wavelength infrared rays near the 10-micron wavelength—the wavelength of peak radiation that Earth gives off. To be a greenhouse gas, the gas must do both of these things.

Greenhouse Gases

43. The graphs in Figure 15.18b show the wavelengths that are absorbed by different gases in the atmosphere. Using the graphs, decide which gases in Table 15.4 have any substantial absorption (tall peaks) in the wavelengths indicated. Circle either *yes* or *no* for the appropriate wavelengths. For example, does methane have any tall peaks anywhere within Earth's reradiation wavelengths (from 3 to 60 microns)? Looking at the top part of Figure 15.18b, you see methane (CH_4) has two tall peaks, one near 3 microns and one near 7 microns, so for methane the answer is yes. Circle *yes* for the appropriate column for methane in Table 15.4. Continue with all the other gases and other wavelengths requested.

Table 15.4 Absorption Characteristics of Some Gases

Gas	ABSORPTION AT DIFFERENT WAVELENGTHS		
	Infrared (anywhere within Earth's reradiation wavelengths)	Infrared, Close to Earth's Peak (7–12 microns)	Visible Light (0.4–0.7 microns)
Methane (CH_4)	yes / no	yes / no	yes / no
Nitrous oxide (N_2O)	yes / no	yes / no	yes / no
Oxygen (O_2) and ozone (O_3)	yes / no	yes / no	yes / no
Carbon dioxide (CO_2)	yes / no	yes / no	yes / no
Water vapor (H_2O)	yes / no	yes / no	yes / no

44. Draw conclusions from Table 15.4.
   a. Which of these gases allow visible sunlight through?

   _____

   _____

   b. Which absorb infrared?

   _____

   _____

   c. Which are greenhouse gases?

   _____

   _____

45. Which of the three gases is the weakest greenhouse gas?

   _____

   Which the strongest? _____ Why?

   Hint: You need to think about Earth's wavelengths and which gases come closest to absorbing the most intense of Earth's radiation.

   _____

   _____

   _____

   Note: The diagram does not show chlorofluorocarbons (CFCs), which are even stronger greenhouse gases.

46. Which greenhouse gases contain carbon? _____

   _____

Although weak as a greenhouse gas, carbon dioxide is the most important, because, at 0.037%, it is the most abundant. Two of the most important greenhouse gases are naturally occurring compounds that contain carbon: carbon dioxide and methane. In Lab 17, we will look at their natural and artificial production as part of the carbon cycle; in the next exercises, we look at how the quantities of these compounds in the atmosphere have changed in the past in relation to temperature.

Greenhouse Gases and Temperatures of the Past

If we had a time machine, we could hop in and go back to a time in the past and collect samples of the atmosphere to see how it has changed. Time machines are science fiction, but we do have something that was in the past and did sample the atmosphere—ice. When ice forms in glaciers, it traps small bubbles of air—small air samples. The ice itself has hydrogen and oxygen molecules that can tell us about the past temperatures. Here is how it works.

Recall from Lab 8 that an *isotope* is an atom with a different weight than other atoms of the same element. For example, hydrogen normally has only one proton in its nucleus, weight of 1, but it can have one proton and one neutron, giving it a weight of 2. This type of hydrogen is called **deuterium.** Oxygen normally has 8 protons and 8 neutrons, with a weight of 16, but occasionally it has 8 protons and 10 neutrons, giving it a weight of 18. The normal oxygen is called *oxygen-16,* or ^{16}O, and the heavier one is called *oxygen-18,* or ^{18}O. These isotopes are not radioactive; they are stable.

These hydrogen and oxygen isotopes are not useful in radiometric dating, but can tell us the average temperature of the atmosphere. The heavier atoms, deuterium and oxygen-18, make water heavier than normal. Heavier water is less likely to evaporate or, if already in the vapor form, is more likely to condense than lighter water. The warmer the temperature, however, the faster the molecules are moving and the more of the heavier water gets vaporized. The proportions of these isotopes if locked in ice can tell us the past temperatures of Earth's atmosphere.

Figure 15.19a, b, and c shows the relationships among age, temperature, carbon dioxide, and methane in the Vostok ice core extracted from Antarctica. Table 15.5 gives temperatures, carbon dioxide, and methane values at depths every 200 m in the same ice core. Temperatures were determined from deuterium values in different layers of the ice.

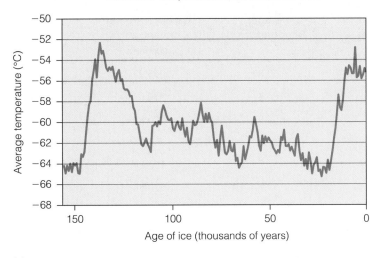

(a)

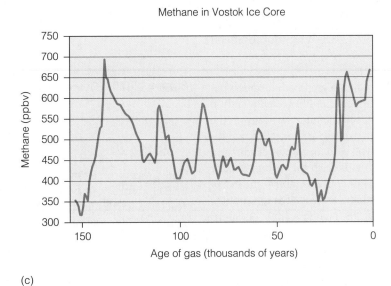

(b)

(c)

Figure 15.19 Vostok Ice core extracted from Antarctica. (a) Temperature versus age. Temperatures are determined from deuterium values in different layers of ice in the Vostok ice core. The age of the ice is calibrated at the top by analyzing annual layers and by theoretical model of how the ice flows for the bottom layers. (b) Carbon dioxide versus gas age. The age of the gas is based on a theoretical estimate of how deep the ice must be to close up holes and trap the gas. (c) Methane versus gas age.

Table 15.5 Measurements from the Vostok Ice Core of Antarctica

Depth in Ice Core (m)	Average Temperature (°C)	Carbon Dioxide Concentration (ppmv[1])	Methane Concentration (ppbv[2])
200	−52.8	263.73	591.92
400	−64.4	220.02	424.48
600	−62.9	220.81	430.79
800	−61.3	207.7	484.84
1000	−62.7	192.84	425.41
1200	−59.1	220.98	524.93
1400	−59.7	227.18	406.5
1600	−60.2	242.53	457.25
1800	−53.8	267.22	594.07
2000	−63.9	191.69	369.66

[1]ppmv = parts per million by volume
[2]ppbv = parts per billion by volume

47. Examine the graph in Figure 15.19a, looking at broad changes, not tiny peaks. Why do you think all the temperature values are so low? _____

_____What two times had the warmest climate? _____ _____ What two times was the climate the coldest? _____ _____ Which of these four times do you think might have corresponded to ice ages?

_____

48. Compare the graphs in Figure 15.19a, b, and c. Does there appear to be a relationship between carbon dioxide and temperature? _____ Between methane and temperature? _____ Does this relationship match or contradict what you know about the greenhouse effect?

_____

49. Use the data in Table 15.5 to draw a graph on the graph paper in Figure 15.20. Plot temperature horizontally, carbon dioxide vertically on the left, and methane vertically on the right. Label your graph appropriately, and make a key indicating what symbols/colors you used for carbon dioxide and methane. Draw a straight line through the carbon dioxide points so that equal numbers of points are above and below the line. Do the same for methane. Does your graph confirm or contradict your observations in the previous question?

_____ Explain your results.

_____

_____

_____

_____

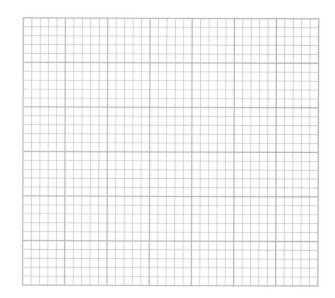

Figure 15.20 Graph paper

Ozone Layer

The process of ozone formation in the stratosphere (Fig. 15.21) absorbs ultraviolet (UV) rays, preventing them from reaching Earth's surface where they could do considerable damage. Ultraviolet radiation damages chromosomes (RNA and DNA), causes sunburns, and can cause skin cancer.

People often talk about the ozone layer being "thin," as a result, many students think ozone is contained within a layer that is small in its vertical dimension. The layer containing 97% of atmospheric ozone, the stratosphere, is about 30 km thick. "Thin" means that ozone is in a low concentration, that

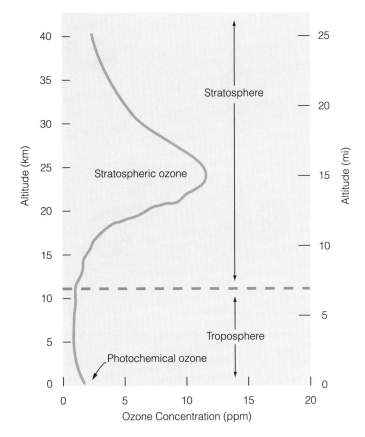

Figure 15.21 Average distribution of ozone in mid-latitudes above the Earth's surface. The ozone layer is the region of the stratosphere where ozone concentrations are higher. Even the highest concentrations of ozone are quite low—only about 12 mg of ozone for every 1 million mg of air (12 ppm). Ozone in the troposphere is a component of smog.

is, it is dilute. Even where ozone is most concentrated, at about 25 km up, it is only 12 ppm or 0.0012% of the atmosphere (Fig. 15.21). Nonetheless, just this little bit of ozone prevents most ultraviolet rays from reaching Earth. However, a decrease of ozone concentration of 1% allows enough increase in UV rays to result in a 2% to 5% increase in incidents of skin cancer.

Formation of the Ozone Layer

How does the ozone layer form? **Ozone** is just an oxygen gas with an extra oxygen atom: (O_3) compared to normal oxygen (O_2). Both oxygen and ozone absorb ultraviolet rays; look at the absorption of UV rays in Fig. 15.18b. Oxygen molecules (O_2) absorb UV from the sun and form ozone in the process. In the stratosphere, UV rays hit oxygen gas molecules and split them apart. The resulting single oxygen atoms (O) join with oxygen molecules to make ozone. Ozone is naturally unstable and will break down if it hits another oxygen atom or another ozone, in both cases producing ordinary oxygen molecules. But the interesting and useful aspect of ozone is that it naturally absorbs UV rays and in doing so breaks back into an oxygen molecule and an oxygen atom.

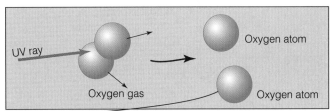
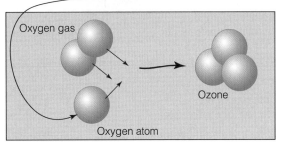

Figure 15.22 Chemical reactions that form ozone in the ozone layer

Figure 15.22 and the following reactions depict the formation of ozone:

$$O_2 \xrightarrow{\text{UV}} O + O \quad (1)$$ This reaction requires energy from the ultraviolet radiation to occur.

$$O + O_2 \longrightarrow O_3 \quad (2)$$ This reaction takes place spontaneously, needing no additional energy.

Once ozone has formed, it does not just sit around but breaks down as it absorbs ultraviolet radiation by Reaction 3 (see also Fig. 15.23). Although this reaction destroys ozone, this is not the ozone-destroying reaction that people make such a fuss about. This is just part of the natural cycle of ozone formation and destruction that takes place continuously in the ozone layer.

$$O_3 \xrightarrow{\text{UV}} O + O_2 \quad (3)$$ This reaction absorbs ultraviolet radiation, protecting us from UV's effects.

Ozone Formation Activity

50. Your lab instructor will provide you with materials for making molecular models, such as Styrofoam balls and toothpicks. Work in a group of three or more people. First, simulate the processes of plants or algae, and build about 10 oxygen molecules, putting aside any extra or leftover balls. Next, one person should simulate ultraviolet radiation from the sun that breaks apart oxygen molecules. One person should react oxygen atoms with oxygen molecules. Use Figure 15.22 and Reactions 1 and 2 to help you. One person should simulate ultraviolet radiation from the sun that breaks apart ozone. Figure 15.23 shows this Reaction 3. Which reactions require molecules from other reactions before they can proceed?

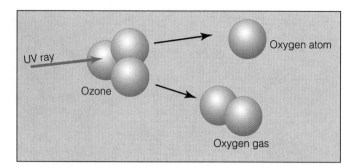

Figure 15.23 Natural destruction of ozone and absorption of ultraviolet radiation

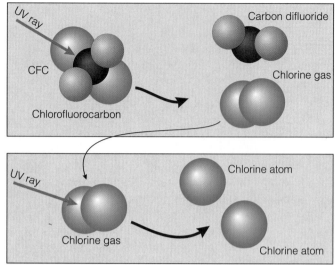

Figure 15.24 Chemical reactions that turn CFCs into chlorine atoms in the ozone layer

Summarize the natural cycle of ozone production and breakdown in your own words. Save your model molecules for the next activity.

_____

_____

_____

_____

For every ozone molecule you produced in the ozone formation activity, there would be about 16,700 oxygen molecules. It is not very likely that an ozone molecule would run into another ozone molecule and thereby destroy both to make three oxygen molecules, although this does happen to a few of them.

Ozone Destruction

The breakdown of CFCs in the stratosphere ultimately causes ozone destruction. Chlorofluorocarbons are manufactured molecules containing chlorine, fluorine, and carbon. A variety of types have been produced, with different proportions of the elements. We will use a CFC that has two chlorines, two fluorines, and one carbon as our example. CFCs gradually diffuse upward in the atmosphere. They break apart into chlorine gas (Cl_2) and the remaining fluorines and carbon when hit by ultraviolet radiation at about 30 km altitude. Figure 15.24 and Reaction 4 show this reaction.

$$CF_2Cl_2 \xrightarrow{UV} Cl_2 + CF_2 \quad (4)$$

The chlorine gas molecules are further broken down into chlorine atoms by more ultraviolet radiation (Fig. 15.24).

$$Cl_2 \xrightarrow{UV} Cl + Cl \quad (5)$$

If a chlorine atom bumps into an ozone molecule, it snatches an oxygen atom from the ozone and becomes chlorine monoxide (ClO), and the ozone turns into an ordinary oxygen molecule (Fig. 15.25, darker gray ellipses).

$$Cl + O_3 \longrightarrow ClO + O_2 \quad (6) \text{ One ozone molecule is destroyed.}$$

But it does not stop there. The chlorine monoxide is highly reactive, and if it encounters a free-floating oxygen atom (one that would have made an ozone molecule in Reaction 2), the oxygen atom combines with the oxygen in the chlorine monoxide to make an ordinary oxygen molecule and releases the chlorine atom (Fig. 15.25, lighter gray ellipses).

$$ClO + O \longrightarrow Cl + O_2 \quad (7) \text{ One ozone molecule is not made.}$$

Now the chlorine atom is free to go back to Reaction 6 and destroy another ozone molecule. The chlorine is used again and again until it combines into a different molecule. One Cl atom destroys on average 100,000 ozone molecules!

As the ozone layer has diminished in the last 45 yr, the incidence rate of skin cancer has increased 200% in the United States, and the increase in mortality from skin cancer is up 150%. As more ozone has been destroyed, very low ozone concentrations over Antarctica, known as the _ozone hole,_ occur in the spring. Conditions over Antarctica are complicated by factors such as stratospheric winds, ice crystals, and the absence and return of sunlight, all of which play a role in the reduction of ozone levels. There is no ozone hole over the Arctic, but values there have declined as well.

Ozone Destruction Activity

51. Again work in a group of three or more. Use your oxygen and ozone molecules from the previous activity, and get new balls to represent chlorine. One person (different than before) should continue to make oxygen atoms and ozone molecules. One person should break down ozone using a chlorine atom (Reaction 6 and Fig. 15.25). The third person should use up some of the oxygen atoms to break apart chlorine monoxide as in Reaction 7 and Figure 15.25. Which reactions depended on the products

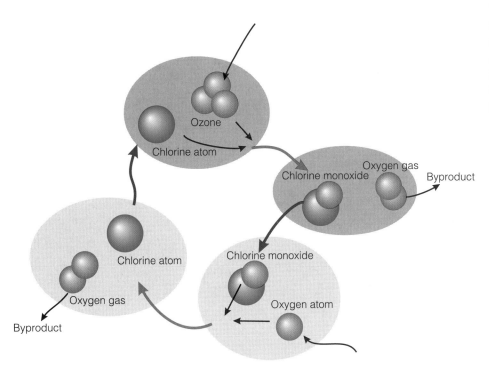

Figure 15.25 Chemical reactions that destroy ozone in the ozone layer. Red arrows signify the chemical reaction; blue and black arrows signify movement of molecules.

Ozone

Chlorine atom

Chlorine monoxide

Oxygen gas

Byproduct

Chlorine atom

Oxygen gas

Byproduct

Chlorine monoxide

Oxygen atom

from other reactions? _____

_____

How many chlorine atoms does it take to destroy, say, 10 ozone molecules and prevent another 10 ozones from forming? _____

Summarize the cycle of ozone destruction in your own words.

_____

_____

_____

_____

Geologic Maps

In this chapter we return to the study of the solid Earth to combine and extend the concepts introduced in Lab 7 (rock masses and their deformation) and Lab 11 (topographic maps and aerial photographs). A geologic map exhibits the underlying rocks and the geologic relationships between them, and it is usually superimposed on a topographic map of the region. Geologic maps combine landforms with characteristics of the rock bodies below. Many nations, and most states and provinces within countries, support a geologic survey that has the directive to produce and distribute geologic maps to aid in research and land development. Such maps are essential in mineral exploration, environmental planning, and civil engineering projects.

What Is Shown on a Geologic Map

Geologic maps symbolically represent the geology at or just below the Earth's surface by showing the distribution of the rock bodies and structural features. For sedimentary and volcanic rocks, the main bodies shown are formations. Any metamorphic and intrusive igneous rock bodies are also shown. Different rock bodies are usually distinguished by representing them as different colors or patterns and letter codes on the map. In the case of maps not printed in color, different

black-and-white patterns may be used (see Fig. 7.2 and an example map using this system in Fig. 7.31). In the letter system employed by the U.S. Geological Survey, the first letter in uppercase for a sedimentary formation indicates the geologic period in which the formation was deposited; subsequent letters in lowercase are an abbreviation of the formation's name (Table 16.1). Some of the other typical types of symbols used on geologic maps are also illustrated in Table 16.1.

An important part of any geologic map is the legend. This gives the reader of the map both geographic information (scale, contour interval, north arrow, etc.) and geological information. The colors and symbols used for formations and other rock bodies in the map are shown as a series of boxes arranged from oldest at the bottom to youngest at the top (the same way they were originally deposited).

Many geologic maps also include geologic cross sections. A cross section is like a topographic profile, but it also shows the underground arrangement of the rock masses, how much they are dipping, and whether they are folded, faulted, or intruded. In other words, it shows how the geology would look if it were possible to make a huge vertical cut in the area. Many areas of high relief—mountains, gorges, and cliffs—display the rocks as if they were sliced in the same manner that a cross section attempts to show. Mountainous areas may have road cuts where the terrain has been sliced through for highways, and these often are actual cross sections of the geology of the region (Fig. 16.1). The line of a cross section is usually drawn on the map, and letters such as A, A′ indicate the ends of the section. Geologic cross sections are actually constructed from the information on the map.

Relationship between Contour Lines and Contact Lines

In layered rock bodies, such as sedimentary or volcanic formations, the contact between any two formations is an approximate plane. Erosion lowers the Earth's surface to form an irregular surface that eventually cuts through a contact. Since two surfaces, the Earth's surface and the contact in this case, intersect to form a line, the contact appears on a map as a line that separates the two formations. This line is called the **trace** of the contact. Any layered rock body has two contacts: the lower bounding surface and the upper bounding surface.

Table 16.1 Geologic Map Codes and Symbols

Category	Item to Appear on Map	Colors, Codes, and Symbols	Example Symbols
Rock units	Rock formations, intrusions and other rock units	Color or pattern fills	See example pattern fills in Fig. 7.2
		Letter code: Geologic period and formation abbreviation Example: Permian age Kaibab Formation	Pk
Contacts	Contacts between rock bodies	Fine black lines	
	Uncertain contact	Dashed black lines	
	Covered or obscured contact	Dotted black lines	
Faults[1]	Faults, undifferentiated	Thick black lines	
	Normal faults	Tick on down-dip side, which is also the down-dropped side U on the up-faulted side and D on the down-dropped side	
	Reverse faults	Tick on down-dip side, which is also the up-faulted side U on the up-faulted side and D on the down-dropped side	
	Thrust faults	Teeth on the upthrust (hanging wall) side	
	Strike-slip faults	Half arrows indicate the sense of motion	
Strike and dip	Strike and dip of bedding	Longer line showing bedding strike with perpendicular short tick in center showing dip (down direction)	
	Strike and dip of vertical bedding	90° dip. Dip line symmetrically crosses the strike line	
	Strike and dip of overturned bedding	Dip direction indicated with a J. The down-dip direction is the straight part of the J.	
	Strike and dip of foliation	Longer line showing foliation strike with triangle at center showing dip	
Folds[1]	Upright anticline fold axis	Long line along the axis, crossing symbol shows direction of the dip away from the axis	
	Upright syncline fold axis	Long line along the axis, crossing symbol shows direction of the dip toward the axis	
	Plunging anticline and syncline	Arrows on the end of the axis show the direction of the plunge (down tilt of the axis)	
	Overturned folds	U shape with arrows pointing in the dip direction; toward the axis for synclines and away from the axis for anticlines	

[1]Fault and fold axis symbols may also be dashed where uncertain and dotted where covered.

Figure 16.1 Road cut through a syncline along Interstate 68 at Sideling Hill, Maryland

Horizontal Surfaces   The shapes of the contact traces of horizontal formations follow the topographic contours (Fig. 16.2a). Since the contact between any two formations is horizontal, it acts like a "natural" contour where it intersects the ground surface. The contact traces are therefore approximately parallel to the contours. Both contours and contact traces have a winding, or *sinuous,* pattern.

The thickness of a formation is the distance between the bottom and top boundaries of the formation, measured perpendicular to the contact surfaces. Thus, in areas of horizontal strata, it is easy to determine the thickness of a formation. The thickness is simply the difference in height, determined from the contours, between the trace of the upper and lower boundaries of the formation. Both contacts may not appear on the map, in which case only a minimum thickness can be determined.

Dipping Surfaces   In beds or other planar features that are inclined, or dipping,[1] the contacts cross the topographic contours because the feature is at different elevations in different areas. In Figure 16.2b and c, the map pattern of the beds form a V or arrowhead that points toward the down-dip side. This is the rule of V's. The **rule of V's** states that the V formed by the surface trace of a planar feature *in a valley* points in the direction of dip of the feature[2] (Fig. 16.2). Thus, if you determine where valleys are, you can determine the dip direction. This is true whether the feature is a sedimentary bed, a volcanic layer, a sill, a dike, or a fault. Even if the map does not have topographic contours, valleys can still be recognized because rivers and streams occupy them.

··············
[1]Review strike and dip from Lab 7 if necessary.

[2]A corollary to the rule of V's is that the contact traces of beds crossing a sharp *ridge* will V in the direction *opposite* to the dip.

In map view, if the beds are vertical planes (Fig. 16.2d), the contact is observed end on and so appears as a straight line *independent of the topography.* The steeper the dip, the straighter (or less sinuous) is the map pattern crossing a valley or ridge. In general, more sinuous map patterns indicate low dips; straighter map patterns indicate steeper dips.

1. Examine the Geologic Map of the Bright Angel Quadrangle in Figure 7.1.
   a. Study the contact between the Muav Formation (€m) and the Redwall Limestone (Mr). What is the relationship between the contact and the contour lines? _____

   b. Are the Redwall Limestone and the Muav Formation approximately horizontal or vertical, or are they dipping at a significant angle? _____ If dipping, what is the dip direction? _____

   c. What is the approximate elevation at the contact between the Muav Formation and the Redwall Limestone (round to nearest 100 ft)? _____ The Redwall Limestone is a shallow marine sedimentary rock; how do you think these rocks arrived at this elevation? _____
   _____

   d. What is the elevation of the contact between the Bright Angel (€ba) and Muav Formation?

   _____

   e. From the information in **c** and **d,** what is the thickness of the Muav Formation? _____

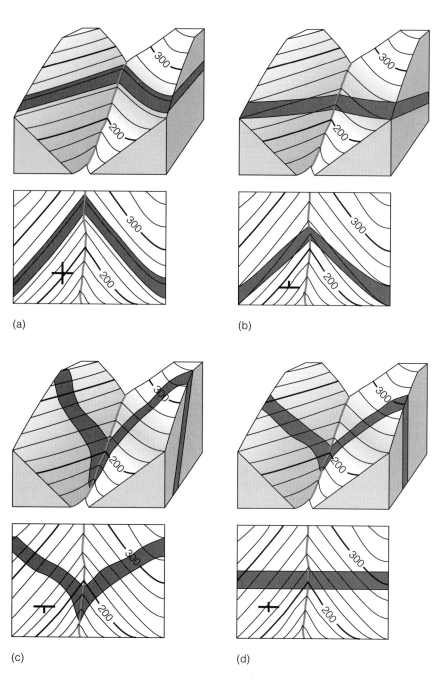

(a)

(b)

(c)

(d)

Figure 16.2 Rule of V's. General rule: the V points in the direction of the dip. All of these block diagrams and maps illustrate areas of identical surface topography. Superimposed on each are the traces of the contacts of a bed, dike, or sill. (a) Horizontal features: the contacts parallel the contours. Observe the symbol for horizontal beds on the map. (b) Feature dipping upstream. Notice the strike and dip symbol on the map. The V made by the contacts points upstream. (c) Feature dipping moderately to steeply downstream. Observe the strike and dip symbol on the map. The V made by the contacts points downstream. (d) Vertical feature. Look at the vertical strike and dip symbol on the map. Contacts of vertical beds, dikes, or other features are not deflected as they cross the stream valley.

f. Is the contact between the Vishnu Schist (p€v) and the Brahma Schist (p€bs) gently or steeply dipping?

_____ Is the foliation in these schists parallel or at a noticeable angle to the contact?

_____

2. Now look at the Geologic Map of Glen Creek, Montana, in Figure 16.3. Using the rule of V's, determine the dip direction of the contacts. _____

3. Compare the northwestern and southeastern parts of the Geologic Map of Pennsylvania in Figure 16.4. In the northwest, are the rocks steeply or gently dipping?

_____ In the southeast? _____

What leads you to these conclusions? _____

_____

_____

_____

Folds on Geologic Maps

In Lab 7, we saw that regional compression or shortening of an area results in folding of the originally horizontal beds (Fig. 7.29). The "pleat" of the fold, or its **hinge** or **axis,** is the line around

Figure 16.3 (*Opposite*) Part of the Geologic Map of Glen Creek, Montana

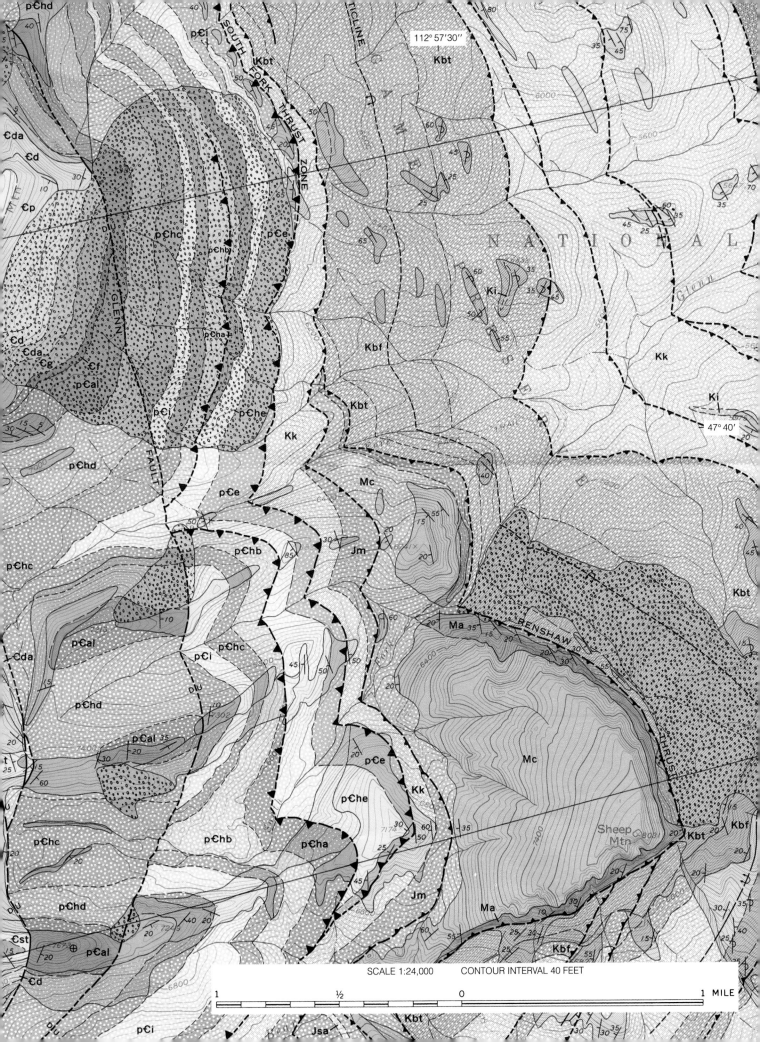

SCALE 1:24,000   CONTOUR INTERVAL 40 FEET

GEOLOGIC MAP OF GLEN CREEK, MONTANA

EXPLANATION

LOWER PLATE

Ktm — Two Medicine Formation

Kv — Virgelle Sandstone

Kt — Telegraph Creek Formation

(Montana Group)

Kmk / Kmsf / Kmf / Kmcf — Marias River Shale

(Colorado Group)

UNCONFORMITY

Kbv / Kbf / Kbf — Blackleaf Formation

UNCONFORMITY(?)

Kk — Kootenai Formation

UNCONFORMITY(?)

Jm — Morrison Formation

Js — Swift Formation

UNCONFORMITY

Jr — Rierdon Formation

(Ellis Group)

Jsa — Sawtooth Formation

UNCONFORMITY

Mc — Castle Reef Dolomite

Ma — Allan Mountain Limestone

(Madison Group)

Ki — Sills

Upper Cretaceous

Lower Cretaceous

Upper Jurassic

Middle Jurassic

U. Miss.

L. Miss.

UPPER PLATE

Єs — Switchback Shale

Єst — Steamboat Limestone

Єp — Pagoda Limestone

Єd — Dearborn Limestone

Єda — Damnation Limestone

Єg — Gordon Shale

Єf — Flathead Sandstone

Upper Cambrian

Middle Cambrian

UNCONFORMITY

pЄau / pЄal — Ahorn Sandstone*

pЄhd / pЄhc / pЄhb / pЄha — Hoadley Formation*

pЄhe — Helena Dolomite*

pЄe — Empire and Spokane Formations

pЄi — Sills

LOCATION OF AREA

From: U.S.G.S.–G.Q. 499 Geology by Melville R. Mudge

Base by U.S. Geological Survey

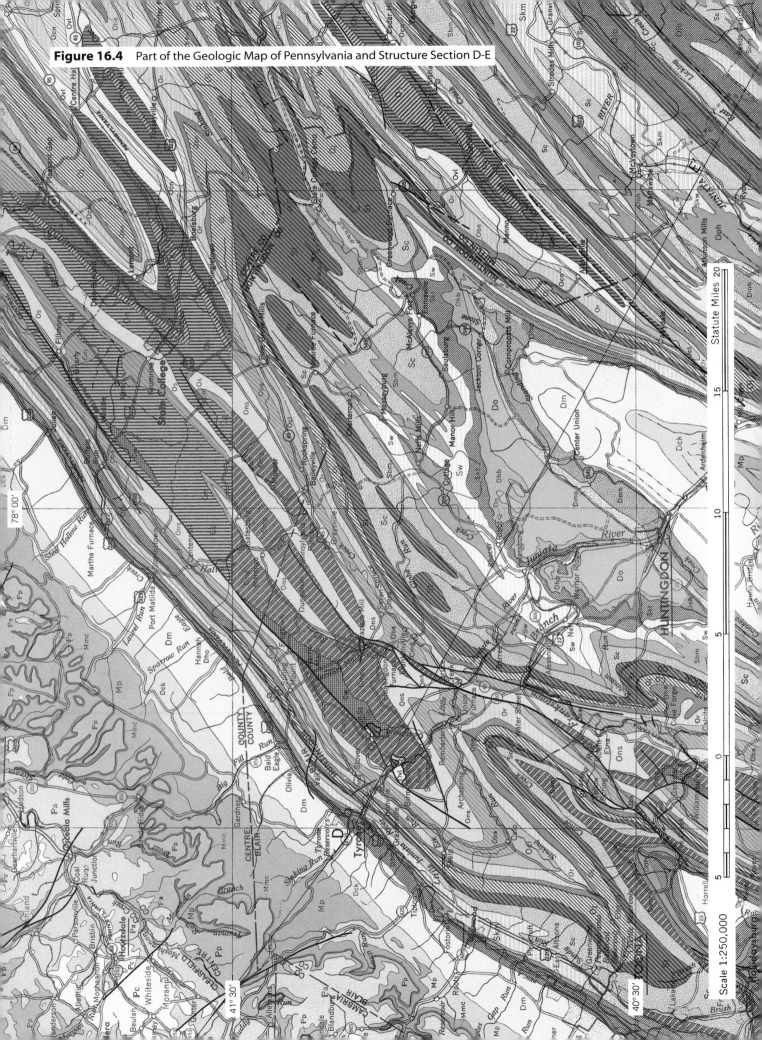

Figure 16.4 Part of the Geologic Map of Pennsylvania and Structure Section D-E

Statute Miles

Scale 1:250,000

Geologic Map of Pennsylvania Structure Section D–E: Horizontal scale same as map. No vertical exaggeration.

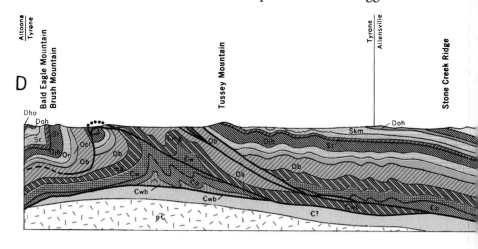

PENNSYLVANIAN

Conemaugh Formation
Cyclic sequences of red and gray shales and siltstones with thin limestones and coals; massive Mahoning Sandstone commonly present at base; Ames Limestone present in middle of sections; Brush Creek Limestone in lower part of section.

Allegheny Group
Cyclic sequences of sandstone, shale, limestone and coal; numerous commercial coals; limestones thicken westward; Vanport Limestone in lower part of section; includes Freeport, Kittanning, and Clarion Formations.

Pottsville Group
Predominantly sandstones and conglomerates with thin shales and coals; some coals mineable locally.

MISSISSIPPIAN

Mauch Chunk Formation
Red shales with brown to greenish gray flaggy sandstones; includes Greenbrier Limestone in Fayette, Westmoreland, and Somerset counties; Loyalhanna Limestone at the base in southwestern Pennsylvania.

Pocono Group
Predominantly gray, hard, massive, cross-bedded conglomerate and sandstone with some shale; includes in the Appalachian Plateau Burgoon, Shenango, Cuyahoga, Cussewago, Corry, and Knapp Formations; includes part of "Oswayo" of M.L. Fuller in Potter and Tioga counties.

DEVONIAN

UPPER

Oswayo Formation
Brownish and greenish gray, fine and medium grained sandstones with some shales and scattered calcareous lenses; includes red shales which become more numerous eastward. Relation to type Oswayo not proved.

Catskill Formation
Chiefly red to brownish shales and sandstones; includes gray and greenish sandstone tongues named Elk Mountain, Honesdale, Shohola, and Delaware River in the east.

Marine beds
Gray to olive brown shales, graywackes, and sandstones; contains "Chemung" beds and "Portage" beds including Burket, Brallier, Harrell, and Trimmers Rock; Tully Limestone at base.

MIDDLE AND LOWER

Hamilton Group

Mahantango Formation
Brown to olive shale with interbedded sandstones which are dominant in places (Montebello), highly fossiliferous in upper part; contains "Centerfield coral bed" in eastern Pennsylvania.

Marcellus Formation
Black, fissile, carbonaceous shale with thick, brown sandstone (Turkey Ridge) in parts of central Pennsylvania.

Onondaga Formation
Greenish blue, thin bedded shale and dark blue to black, medium bedded limestone with shale predominant in most places; includes Selinsgrove Limestone and Needmore Shale in central Pennsylvania and Buttermilk Falls Limestone and Esopus Shale in easternmost Pennsylvania; in Lehigh Gap area includes Palmerton Sandstone and Bowmanstown Chert.

Oriskany Formation
White to brown, fine to coarse grained, partly calcareous, locally conglomeratic, fossiliferous sandstone (Ridgeley) at the top; dark gray, cherty limestone with some interbedded shales and sandstones below (Shriver).

Helderberg Formation
Dark gray, calcareous, thin bedded shale (Mandata) at the top, equivalent to Port Ewen Shale and Becraft Limestone in the east; dark gray, cherty, thin bedded, fossiliferous limestone (New Scotland) with some local sandstones in the middle; and, at the base, dark gray, medium to thick bedded, crystalline limestone (Coeymans), sandy and shaly in places with some chert nodules.

SILURIAN

Keyser Formation
Dark gray, highly fossiliferous, thick bedded, crystalline to nodular limestone; passes into Manlius, Rondout, and Decker Formations in the east.

Tonoloway Formation
Gray, highly laminated, thin bedded, argillaceous limestone; passes into Bossardville and Poxono Island beds in the east.

Wills Creek Formation
Greenish gray, thin bedded, fissile shale with local limestone and sandstone zones; contains red shale and siltstone in the lower part.

Bloomsburg Formation
Red, thin and thick bedded shale and siltstone with local units of sandstone and thin impure limestone; some green shale in places.

McKenzie Formation
Greenish gray, thin bedded shale interbedded with gray, thin bedded, fossiliferous limestone; shale predominant at the base; intraformational breccia in the lower part. Absent in Harrisburg quadrangle and to the east.

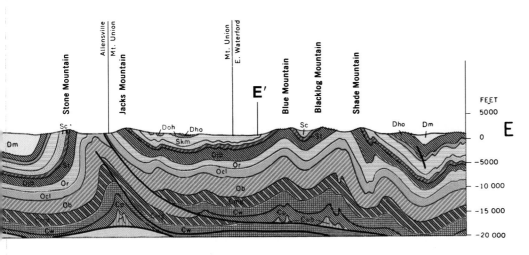

E' ... E

FEET
5000
0
-5000
-10 000
-15 000
-20 000

Stone Mountain / Allensville / Mt. Union / Jacks Mountain / Mt. Union / E. Waterford / Blue Mountain / Blacklog Mountain / Shade Mountain

SILURIAN (CONT)

Clinton Group
Predominantly Rose Hill Formation - Reddish purple to greenish gray, thin to medium bedded, fossiliferous shale with intertonguing "iron sandstones" and local gray, fossiliferous limestone; above the Rose Hill is brown to white quartzitic sandstone (Keefer) interbedded upward with dark gray shale (Rochester).

Tuscarora Formation
White to gray, medium to thick bedded; fine grained, quartzitic sandstone, conglomeratic in part.

ORDOVICIAN

Juniata Formation
Red, fine grained to conglomeratic, quartzitic sandstone with well developed cross-bedding and with interbedded red shale in places.

Bald Eagle Formation
Gray to greenish gray, fine grained to conglomeratic, thick bedded sandstone; often iron-speckled and cross-bedded; some greenish gray shale in places.

Reedsville Formation
Dark gray, olive weathering shale with thin silty to sandy interbeds; black shale of Antes Formation at the base.

Coburn Formation
Dark gray to black, thin bedded limestone with black shale interbeds.

Salona Formation
Dark gray, thin bedded, dense limestone.

Nealmont Formation
Bluish gray, finely crystalline, fossiliferous limestone; lower part grades laterally into Curtin Formation.

Curtin Formation
Gray, impure limestone; bluish gray, fine grained, high calcium limestone with some larger calcite grains (Valentine Member, Ov) at the top.

Ov
Ocl
Ovl

Benner Formation
Gray, mottled, dolomitic limestone and coarse granular limestone.

Hatter Formation
Dark gray, impure, fossiliferous limestone.

Loysburg Formation
Dense limestone over irregularly banded dolomitic limestone.

ORDOVICIAN (CONT)

Bellefonte Formation
Gray, cream to tan weathering, medium bedded dense dolomite.

Axemann Formation
Bluish gray, medium bedded, impure limestone.

Oba

Nittany Formation
Gray, thick bedded, coarsely crystalline dolomite.

Stonehenge-Larke Formation
Stonehenge- Bluish gray, finely crystalline limestone and dark gray, laminated limestone with abundant "edgewise" conglomerate. Larke- Dark, coarsely crystalline dolomite; equivalent to Stonehenge.

Ons   Ob

Beekmantown Group

CAMBRIAN

Mines Formation
Bluish gray crystalline dolomite; largely oolitic with much oolitic chert.

Cm

Gatesburg Formation
Bluish gray, coarse crystalline dolomite with many sandstone interbeds; cryptozoon reefs common.

Warrior Formation
Bluish gray, fine grained dolomite with shaly partings.

Pleasant Hill Formation
Dark gray, thick bedded limestone and thin bedded, shaly limestone.

Cp

Waynesboro Formation
Green and red shale, sandstone, and conglomerate.

Cwb

which the beds are bent. If the beds on either side of the hinge dip away from the axis, the fold is an anticline; if the beds dip toward the hinge, it is a syncline. In nature, folds rarely exist in isolation but rather in trains of alternating anticlines and synclines.

Geologic maps may show the trends of the axes of anticlines and synclines as a dashed or solid line. The symbols on the axes are arrows that point in the direction of the dip, away from the fold axis for anticlines, or toward the axis for synclines (Table 16.1). Even if the fold axis symbols are not plotted on the map, the axes can be easily determined by using the strike and dip symbols on the map or using the rule of V's to determine the dip direction. Simply look for areas where formations seem to form a "mirror image" repetition of each other. The axis will be approximately where the "mirror" plane would be. Then use either strike and dip symbols or the rule of V's to determine whether the beds dip away from this region for anticlines or toward it for synclines. A corollary of the principle of stratigraphic superposition, discussed in Lab 7, states that tilted beds dip toward the outcrop of younger sedimentary layers, because the younger beds are on top of all the older beds. This principle applies to folded beds as well. In synclines the youngest rocks will occur in the center of the fold; in anticlines the oldest rocks will occur in the core or center of the fold. To aid in visualizing the folds, it helps to draw the fold axis symbol on your copy of the map (in pencil). Draw them freehand, not with a ruler, as the fold axes are frequently somewhat curved.

Upright Folds

Upright folds are anticlines and synclines that have horizontal fold axes and vertical axial planes (Fig. 7.29). The strikes of sedimentary beds in an upright fold are parallel to each other on opposite sides of the axis (Fig. 7.31 and Fig. 16.5).

4. **a.** Examine a block or clay model of an upright anticline and syncline. Try to visualize how the rock layers would penetrate through the block. Locate the axis of the fold. If you place the model on the table, is the axis horizontal? _____

   b. Look at the top of the model (map view) and describe the pattern the rock layers make on this surface.

   _____

   _____

   _____

   c. Determine which rocks are oldest and which are youngest using stratigraphic superposition. Locate the anticline. Are older or younger rocks found in the core or axial area of an anticline? _____

   d. Are older or younger rocks found in the core of a syncline? _____

5. Examine the map in Figure 16.6. Determine the position of the fold axes, and draw in the appropriate fold axis symbols using strike and dip symbols and rule of V's. Are the ages of the formations, the dip shown by strike and dip symbols, and the rule of V's all consistent? _____

If not, correct the ages and/or the strike and dip symbols to make them consistent with the rule of V's.

The folds in the previous exercise are upright folds. Except for deviations caused by valleys and ridges, upright folds tend to produce an approximately parallel-striped pattern on maps (Figs. 7.29 and 7.31).

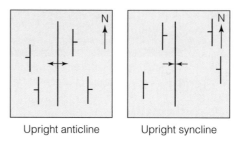

Upright anticline   Upright syncline

Figure 16.5   Strike and dip of beds in upright folds.

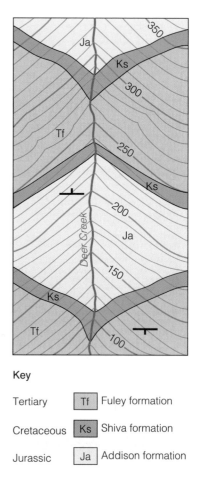

Key

Tertiary	Tf	Fuley formation
Cretaceous	Ks	Shiva formation
Jurassic	Ja	Addison formation

Figure 16.6   Geologic map of a region with folds. Locate and draw in the fold axes using the age of rock layers, the dip of the beds, and the rule of V's.

Plunging Folds

Most folds have axes that are not horizontal because the compression causing the beds to fold and the beds themselves are irregular. Such folds are called **plunging folds** because their axes plunge into the Earth at an angle (see the lower block diagram in Fig. 16.7). As Figure 16.7 shows, where there are plunging anticlines and synclines, erosion will expose the formations as a series of zigzag patterns. When seen on geologic maps, such a map pattern indicates that plunging folds are present. Beds in plunging folds still dip away from the axis of an anticline and toward the axis of a syncline, but the strikes are not parallel (Figs. 16.7 and 16.8) as they were for upright folds. The dip of beds near the fold axes will point in the general direction of the plunge.

Plunging folds tend to occur in sets of alternating anticlines and synclines plunging in the same direction. Look more closely at Figure 16.7. Arrowheads on the ends of the fold axes portray the plunge direction. Also, you can see that for anticlines the bend in rock layers points in the direction of the plunge of the fold axis, and the opposite is the case for synclines—the "nose" of a syncline points away from the plunge.

When drawing fold axes on a map of plunging folds, as for upright folds, you should draw the axis where a mirror would be to produce the image of repeated rock layers. For plunging folds the axial plane (mirror) also connects the "kinks," bends, or "noses" of the folds. The axial planes in Figure 16.7 correspond to the position of the mirror planes for the folds shown in the block diagram. On a map, the fold axis coincides with the trace of the axial plane.

6. Examine a block or clay model of a *plunging* anticline and syncline.
   a. Try to visualize how the rock layers would penetrate through the block. Locate the axis of each fold. If you place the model on the table, is the axis horizontal?

   _____

   b. In your own words, describe the pattern the rock layers make on the top (map view) of the model.

   _____

   _____

   _____

   _____

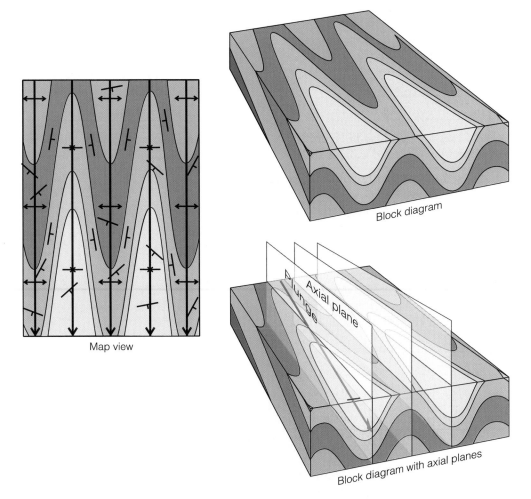

Map view

Block diagram

Axial plane

Plunge

Block diagram with axial planes

Figure 16.7   Plunging anticlines and synclines. The red arrow (pink underground) on the lower block diagram is the fold axis and shows the plunge of the folds. Observe the arrow heads on the map at the end of the fold axes on the downward (plunging) end.

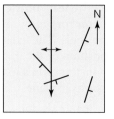 

Plunging anticline   Plunging syncline

Figure 16.8   Strike and dip of beds in plunging folds.

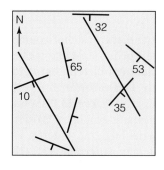

Figure 16.9   Folds marked by strike and dip symbols. The fold axes need appropriate arrows added.

7. Figure 16.9 shows strikes and dips of beds in an area of folding. Fold axis traces are drawn on the figure for you (without the arrows). Determine the type of folds, upright or plunging. _____ Also determine which is an anticline or a syncline, and finish drawing the fold axis symbols on the map. If the folds are plunging, what is the direction of the plunge? _____

8. Look at the following figures in Lab 7, and decide whether the folds are upright or plunging:

   Figure 7.4 (p. 125) _____

   Figure 7.5 (p. 125) _____

   Figure 7.27 (p. 137) _____

   Figure 7.31 (p. 140) _____

9. Draw complete fold axes for each map in Figure 16.10. Also, fill in the information requested in Table 16.2.

10. Examine the southeastern three-fourths of the Geologic Map of Pennsylvania in Figure 16.4.
    a. What geologic structures are suggested by the overall pattern or arrangement of rocks? _____
    b. Describe the geologic structure east of Huntingdon and Center Union in the south-central part of the map. _____

    What are three pieces of evidence that support your answer? _____

    _____

    _____

    c. To the northeast of this area are some dark blue ruled rocks, labeled Oa and Obf, near Menno in Mifflin County. These rocks occur in the core of a(n) _____. What is the direction of the plunge in this area? _____

11. Carefully examine the Geologic Map of the Southern Half of Somerset County, Pennsylvania, in Figure 17.11 to help you answer the following questions:
    a. Use the rule of V's to determine whether the beds are nearly horizontal or dipping in each of the following locations. If dipping, give the dip direction.

       ■ Whites Creek near Unamis in the southern part of map west of the areas of Dck (orange)

       _____

       ■ Pak (green) near 39°44′45″N and 79°10′30″W

       _____

    b. Are the beds of Dck (orange) and Mp (pale blue) near Packhorse Mtn steeply or gently dipping?

       _____

    c. Using your answers in **a** and **b,** visualize and draw a rough draft of a slice from Boardman Ridge to Maust Hill through the structure.

    d. What structure is present in this area?

       _____

    e. What is the direction of plunge of this fold?

       _____

    The mountains of Montana were formed by the convergence of the North American and Farallon Plates. The Farallon Plate was an oceanic plate and was largely subducted; only a small fragment exists today, known as the Juan de Fuca

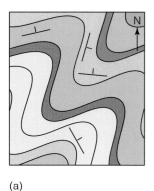 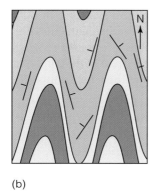 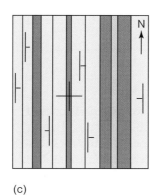

(a)   (b)   (c)

Figure 16.10 Simplified Geologic Maps of Folds

Table 16.2 Practice with Folds

Fold	MAP		
	16.10a	16.10b	16.10c
Upright or plunging?			
Plunge compass direction			
Compression compass directions			

Plate (Fig. 16.11). Let's look at structures that formed there as a consequence.

12. Carefully examine the Geologic Map of Devils Fence Quadrangle, Montana, in Figure 8.12 to help you answer the following questions:

  a. Use the age of the rocks to determine whether the large fold in the eastern half of the map is a syncline or anticline. _____ What other pieces of evidence support your answer? _____

_____

  b. What is the direction of plunge? _____

  c. Draw an appropriate fold axis on the map.

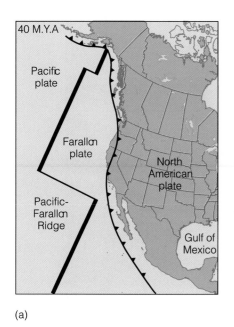 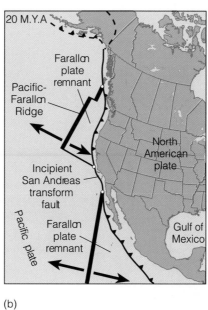 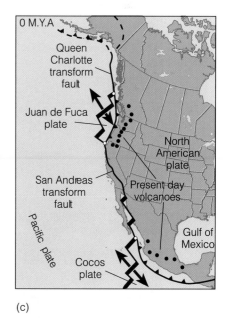

(a)   (b)   (c)

Figure 16.11 Convergence of the Farallon Plate and development of strike-slip movement along the West Coast of North America. As North America drifted west, it overrode and subducted the Farallon Plate, whose remnants are the present Juan de Fuca and Cocos Plates. (a) 40 Myr ago. (b) 20 Myr ago. (c) Today. Black dots are present-day volcanoes.

d. What type and compass directions of force caused these folds? _____ _____

Folds and Differential Erosion

Because different formations may be composed of rocks of different hardness, they erode at different rates. Harder rocks tend to stand out as hills or ridges and softer rocks become valleys. This is known as differential erosion. The zigzag pattern of an area with plunging folds may be seen as a pattern of zigzag ridges in the landscape. This is particularly common in the Appalachian region of Tennessee, Virginia, and Pennsylvania. In some locations, an anticline may be a hill or mountain where the core is made up of resistant rock, such as quartz sandstone. In other places along its axis, the same anticline may be a valley, where it has an easily erodable core of mudstone or limestone (limestone is easily erodable in wet climates but not in dry climates).

Folds on Air Photos Air photos are often used to assist in making geologic maps from above, because rock types and structures are often visible due to differential erosion or vegetation types. Uneroded folds can be mapped with air photos quite easily, but where erosion has cut into the folds, strike and dip measurements may be needed to distinguish anti-clines and synclines in air photos. This is because, as discussed earlier, anticlines are not necessarily mountains and synclines are not necessarily valleys after erosion. Some of the valleys occur at the cores of anticlines, where the core is shale or limestone, for example.

13. The Appalachian Mountains, including the part of the Valley and Ridge Province in Pennsylvania shown in Figure 16.12, formed by the collision of Africa into North America.

   a. Use the stereoscope to examine the stereo air photos (Fig. 16.12) of this area of Pennsylvania. What makes the pattern of hills and valleys in this area?

   _____

   _____

   _____

   _____

   b. All mountains in the air photo pair are formed by erosion-resistant sandstone, and the valleys are of the less resistant shale or limestone. Locate the sandstone in the air photo stereo pair in Figure 16.12, and then label the areas of sandstone, ss, on Figure 16.13. Shade or color the sandstone layers.

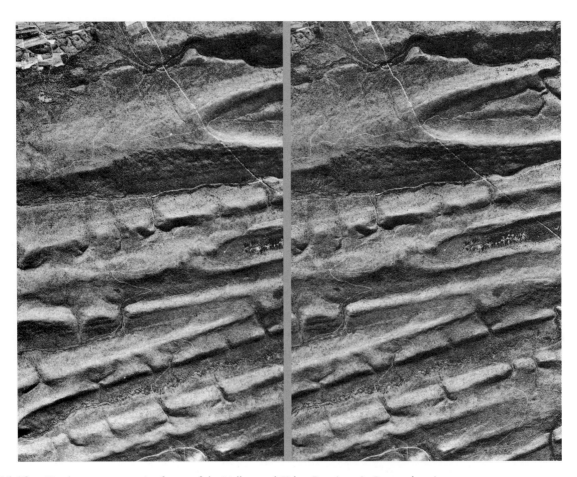

Figure 16.12 Air photo stereo pair of part of the Valley and Ridge Province in Pennsylvania

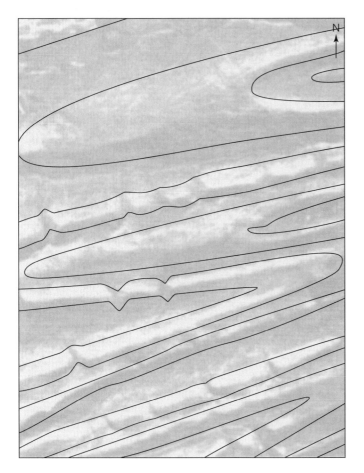

Figure 16.13 When making geologic maps, sometimes air photos may show the location of certain rock types or layers because of differential erosion. This figure is a preliminary geologic map made by tracing approximate contact locations on one of the air photos in Figure 16.12 of part of the Valley and Ridge Province in Pennsylvania.

c. See whether you can tell which areas are anticlines or synclines without having strike and dip symbols. Sketch in fold axes on Figure 16.13 using the appropriate symbols.

d. Are the folds plunging or upright? _____

e. Can you tell the direction of plunge? If so, what is it?

_____

f. What is the direction of shortening that formed the folds? _____

Faults on Geologic Maps

Remember that faults are breaks in the Earth's crust where one side has slipped relative to the other side (Fig. 16.14). Fault contacts on geologic maps are shown as thicker lines than other types of contacts (Table 16.1). Contacts between rock units on one side of a fault commonly end abruptly at the fault and are displaced relative to the other side (Fig. 16.14d and e). Occasionally, faults—especially thrust faults (Fig. 16.14c)—occur parallel to bedding planes, in which case offsets may not be visible.

Faults are planar features, in the same way that contacts between formations are. Consequently, they obey the rule of V's in exactly the same way. Additionally, steeply dipping faults have straighter map patterns than more shallowly dipping faults.

As discussed in Lab 7, there are three basic kinds of faults: dip-slip, strike-slip, and oblique slip. Within these are subcategories. *Dip-slip* faults consist of **normal** or **reverse faults,** with **thrust faults** a subcategory of reverse faults. Dip-slip faults are often defined in terms of a hanging wall and a footwall. The hanging wall is the block of rock physically above the fault plane, whereas the footwall is below the fault plane. You can also think of the footwall as a ramp and the hanging wall as a block sitting on the ramp. Figure 16.14 and its caption define these faults and how to distinguish them. In addition to their defining characteristics, normal faults (Fig. 16.14a) are usually steeply dipping and so generally have fairly straight map patterns. Reverse faults either may dip steeply (Fig. 16.14b) and have a fairly straight map pattern or may dip shallowly, called *thrusts* (Fig. 16.14c), resulting in a more sinuous map pattern. Thrusts are often found in association with major folds whose axes run approximately parallel to the faults (Fig. 16.3).

Maps can be used for different purposes; therefore, depending on the map maker's interests and intent, the symbols may vary or details may be included or left out. For dip-slip faults of all kinds, if no U and D symbols or teeth are marked on the map, it is generally possible to tell the direction of slip using the age of beds. Remember that younger rocks occur on the down-dropped side after erosion has worn away some of the up-thrown rocks (Fig. 7.33).

Strike-slip faults also have different types. Recall that the slip on such faults is approximately horizontal, parallel to strike. Figure 16.14d and e show the two possible directions of slip. Imagine that you are standing on one side of the fault and look across to the other side. If the arrow on the *other* side points to your right, then the slip is **right-lateral** (Fig. 16.14d); if the arrow points to the left, then it is **left-lateral** (Fig. 16.14e). Strike-slip faults are usually very steeply dipping and so tend to have the straightest map patterns of any fault.

14. Examine and name each fault in Figure 16.15. Label them on the diagram.

Normal faults result from extension; reverse and thrust faults result from shortening; strike-slip faults result from horizontal shear. Although different types of plate boundaries commonly create more than one type of stress, a general association exists among the three types of plate boundaries, the three types of stress, and these three types of faults. Divergent plate boundaries have normal faults, convergent boundaries have reverse and thrust faults, and transform plate boundaries are associated with strike-slip faults (Fig. 16.14). The converse is not necessarily true, because forces at plate boundaries can

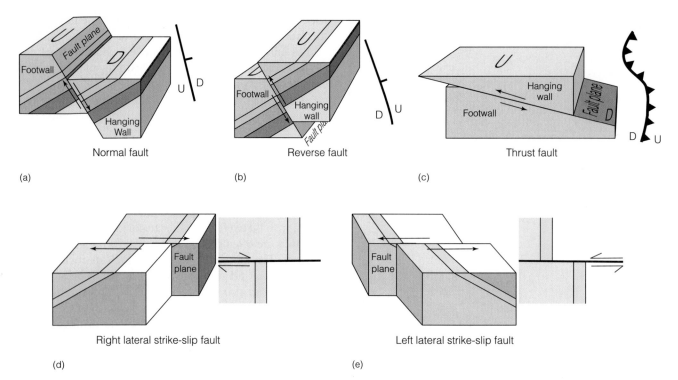

Figure 16.14 Types of faults and their map symbols (to the right). (a) **Normal faults** are dip-slip faults for which the hanging wall has slipped *down* the dip of the fault plane and/or the footwall moved up. The block moved down the ramp. On geologic maps, a small tick or arrow shows the down-dip side of the fault trace. The up-thrown side may also be marked with a U and the down-thrown side with a D symbol. The down-drop side (D) is in the direction of dip of the fault plane. (b) **Reverse faults** are also dip-slip faults but with movement in the opposite direction of normal faults. The slip of the hanging wall is *up* the dip of the fault plane and/or the footwall moved down. The block slid up the ramp. For reverse faults, the small tick in the down-dip direction will occur on the same side as the U. (c) **Thrust faults** are low-angle reverse faults. As for other reverse faults, the slip of the hanging wall or upper plate of a thrust fault is up the dip of the fault plane. The up-thrown side of the fault (hanging wall) is decorated with triangular, teethlike symbols pointing in the direction of dip of the fault. (d) A **right lateral** *strike-slip fault* is one where the opposite side of the fault moved horizontally to the right. (e) For a **left lateral** *strike-slip fault,* the opposite side moved horizontally to the left.

be complex (Fig. 16.16) and deformational forces outside plate boundaries may also produce faults (Fig. 16.17).

15. Examine the Geologic Map of the Glen Creek Quadrangle (Fig. 16.3). Compare the Glenn Fault with the South Fork Thrust Zone.
    a. What is the dip direction of the Glenn Fault?

    _____

    b. What is the dip direction of the South Fork Thrust Zone? _____

    c. Which of the two is the most steeply dipping?

    _____

16. On the Geologic Map and Structure Section of Pennsylvania (Fig. 16.4), find the cross section line (structure section line) labeled D to E. Find the faults in the cross section for this line.

    a. Are the faults mostly low or high angle?

    _____

    b. Were older rocks shifted up over younger rocks or younger shifted down over older? _____

    c. What type are most of the faults in this cross section?

    _____

17. Examine Structure Section B-B′ and the Geologic Map of Devils Fence Quadrangle, Montana (Fig. 8.12), then answer the following questions:
    a. What kind of fault is the fault near €w on section B-B′? _____

    b. Add the appropriate symbols to the fault on the map.

    Observe that the contacts end abruptly at the fault on both the map and the cross section and how the same contacts are

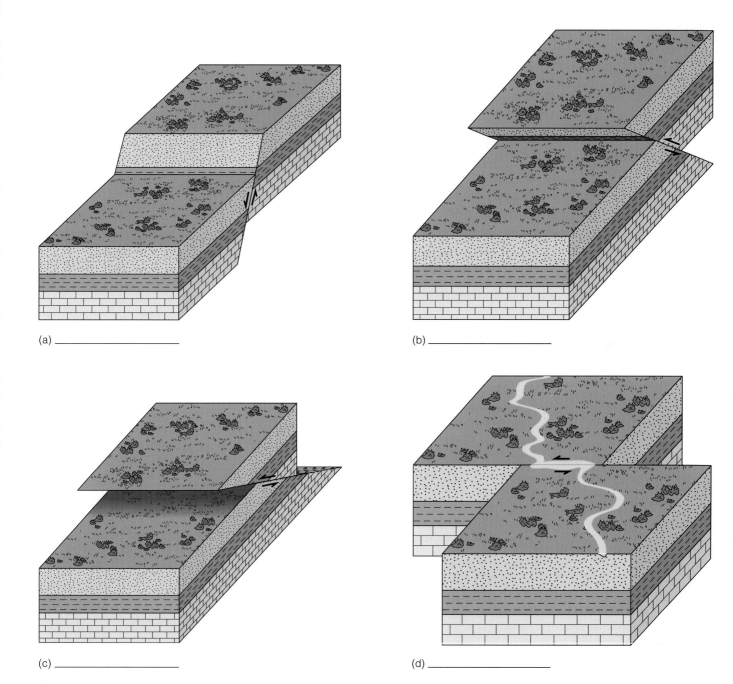

(a) _____

(b) _____

(c) _____

(d) _____

Figure 16.15 Label the correct fault name on each diagram.

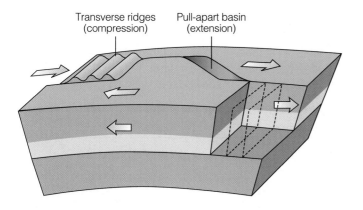

Transverse ridges (compression)

Pull-apart basin (extension)

Figure 16.16 Model of bends in a right lateral strike-slip fault shows shortening and extentional features at a transform plate boundary. A bend to the left results in shortening, producing transverse ridges with folding and thrust faulting. A bend to the right generates extension, resulting in a pull-apart basin and normal faults. For left-lateral faults, view this figure in a mirror; the left bends would produce extension and right bends, shortening.

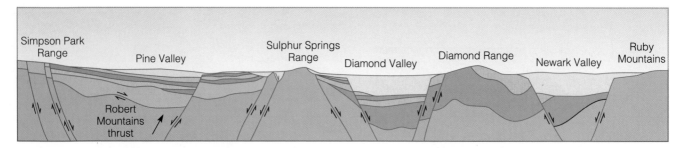

Figure 16.17 Basin and Range Province. This cross section of part of the Basin and Range Province in Nevada show normal faulting by regional extension not at a plate boundary.

offset across the fault. Sometimes, although not in this case, the matching contacts have been moved so far along the fault that both sides do not appear on the map or cross section.

c. Locate the Horse Gulch Fault on the western edge of the map. What type of movement occurred on this fault? _____ What is your evidence?

_____

_____

d. Draw correct map symbols on the Horse Gulch Fault. Refer to Table 16.1 if needed.

18. Examine the Geologic Map of the White Pine District, Nevada, in Figure 16.18.

a. What is the dip direction of the Belmont Fault (just west of center)? _____

What is the dip angle? _____

b. Which side moved down? _____

c. What type of fault is this? _____

d. Use the rule of V's to determine the dip direction of the Monte Cristo Thrust (SE quarter of the NW quarter of the map). _____

e. Did both types of faults form by the same deformational forces? _____ Do you think it is likely that they formed at the same time? _____

19. Examine the fault in the northeastern section of the Geologic Map of Washington in Figure 1.3, between the Entiat and Wenatchee Rivers.

a. Which side moved up? _____ How can you tell? _____

_____

b. Without knowing the dip of the fault plane, can you tell which kind of fault this is? Yes / No.

You may want to experiment with block models of faults before you answer this question.

c. Although it is not shown on the map, the fault dips to the west-southwest. What type of fault is this?

_____

d. Add the correct map symbols to the fault.
e. Locate a thrust fault on this map. What features, symbols, or aspects of this fault help suggest it is a thrust?

_____

_____

_____

Faults Computer Exercise

■ Start **TASA Plate tectonics.** Click stop (■).
■ Click **13 Transform Fault Boundaries.**

20. What type of fault is shown in the fault blocks near the beginning of this section?
a. Normal fault
b. Reverse fault
c. Thrust fault
d. Right-lateral strike-slip fault
e. Left-lateral strike-slip fault

21. What type of fault is shown in the detail of the southern part of the Mid-Atlantic Ridge?
Hint: Answer the question after you have seen the fault in motion.
a. Normal fault
b. Reverse fault
c. Thrust fault
d. Right-lateral strike-slip fault
e. Left-lateral strike-slip fault

22. What is the name of the major fault running through California? _____

23. What type of fault is it?
a. Normal fault
b. Reverse fault
c. Thrust fault

d. Right-lateral strike-slip fault
e. Left-lateral strike-slip fault

■ Click the stop (■) button, then the **QUIT** button.

■

Faults on Topographic Maps and Air Photos

Faults may sometimes be visible on topographic maps and air photos for several reasons. Rocks of very different hardness and resistance to weathering may be juxtaposed by fault movement; one side of the fault may therefore erode faster than the other, creating a steep slope on the site of the fault. Faults also commonly occur as valleys because rocks in the immediate vicinity of the fault may be fractured and crushed and hence may be more easily eroded than rocks away from the fault. Streams often preferentially follow faults for at least part of their course as a result. Finally, if the fault is active (i.e. moving at the present or within geologically recent time, possibly with some seismic activity), very recent features such as streams and ridges and even manmade features such as fences or orange groves that crossed the fault may be offset by the fault.

San Andreas Fault, California The San Andreas is the large active fault in California; its movement caused the great San Francisco earthquake of 1906 and numerous earthquakes since then. The movement of the fault in the 1906 earthquake was more horizontal than vertical. The same seems to be true of many thousands of earthquakes that have occurred in the region in the last few million years. As a result, geologic features on one side of the fault may have been displaced over time tens of miles or more with respect to the same features on the other side of the fault.

The fault trace (often erroneously called the "fault line") can be seen on photos.

24. Locate the San Andreas Fault on the oblique air photo in Figure 16.19.
   a. What is the approximate strike of the fault, keeping in mind that the view is northeastward? _____
   Draw the fault on the air photo with a thick black pen or in color.
   b. Describe features that indicate the location of the fault. _____
   _____
   _____
   c. Describe features that indicate the amount of displacement along the fault trace. _____
   _____
   _____
   d. What type of fault is this? _____

e. If you had to live somewhere in the area shown on the air photo, where would you choose to live? Give a general description, such as northwest corner or southwest section of the northeast quarter of the air photo. _____
_____
_____

f. In what way would this close proximity to the fault affect your everyday life? Consider day-to-day inconveniences and hazards. _____
_____
_____
_____

g. Now check the fault movements you saw with the plate map of the San Andreas Fault in Figures 9.13 and 16.11c. Does your answer in **d** match the movement shown in both figures? _____ Use Figure 16.11 to help you determine how long the San Andreas Fault has been moving. _____

Unconformities on Geologic Maps

The essential characteristic of an angular unconformity is that the base of the unconformity cuts across the contacts of all of the beds underneath it. This characteristic is easiest to see in cross section (Fig. 8.9), but it is also visible on a geologic map (Fig 16.20) where it covers the beds below and their contacts. Often the unconformity surface dips more gently than the underlying beds. As a result, the map pattern of the unconformity may be more sinuous than the contacts underneath it.

25. Find the angular unconformities on the Geologic Map of the Southern States in Figure 16.21. (One is easy to spot; the other is much more difficult.)
   a. Name the formation that is clearly lying unconformably on other rocks. _____
   b. What is the direction of dip of the beds above the unconformity (use the rule of V's)? _____
   c. Compare the strikes and/or dips of the rocks below the unconformity with those of the rocks above the unconformity. _____
   _____
   _____
   d. Name the group that occurs just above the less obvious unconformity. _____

39° 15'

39° 10'

Scale $\frac{1}{48,000}$

Contour interval 40 feet

2000   0   2000   4000   6000   8000   10000   12000   14000   16000   18000   20000 Feet

TRUE NORTH

MAGNETIC NORTH

17°

APPROXIMATE MEAN
DECLINATION, 1951

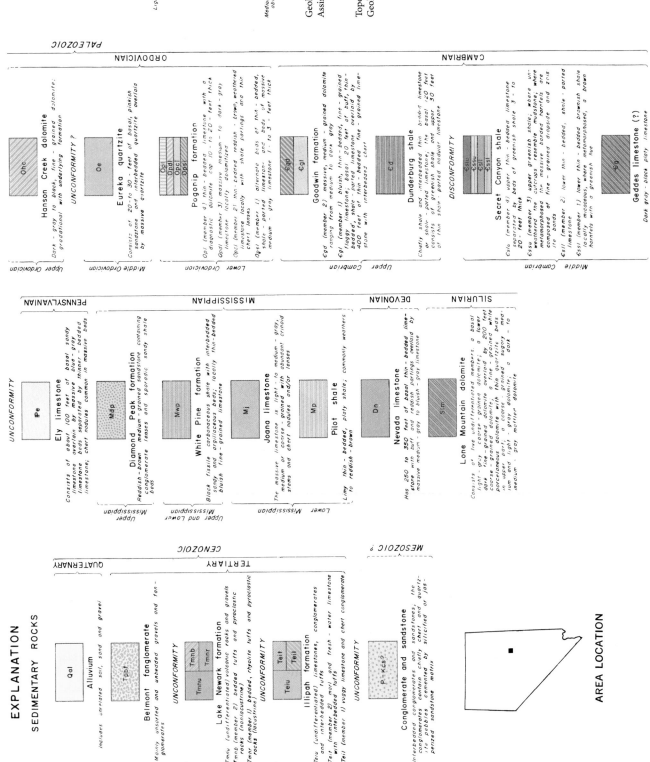

GEOLOGIC MAP OF THE WHITE PINE DISTRICT, NEVADA

From: Nevada Bureau of Mines Bulletin 57

Figure 16.18 Part of the Geologic Map of the White Pine District, Nevada

Figure 16.19 Oblique air photo of the San Andreas Fault, looking northeastward from the Carrizo Plain. Downstream channels have been offset 130 m and 380 m from their upstream source. The greater offset for one channel has disconnected it from its upstream source (known as *beheaded*). Sec. 33, T30S, R20E, San Luis Obispo County, California.

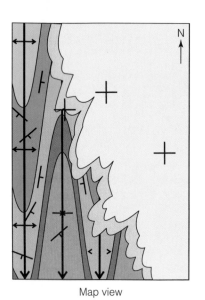

Map view

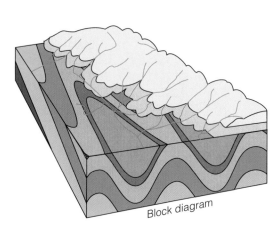

Block diagram

Figure 16.20 Map and block diagram showing the appearance of an unconformity. The unconformity occurs at the bottom of the yellow layers. The trace of the unconformity cuts across the contacts of the formations beneath the unconformity.

e. What is the evidence for this second unconformity?

_____

_____

_____

26. Review your answers to Exercises 1b and 1f in this lab for the Bright Angel Quadrangle map in Figure 7.1. What is the rock unit just above the unconformity (basal unit)

near the location for Exercise 1f? _____

There are actually two unconformities near here; see whether you can find the other one and name its basal

unit. _____

Igneous Contacts on Geologic Maps

The map pattern of the contacts of an intrusive igneous rock depends on whether the magma is **concordant** (intrudes *along* existing bedding or foliation) or **discordant** (intrudes *across* existing bedding or foliation).

In concordant intrusions, of which sills are the most common, the intrusive contacts are parallel to the contacts of the beds the sill intrudes (Fig. 7.16). Examine p€r in cross sections B-B′ and C-C′ in Figure 7.1b. Notice how the top and bottom contacts of the sill are parallel to the bedding contacts. Except at the end of an intrusion where it pinches out, a sill appears to be just another formation within the sequence. Unlike beds, however, sills are not laterally continuous and often end in a wedgelike form or turn into dikes, cutting across the layers. Lava flows, too, are parallel to sedimentary layers unless they lie unconformably on tilted layers.

In discordant intrusions such as dikes, stocks or batholiths (Fig. 7.16), the intrusive contact cuts across the existing beds and formation boundaries. Thus, on a map the trace of such intrusive boundaries cuts through the traces of the formation boundaries.

The map views of concordant intrusive contacts are similar in appearance to sedimentary unconformities. This similarity of appearance may initially cause confusion in differentiating the two rock bodies on a geologic map. Fortunately, they are easy to identify if a map legend is available. If the rock unit that cuts through the other contacts is an igneous intrusive rock, such as granite, diorite or gabbro, then the contact must be intrusive. If, on the other hand, the unit is sedimentary, then the contact must be an unconformity. Remember that cross-cutting rocks are younger than the formations being cut off.

27. Examine the Geologic Map of the Cross Mountain, California in Figure 7.22.
   a. Is the contact between Qal (areas of yellow) and other rocks an intrusive or unconformable one?

   _____

   What is your evidence?

   _____

   _____

   b. Describe the contact of gr with gd using terms such as *intrusive* and *unconformable.*

   _____

   How about ga and gd? _____

   c. How about Qts and gd? _____

   Tsv and gr where the contact is not faulted?

   _____

28. Examine the Geologic Map of Devils Fence Quadrangle, Nevada, in Figure 8.12.
   a. Find ad (orange color) in the southeastern part of the map in Sections 23, 24, and 25. Use the rule of V's to determine its dip direction in Section 23.

   _____

   b. Judging from its shape, its dip compared with surrounding rocks, and other factors discussed, which of the following best describes ad? Circle one.

   ■ Sediments unconformably overlying other rocks
   ■ A discordant intrusive
   ■ A concordant intrusive

Drawing a Cross Section

By now you know what a cross section is, and you have worked with a number of them that were already constructed. Here you will learn how cross sections are constructed and will draw a simple one. Figure 16.22 shows the method of constructing a cross section. Read through the instructions in the figure caption completely before starting. Ask your instructor to help you or to demonstrate the technique if you have questions or do not understand. You may want to try drawing a small part of the cross section in the figure as practice to see whether you produce the same results.

29. Use the topographic profile in Figure 16.23b to draw a cross section along the line from A to A′ in the map in Figure 16.23a. What structures do you see in your cross

   section when you are done? _____

   _____

Making Geologic Maps from Field Investigations: A Simulation

Geologists construct geologic maps by making observations at natural outcroppings of rocks and plotting this information on topographic maps and by taking careful notes. Observations include noting rock type, identifying the formation this rock type probably belongs to, and measuring the dip and strike of beds or foliation. If deformational features or structures are present, they, too, are measured. For example, the strike and dip of a fault zone can be directly measured in the field, and the type of fault determined, if the fault is sufficiently well exposed. A fold, too, may be sufficiently exposed to determine whether it is a syncline or an anticline, upright or plunging.

While in the field and later the geologists compile all this information on a topographic map using appropriate symbols. Then they interpret the observations and try to suggest where the contacts of all the rock bodies are. Often in desert areas, a geologist can easily follow a contact across the landscape, but in vegetated areas the contact is usually hidden, and the mapper has to suggest where it might be. Often aerial photographs can be used to aid in this interpretation.

Your instructor has prepared a simulation of the process of geologic mapping in a sandbox model. He or she has sculpted the sand to form a landscape and inserted small cards, flags and possibly some actual rocks into the sand. These simulate outcrops, and each one is numbered to correspond with observations written in a geologist's notebook. The cards show the orientation of the layers of rock

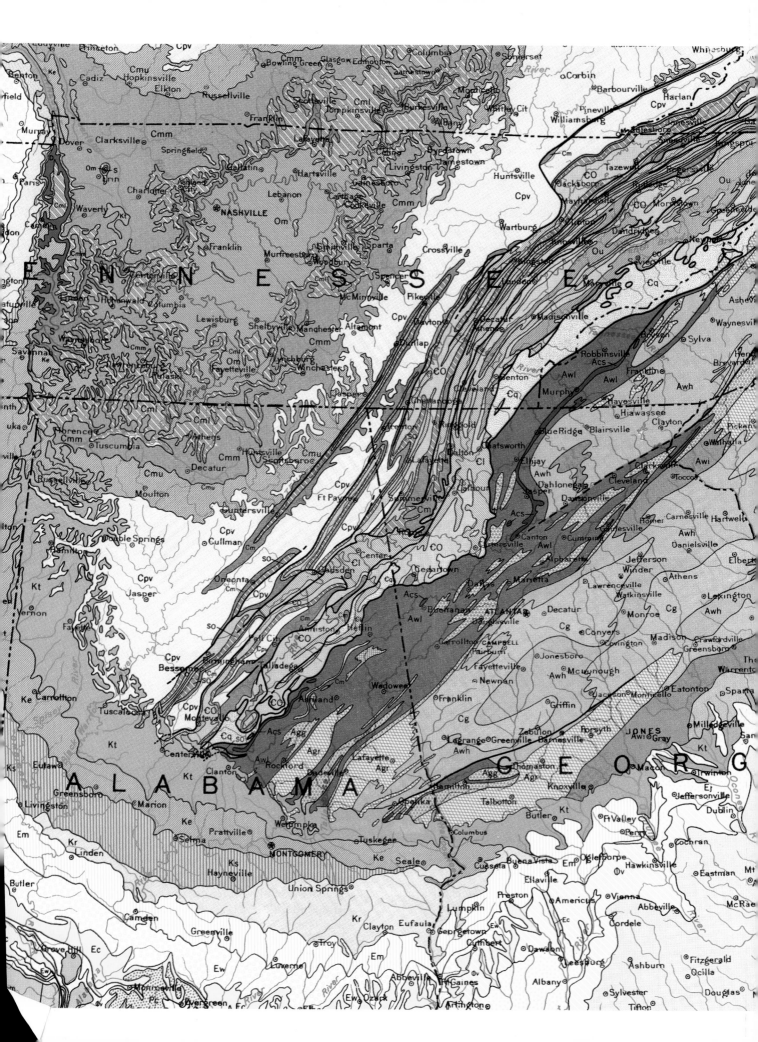

EXPLANATION

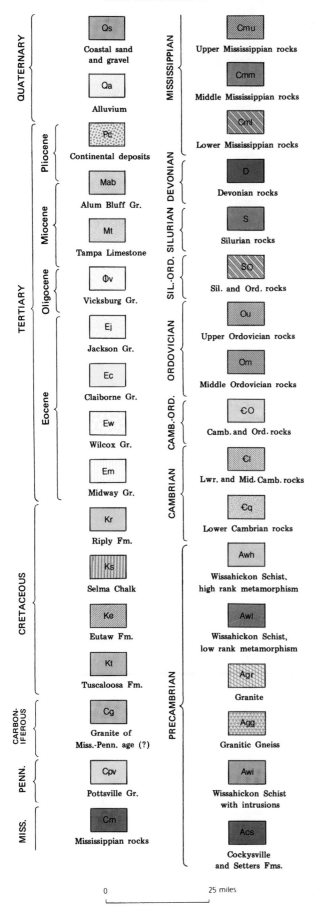

From: U.S.G.S.

Figure 16.21 (*opposite*) Part of the Geologic Map of the Southern States

in the field. A north arrow and scale are indicated on the model as well.

30. With your lab partner(s), using the grid in Figure 16.24, measure the sides of the sandbox with a ruler, and then draw the boundaries of the sandbox to scale onto the grid. To do this you will need to decide on a scale to represent the sandbox that is small enough to fit and large enough to fill most of the grid. What is the scale comparing your grid map to the sandbox? _____ The sandbox model also has a scale: what is it? _____ Your map represents both the sandbox and the part of the Earth the sandbox represents. On your map write down the scale of your map compared to the Earth's surface. Show the calculations here.

31. Place the acrylic sheet over the sandbox. Decide on a reasonable contour interval and with a washable overhead-marking pen, carefully draw topographic contour lines on the acrylic. Take turns doing this so each person has an opportunity to try it. Draw these contour lines to scale on your grid map of the sandbox. When completed, you will have prepared the topographic base map for the geologic map.

32. Mark the position of each outcrop on your base map, and write down its identifying number. Transfer the field observations for each location from the geologist's notebook[3] and your observations of any rocks in the sand to your map. Use colors or appropriate rock symbol patterns for each rock type (refer to Fig. 7.2 if needed).

33. Place the strike and dip symbols at each outcrop location where bedding or other planar structure is present. Be careful to make sure that the strike is correctly oriented with respect to north and that the tick on each strike symbol is on the downward-dipping side of the bed.

When these steps are completed, you have an outcrop map. Here is where you make an interpretation of the geologic data.

34. Try to evaluate where the contact traces should be drawn and what kind of contacts they are. Use the rule of **V**'s to help you decide where uncertain contacts should cross valleys. Color the areas on the map between the contacts or use appropriate rock symbols. Which of the following

···········

[3]The "geologist's notebook" here may be in the form of a handout, notes written on the board or an overhead, or an actual notebook. Consult your lab instructor for the form these take.

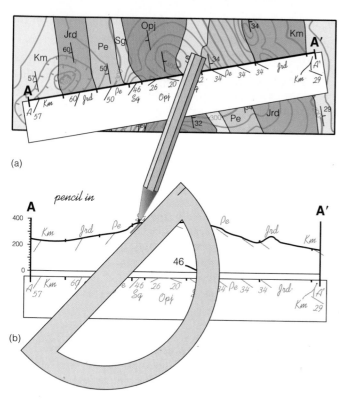

(a) Line up a thin strip of paper along the cross section line, and mark the contacts and dip angles.

(a)

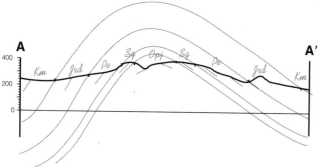

(b) Use pencil to transfer the information from the strip of paper to a topographic profile for the same line. Place the information at the land surface. Use a protractor to draw the dip angles accurately.

(b)

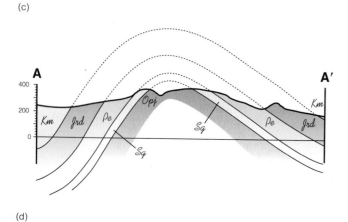

(c) Continue using pencil while you sketch in the inferred location of the contacts in the cross section. Maintain constant bed thickness if possible, except where the beds are known to be eroded or to change thickness.

(c)

(d) When satisfied with the results, ink in the correct lines and formation names, and erase extraneous marks. Dash in contact lines inferred to have been present above the current land surface before erosion. Color the units below ground level, leaving areas uncolored where information is unavailable or sketchy.

(d)

Figure 16.22   Method of drawing a cross section on a previously constructed topographic profile. CI = 20.

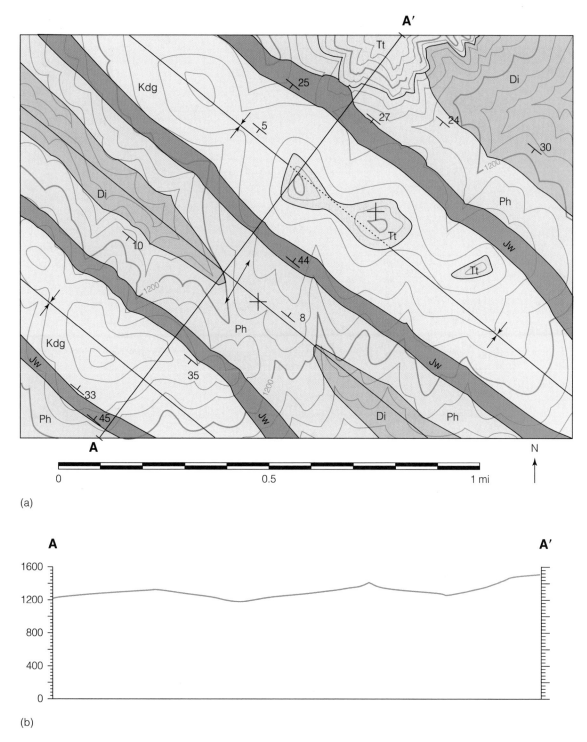

(a)

(b)

Figure 16.23 (a) A geologic map of a hypothetical area. CI = 40. Draw a geologic cross section along the line from A to A'. (b) Topographic profile base for drawing a geologic cross section for the map above.

features—what type and how many—occur in the area represented by the map?

a. Anticline: Type? _____ How many? _____

b. Syncline: Type? _____ How many? _____

c. Fault: Type? _____ How many? _____

d. Unconformity: Type? _____

How many? _____

e. Intrusion: Type? _____ How many? _____

Discuss your interpretation with your instructor.

Figure 16.24 Grid for Drawing a Geologic Map of the Sandbox

Resources

OBJECTIVES

1. To understand the importance of resources
2. To understand the origin, migration, and trapping of oil and natural gas
3. To understand the formation of coal, its exploration, and the environmental effects of coal mining
4. To learn the carbon cycle and to understand its relationship to resources and environmental problems
5. To learn the composition of some common materials and the mineral resources used to make them
6. To understand an extraction method for placer minerals

The Earth provides us with many resources, from the air we breathe to the materials and energy used to make and fuel our homes and automobiles. A **resource** is any naturally occurring substance that can be used. Sometimes resources are discussed in terms of the elements used or in terms of the minerals, rocks, biological materials, or naturally occurring liquids or gases that are useful. In this lab we will concentrate on geological resources. Our primary sources of energy are from fossil fuels, in solid (coal), liquid (petroleum), and gaseous (natural gas) forms. In this lab we will look at fossil fuels and where they are found. Common ordinary materials used to build walls, sidewalks, and bridges or make eating utensils, wire, and cans are made from minerals extracted from the ground. We will find out what minerals provide the material to make some common items. We will also consider the impact that the extraction and use of resources has on the planet, with the consequent scarring of the land, pollution of the water, and climate change, since these results have the potential to affect our everyday lives for decades.

Resources may be renewable, nonrenewable, or potentially renewable (Fig. 17.1). A **renewable resource** is one whose quantity is not changed by its use; solar energy is a good example. A nonrenewable resource is one that has a limited quantity that is diminished by use. A **potentially renewable resource** is one that is replenished but may or may not be used faster than the rate of replenishment.

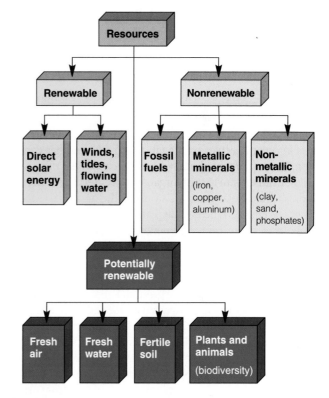

Figure 17.1 Types and examples of resources. Resources may be renewable, nonrenewable, or potentially renewable. Renewable resources do not diminish with use. Nonrenewables have essentially finite supply that is not replenished or is replenished very slowly. Potential renewables may be used more slowly or more rapidly than their rate of replenishment; if more slowly, then they resemble renewable resources, but if more rapidly, then they are nonrenewable resources.

Fossil Fuels as Resources

Fossil fuels are energy resources that come from the remains of organisms preserved in rocks (Fig. 17.2). Remember that a fossil is any evidence in the rock record of preexisting life. The *fossil* in *fossil fuel* is organic matter that once was part of a living organism. The energy in fossil fuels is chemical energy captured from sunlight by photosynthesis. To a chemist, synthesis means putting atoms together to form molecules or recombining

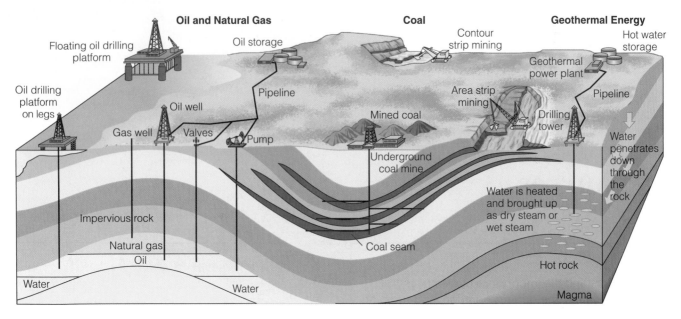

Figure 17.2 Earth's crust provides a number of different energy resources, including fossil fuels, geothermal energy (which provides heat from the Earth's interior), and uranium for nuclear energy.

atoms in one molecular combination to make another combination. Plants perform **photosynthesis,** a chemical synthesis using light from the sun to form organic molecules from water and carbon dioxide. Solar energy is stored in the organic molecules as chemical energy. When we eat plants, these organic molecules give us (and other animals) energy, calories, and also energy stored in muscle and fat tissues. Plants produce more organic molecules and stored chemical energy than are used by all living things; this excess, in the form of organic remains of plants and animals, becomes buried in the ground. Large quantities of organic material, preserved compacted and sealed off by sedimentation over time, result in a potential energy resource. If the buried material has a sufficient concentration of stored energy, we can burn it and use the energy to generate electricity and power our cars and other things. We call this material *fossil fuel,* and in a sense it is stored solar energy. Coal is mainly from plants, and oil and natural gas form from a mixture of organisms, but all fossil fuels are preserved organic materials.

Fossil fuels take millions of years to form so are nonrenewable resources (Fig. 17.1). Humans are using fossil fuels fast enough to exhaust the supply within a few centuries. In the United States, 85% of the energy resources come from fossil fuels (Fig. 17.3). Of that amount, about 63% comes from petroleum and natural gas combined. Because of this rapid rate of use, the United States, which already imports about half of the oil it uses, will exhaust its supply of fossil fuels faster than the rest of the world.

Petroleum and Natural Gas Exploration

Petroleum (oil) and natural gas are fluid fossil fuels. They are *hydrocarbons* that form from the buried organic matter of microscopic marine organisms. Petroleum or crude oil is a liquid mixture of large, sometimes complicated hydrocarbon

molecules. Gasoline—a liquid commonly called "gas" (just to confuse you)—is refined from crude oil and has a high proportion of octane (C_8H_{18}), a substantially smaller and simpler molecule than those making up crude oil. We also get motor oil from crude oil. Natural gas is chemically mostly methane (CH_4) but may contain heavier molecules such as ethane (C_2H_6), propane (C_3H_8), or butane (C_4H_{10}). Gas stoves commonly run off of methane or propane.

1. **a.** Oil or petroleum and natural gas form from what fossilized organisms? _____

    b. What are the two major elements making up natural gas and petroleum? _____

    _____

    c. When burned, each of these elements combines with oxygen. What two substances do you think are produced by burning oil or natural gas? _____

    _____

2. Examine and describe two samples of crude oil. Fill in Table 17.1. From your experience or that of your lab partners, describe gasoline and motor oil in Table 17.1. What characteristics can you describe that differentiate crude oil from gasoline? Crude from motor oil?

    _____

    _____

    _____

The refining process of crude oil results in many different products that separate out during the distillation; not all are fuels. A few examples are the petrochemicals used in plastics, paints, and insecticides.

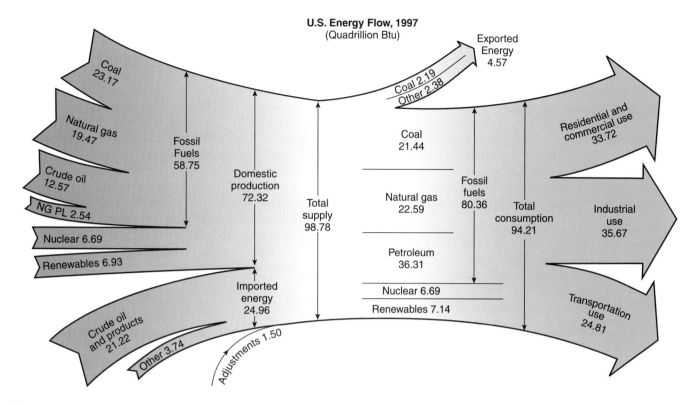

Figure 17.3 Energy flow in the United States for 1997. The U.S. Energy Information Administration provides figures on energy sources and use in the United States.

Table 17.1 Crude Oil and Its Derivatives

Description	Color	Viscosity	Other Description and Comments
Light crude			
Heavy crude			
Gasoline			
Motor oil			

3. List five items in the room that you think used petrochemicals in manufacturing, and discuss what materials could substitute for them if fossil fuels were not available.

_____

_____

_____

_____

_____

The organisms that will eventually become oil and gas settle to the bottom of the sea with other fine-grained material such as clay or fine-grained carbonate. The resulting shale or limestone is known as a *source rock*. Because the source rocks are fine grained, their permeability is low; therefore, they are poor choices for oil or gas extraction. The geologic processes known as *migration* and *trapping* move the oil and natural gas to areas where the oil can be extracted. The oil and natural gas must migrate to a permeable rock but must not migrate all the way to the surface. If they did that, they would be lost to weathering and to the atmosphere. The process of migration is aided by water.

4. Examine the jar containing oil, gas, and water. How can you tell which is more dense or less dense?

_____

_____

List these fluids in order from most dense to least dense.

_____   _____   _____

5. Combining your knowledge from Lab 13 (groundwater) with your observations in Exercise 2, imagine drops of oil and bubbles of natural gas immersed in water in the pores of a permeable limestone or sandstone. What

direction would each migrate? Oil _____

Natural gas _____ If the permeable rock were

sealed from leakage describe the relative final positions of these three fluids in the bed. _____

_____

To keep the oil and natural gas from escaping to the surface, geologic structures or features must be present that trap the fuels. A proper trap must have impermeable rocks above permeable ones so that the fluids can migrate into the permeable rocks, as in Figure 17.4, but not migrate upward out of the ground, as in the oil seep in Figure 17.4a. The impermeable rocks prevent further upward migration. A body of permeable rock containing oil and natural gas is called a **reservoir.**

6. Look at the jars with oil, gas, and water sitting on inclined planes; also examine Figure 17.4a–d. The reservoir rock for the resource is simulated by the jars—they may not look like rock beds, but the jars hold liquids and gases, as does a bed of permeable rock with an impermeable "cap." What do you notice about the relationship of the oil-water interface and the oil–natural gas interface with respect to the dip of the strata or the tilt of the jars?

_____

_____

_____

_____

7. Examine Figure 17.4a–d, and use the diagrams to help you color likely oil and natural gas reservoirs in the cross sections in Figure 17.5. Remember that an oil or gas reservoir (where one can drill and pump it out) must be in permeable rocks. If necessary, review rock symbols from Figure 7.2. First, draw the fluid interfaces (oil-water and gas-oil boundaries) so that you maintain the angular orientation that you described in your answer in Exercise 6. Then color in the key for each fluid and color the cross sections correspondingly where you would expect oil, natural gas, and water.

8. Assuming each well was drilled only as deep as the first trapped fluid (gas or liquid), which wells in Figure 17.5 are extracting natural gas? _____

Oil? _____

Water? _____

Now that you have found oil in cross sections, let's find some in a map. Imagine you are trying to deduce where oil is in an area that already has some oil wells. You want to buy up the oil rights, and you need to know which plot of land will most likely be the most productive.

9. The map in Figure 17.6 shows the locations of holes drilled for oil. Numbers next to each hole are the elevation of the contact between a porous, permeable sandstone unit, below, and a shale layer, above. Which would contain the oil, the shale, or the sandstone?

_____

a. What rock type is present below 3060 ft at the well in Section 21? _____

Above 3060 ft? _____

b. Using a 200-ft contour interval, draw contours on the map in Figure 17.6 showing the elevation of the sandstone-shale contact. Label each contour with its elevation. This is known as a **structure contour map.**

c. These structure contours represent the three-dimensional shape of the sandstone-shale contact underground. As for all contours, each line corresponds to the intersection of a horizontal plane with the contoured surface. Keeping these ideas and your answer to Exercise 6 in mind, use your structure contours on the map to guide you to draw in the approximate location of the oil-water boundary. What elevation corresponds to this boundary? _____

Using a different color than your structure contours, draw in a contour for this elevation. Shade in where oil is present beneath the surface of the map area.

d. Look at Figure 17.4 and decide which one of the traps shown corresponds to the trap on your map.

_____

e. If Section 22 had some acreage for lease, which area would you lease, if any, to drill for oil? _____

How about Section 21? _____

Section 17? _____

What proportion of Section 8 has oil? _____

f. Where (in what quarter section) could you drill to get the most oil from a single well? _____

What is your reasoning? Draw a sketch to help illustrate your reasoning if appropriate.

10. Once the oil and natural gas are extracted, they need to be transported to their destination. Depending on details of the regional geography, this transport may be done by pipeline, tanker trucks, tanker rail cars, or ships along navigable waterways. Can you think of any problems that might arise from any of these methods of transportation related to

a. the environment? _____

_____

_____

_____

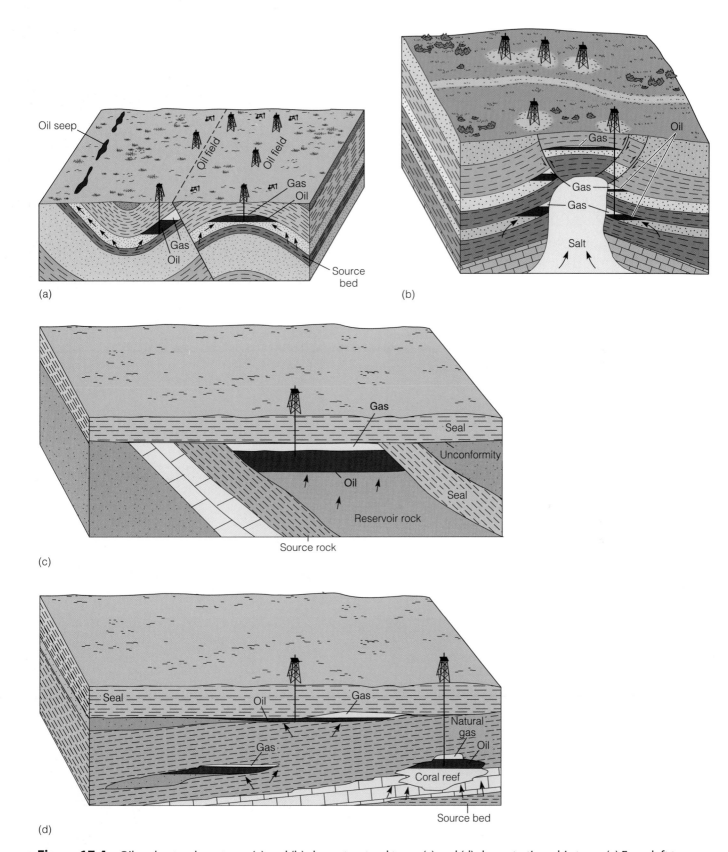

Figure 17.4 Oil and natural gas traps. (a) and (b) show structural traps; (c) and (d) show stratigraphic traps. (a) From left to right shows an oil seep, a fault trap, and an anticline trap. The oil seep is not a trap. (b) Salt dome trap. (c) Unconformity trap. (d) Pinchout and reef traps.

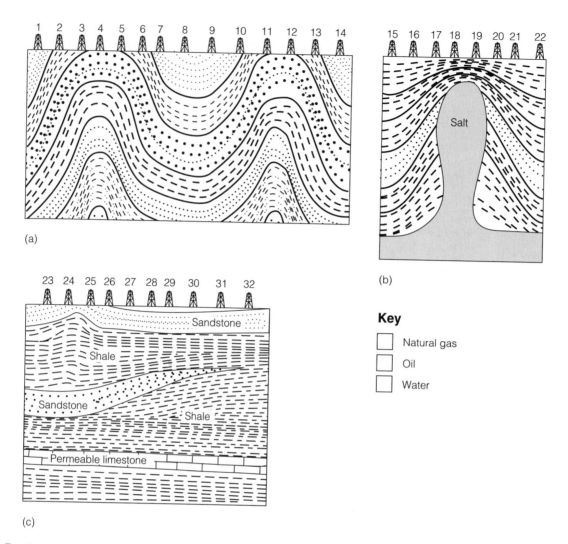

(a)

(b)

(c)

Key

- [ ] Natural gas
- [ ] Oil
- [ ] Water

Sandstone

Shale

Sandstone

Shale

Permeable limestone

Salt

Figure 17.5 Cross sections of regions with oil and natural gas. For each cross section (a, b, and c), locate and color in where you would expect oil and natural gas to collect assuming source beds occur nearby. Fill the appropriate colors in the key. Use (c) as a key for the rock symbols in all three cross sections.

Figure 17.6 Map of a township with the locations of oil wells. Black circles indicate successful oil wells, and open circles indicate "dry holes." "Dry holes" are not actually likely to be dry; they do have water, but they do not have oil. No natural gas has been found in this area. The numbers 15–20 are section numbers. Each well is labeled with the elevation in feet above sea level of the contact between the Eocene Fairchild Sandstone and the Miocene Humphrey Shale as found in that well.

Table 17.2 Solid Fossil Fuels

Sample Number	Type of Coal or Peat	Brief Description of Sample	Water Content	Heat Content Compared to Lignite
	Peat		>~75%	60%–80%
	Lignite		~10%–75%	~100%
	Bituminous coal		~5%–10%	110%–150%
	Anthracite		<~5%	120%–170%

b. safety? _____

_____

_____

_____

11. The burning of oil and natural gas can produce carbon dioxide, water vapor, carbon monoxide, nitrogen oxide gases, and, for oil, sometimes sulfur dioxide gas. Are there any environmental problems with this?

_____

_____

_____

_____

_____

Coal

Another fossil fuel, coal, is a brown or black rock made of organic matter that can burn. It is the most abundant fossil fuel, with especially large quantities in Russia, the United States, and China. It forms by the accumulation, burial, and lithification of plant matter from a swamp or wetland setting. The wet environment and rapid accumulation of organic material help to reduce oxygen and protect the organic matter from further decomposition to carbon dioxide. As overlying sediments bury and compress the organic material, it loses water (Table 17.2 and Fig. 17.7), hydrogen, and nitrogen, causing carbon to increase. At first the material becomes **peat** then turns into a soft brown coal called **lignite** (Fig. 17.7). Continued burial concentrates carbon further, turning the coal into harder, lustrous **bituminous coal.** With low-grade metamorphism even harder, denser and shinier **anthracite coal** forms. (See Lab 5 for more complete descriptions of these types of coal.) The progression from peat to lignite to bituminous to anthracite coal not only is accompanied by an increase in carbon content but also results from increasing temperature of formation and increasing density (Fig. 17.7). The energy extractable per unit mass also increases (Table 17.2).

12. In what environment does the organic matter found in

coal accumulate? _____

What other rocks might you expect to form above or below the coal (what might be deposited in a similar environment)?

_____

13. Examine the samples of coal and peat provided. Match the samples with descriptions in Lab 5, and enter the sample numbers and a brief description of the samples in the appropriate places in Table 17.2. Draw an arrow in Table 17.2 indicating increasing change of the samples

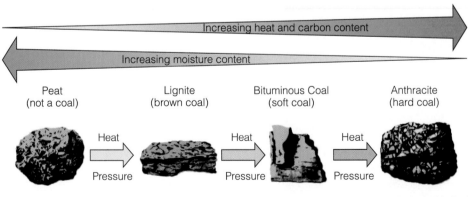

Figure 17.7 Coal types and formation. Over millions of years, coal may form and change with increasing heat and pressure. Moisture content decreases, and carbon content increases. Peat is a soil material, lignite and bituminous coal are both sedimentary rocks, and anthracite is a low-grade metamorphic rock.

Increasing heat and carbon content

Increasing moisture content

| Peat (not a coal) | Lignite (brown coal) | Bituminous Coal (soft coal) | Anthracite (hard coal) |

Heat / Pressure

Partially decayed plant matter in swamps and bogs; low heat content

Low heat content; low sulfur content; limited supplies in most areas

Extensively used as a fuel because of its high heat content and large supplies; normally has a high sulfur content

Highly desirable fuel because of its high heat content and low sulfur content; supplies are limited in most areas

from the original organic matter—from least changed to most changed.

Coal Mine Evaluation In general, extraction of coal or other resources depends on the economic feasibility of the entire operation. The price of the resource on the market needs to exceed the cost of its extraction. Coal mining can be done underground, but it is much safer to remove the unwanted rock above the coal, called the **overburden,** using a surface mining technique such as strip mining (Figs. 17.8 and 17.9). This technique is only feasible if the coal is close enough to the surface. To calculate whether it is economically feasible to remove the overburden, miners use a simple overburden to coal ratio such as 30:1. With a 30:1 ratio, up to 30 ft of overburden could be removed for a 1-ft-thick coal bed. If the bed were 2 ft thick, the removable overburden could be up to 60 ft thick. Miners may need to adjust the ratio to allow for the quality and value of the

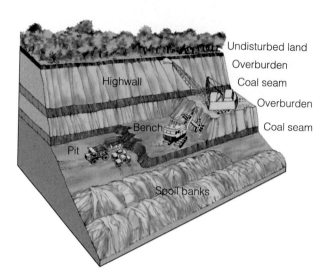

Figure 17.8 Strip-mining coal. Overburden is removed in strips and piled in the spoil banks; then the coal is removed. The next strip of overburden is removed and piled in the previous strip, and mining continues.

(a)

(b)

Figure 17.9 Actual strip mines. (a) Strip mining for a coal-fired power plant in New Mexico. Hummocky areas have been mined; smooth areas are still to be mined. The parallel striping of the hummocks shows the pattern of mining along parallel strips. (b) An abandoned coal strip mine in Belmont County, Ohio. The coal layer is visible (gray) in the highwall (artificial cliff face). The hummocky spoil banks are quite permeable and allow the formation of sulfuric acid and leaching of cadmium and manganese into acid pools.

coal and the difficulty or expense of removing the un-wanted material.

14. Examine the map in Figure 17.10, and review the rule of V's in Lab 16 if necessary.

   a. Is the Mosby Coal bed horizontal or dipping?

   _____ If it is dipping, what is the dip di-

   rection? _____ If it is horizontal, what is

   the elevation of the top of the bed? _____

   b. The Mosby Coal is 3 ft thick in this area. What is the maximum overburden that could be removed eco-nomically if the overburden to coal ratio were 25:1?

   _____

   c. Color the map and make a corresponding key for (1) the area where no coal is present because it has been eroded away, (2) the area where the overburden is thin enough to mine the coal, and (3) the area where the overburden is too thick.

15. Examine the Geologic Map of the Southern Half of Som-erset County, Pennsylvania, in Figure 17.11. The red con-tours (structure contours) mark the elevations of the up-per Kittanning coal bed (*uk*).

   a. What formation directly underlies the structure con-tours of the upper Kittanning coal bed (read the map

   explanation)? _____

   What formation directly overlies the structure con-

   tours (*uk*)? _____

   b. What would you expect to find on the ground where the structure contours equal the topographic contour elevations (e.g., where the 2700 structure contour in-tersects the 2700 topographic contour east of Ringer Hill at longitude 79°11′ W)? Use Figure 17.12 to help you visualize this.

   _____

   c. If the structure contours are lower than the topo-graphic contours, at what level is the upper Kittan-ning coal bed? Circle one.

   - Below the ground
   - Above the ground
   - Lower than the structure contours
   - Higher than the structure contours

   d. If the structure contours are higher than the topo-graphic contours, what has happened to the coal? Since this area has not been mined, think of a natural process that has occurred.

   _____

16. Now apply your answers in Exercise 15 to a situation on the same map (Fig. 17.11). In the center of the anticline trending northeast across the entire map, find the mine

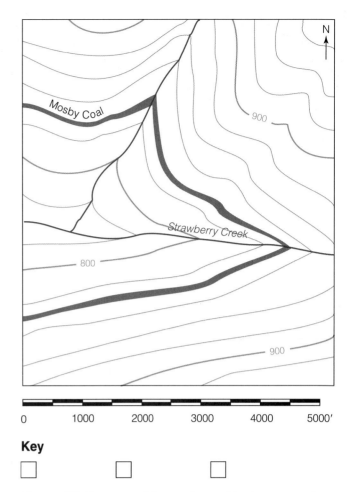

Figure 17.10 Map of the Mosby Coal bed near Strawberry Creek. The 3-ft-thick coal bed is shown in gray. Contour inter-val is 20 ft.

or quarry (the crossed hammer and pick symbol) in the Mn (pink) bed south of the lake.

   a. What is the elevation of the land here? _____

   Of the upper Kittanning coal bed? _____

   b. Would this be a good location to mine the *uk* coal

   bed? _____ If so, how far down would you have to

   mine to hit it? _____ If not, why not?

   _____

   c. Is the pink rock, Mn, in the map key, above or below

   the *uk* coal bed? _____ Is this consistent

   with your answers in **a** and **b**? _____

17. The following questions **a–e** are intended to help you de-termine the economic feasibility of mining in the part of Figure 17.11 shown in Figure 17.13.

   a. Strip mining of Kittanning coal is feasible at an over-burden to coal ratio of 30:1. If the *uk* coal bed is about 5 ft thick, how many feet of overburden could be

   removed economically? _____

SCALE 1:62500
CONTOUR INTERVAL 20 FEET
DATUM IS MEAN SEA LEVEL

NEGRO MOUNTAIN ANTICLINE

YOUGHIOGHENY SYNCLINE

79°10'
79°15'
39°50'
39°45'

4 MILES

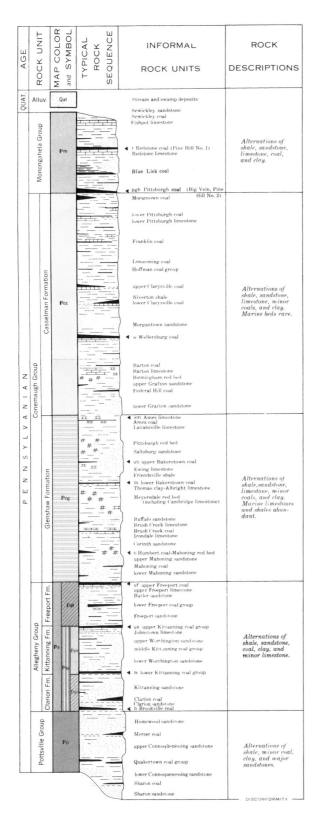

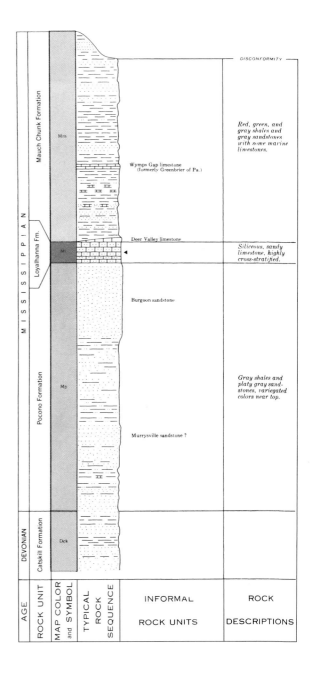

From: Pennsylvania Topographic and Geologic Survey Bulletin C56A

Geology by Norman K. Flint

Base from U.S.G.S. 15′ and 7½′ Quadrangles

Figure 17.11 Part of the Geologic Map of the Southern Half of Somerset County, Pennsylvania. Red lines are structure contours on the *uk* coal bed.

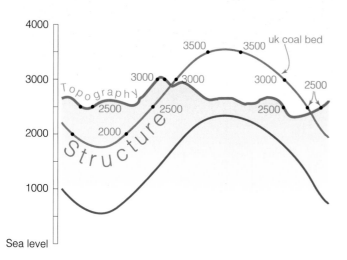

Figure 17.12 Cross section showing the relationship between topographic and structure contours

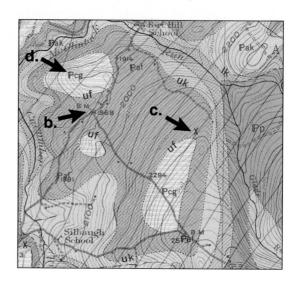

Figure 17.13 Reference map for locating features in Exercise 17 on the map in Figure 17.11. The arrows and the letters, b, c, and d show the location of features mentioned in questions b, c, and d, respectively.

Use Figure 17.13 to pinpoint the locations for Questions **b, c,** and **d** in Figure 17.11.

b. Study the area of ℙaf and ℙcg between Silbaugh School and Fort Hill School to see whether this area could be strip mined. Find the 1968 benchmark (**BM**). What is the elevation of the Kittanning coal bed there? _____ Use the benchmark elevation to determine how much overburden is present there. _____ Could this be strip mined at a profit? _____

c. Find the **X** about a mile to the east of the benchmark on the map. How much overburden is present at the **X**? _____

d. To the northwest of the benchmark is a yellow patch labeled ℙcg. Notice the *uf* bed at the contact between the yellow and green, and refer to the key. Would it be economically feasible to mine down to the *uk* bed in these yellow patches if the *uf* bed is about 6 ft thick? _____

e. Judging from your answers in **a–d**, what area could be strip-mined in this vicinity? Outline the area on the map in Figure 17.13.

Coal and the Environment Using coal has advantages, mainly involving abundance and energy production, and disadvantages which are chiefly environmental. Mining for coal or mineral resources can severely change the shape of the landscape. **Strip mining** is a method usually used for shallow horizontal or nearly horizontal mineral or coal deposits. The overburden is removed from a strip and piled up next to it, and then the coal or mineral is removed. Overburden from the next parallel strip is piled in the previously mined strip, and the process continues. Figure 17.9 shows a coal strip mine in New Mexico (a) and a closer view of a strip mine in Belmont County, Ohio (b).

Surface mining such as strip mining is especially disruptive to the land because so much material is removed from above the coal or other resource. The removed overburden, or **spoil**, is broken and crushed waste rock. The volume of the spoil is around 30% greater than the original volume of the overburden because of the increased porosity of the loosely packed spoil. The volume of coal or ore removed is generally less than the increase created by breaking up the overburden, creating a problem of where to put this large volume of waste.

18. Examine Figure 17.9a.

   a. Where did the miners put the spoil? _____

   _____

   _____

   b. Do you think they should leave it like this? _____

   Why or why not? _____

   _____

   _____

   c. Would your opinion change if (i) the area were forested beforehand? (ii) If it were in sight of your home? State why or why not for each case and consider environmental issues.

   _____

   _____

   _____

   _____

   _____

   _____

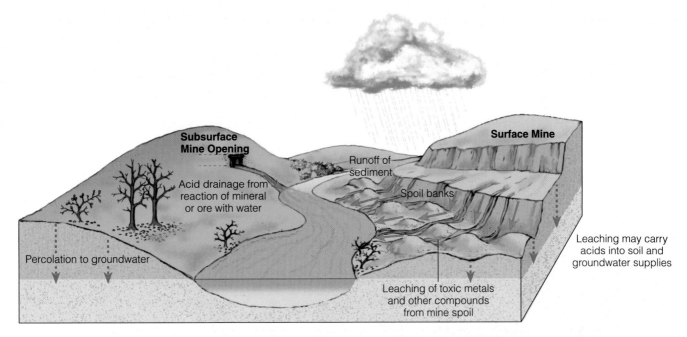

Figure 17.14 Mining can cause pollution in the form of sediment runoff from spoil banks of surface mines, in the form of acid mine drainage from surface and underground mines, and in the form of toxic chemicals such as metals or mining chemicals.

The spoil is readily eroded and tends to increase sediment in streams that harms aquatic life (Fig. 17.14); it also may fill stream channels, increasing the chance of flooding. Coal or spoil piles may become landslide disasters if saturated during heavy rains. In addition, loose spoil has a greater permeability than the original overburden and allows water to percolate through it, which causes additional problems with water pollution.

Of all of the fossil fuels, coal can have the highest sulfur content, from 0.2% to 7%. This is because of the very low oxygen conditions where coal forms. In fact, the organic matter can strip oxygen from substances, a process called **reduction.** Any oxidized sulfur (sulfate, SO_4^{2-}) present in the organic matter or in fluids moving through it will be reduced (stripped of oxygen) to sulfur ions (sulfide, S^{2-}) and combined with iron to form pyrite, an iron sulfide.

Mining coal leads to water pollution when sulfur in the pyrite interacts with rainwater and makes sulfuric acid. This produces **acid mine drainage** (Figure 17.14). If the acid water gets into streams, lakes, and rivers, it may kill fish and other aquatic organisms directly or in lower concentrations destroy their eggs. It also causes toxic metals to leach out of the mine area and surrounding rocks, adding metallic poisons to the water. The water is commonly red to orange-brown due to very fine iron oxides in the water (Fig. 17.15).

Use of coal for energy involves burning, which results in substantial air pollution. Burning of coal oxidizes whatever pyrite is present to sulfur dioxide gas, which becomes air pollution and contributes to acid rain by forming sulfuric acid when it combines with rainwater (Fig. 17.16). Oil has some sulfur but not nearly as much as high-sulfur coal. In addition, compared to other fossil fuels, coal has the highest proportion

Figure 17.15 Acid mine drainage from a gold mine, Bear Trap Creek, Montana. The orange color comes from very fine iron oxide particles suspended in the acidic water resulting from the dissolution of iron from pyrite.

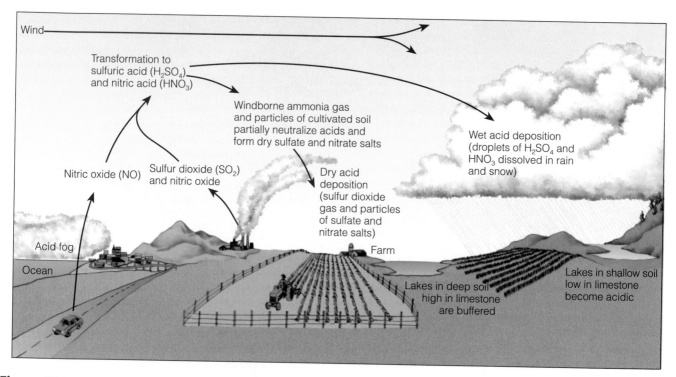

Figure 17.16 Acid deposition, in the form of rain, snow, fog, dust, or gas, comes from air pollution from automobiles, and coal-burning power plants and influences, soil, farming, lakes and rivers, and forests. Limestone neutralizes the acid, but areas without limestone become acidified.

of carbon and therefore releases the most carbon dioxide, a greenhouse gas, per unit of energy produced.

19. If all or part of the area in Exercise 17**e** were strip-mined, what problems would be created for the people in the town of Confluence near the western edge of the map in Figure 17.11, west of the area in Figure 17.13?

_____

_____

_____

_____

_____

_____

20. In summary, list two problems associated with coal for each of the following:

a. Hazards of mining _____

_____

b. Environmental problems with mining _____

_____

_____

c. Environmental problems with using coal _____

_____

_____

Carbon Cycle

Carbon is an important resource element because it combines to make so many useful things. Food, clothing, shelter, plastics, and our fossil fuel energy resources all have carbon or involve resources containing carbon. Organic and inorganic carbon is prevalent in places such as the atmosphere, the oceans (hydrosphere), all living things (biosphere), and even rocks (lithosphere). These places or spheres are sometimes called *carbon reservoirs*. Carbon exists in different chemical forms such as carbon dioxide, methane, carbonate minerals, and organic carbon such as sugar, starch, fat, oil, protein, wax, cellulose, and plant and animal fibers. It moves from one reservoir to another by many different processes, including human activities. This movement of carbon from one form to another and from one place to another is known as the **carbon cycle** (Fig. 17.17). You will be constructing a graphical model of the carbon cycle on the diagram in Figure 17.18. In this diagram, the largest ellipses represent places where carbon exists on or in the Earth. Within these reservoirs, substances containing carbon are in smaller ellipses.

21. Try to think of as many substances that contain carbon as you can, and pencil them into the ellipses in appropriate spheres on the carbon cycle diagram in Figure 17.18. Then check your answers and add any you may have missed using the list of substances (left column) in Table 17.3. Table 17.3 does not list all possible substances containing carbon, so you may have to add ellipses or double up substances if you have thought of different ones than those listed in the table.

22. On the carbon cycle diagram, draw arrows between the ellipses, showing how carbon moves from one reservoir to another. Each arrow is a process. Try to think of as many processes as you can, and pencil in arrows and label each arrow with the name of the process. Then check your answers and add any you may have missed using all of the processes (right column) listed in Table 17.3. Each arrow should be labeled. The descriptions listed in the table after the processes will help you determine where to draw the arrows. Table 17.3 does not list all possible processes involving carbon, so you may have thought of some not listed.

When you have used all the substances and processes in Table 17.3, you have completed the carbon cycle. Next let's look at some Earth materials that contain carbon in your carbon cycle diagram. Some of these are samples you have identified earlier in this lab.

Earth Materials and the Carbon Cycle

Carbon comes in three basic chemical types:

- *Organic unoxidized or reduced carbon*—humans can burn the reduced type of carbon to obtain energy, which turns it into the oxidized form. Solid or liquid organic carbon is reduced and may have a brown to black color. Gaseous reduced carbon is mostly methane, CH_4.
- *Elemental carbon* occurs naturally as graphite or diamond. Graphite and diamonds can also burn although we do not burn them for energy.
- *Oxidized carbon* is carbonate if solid and may effervesce when acid is applied. Gaseous oxidized carbon is carbon dioxide.

Both carbon gases mentioned here are odorless, tasteless, and colorless; but one is flammible and the other is not.

23. Examine the samples provided and fill in Table 17.4. For Type of Carbon, indicate whether carbon is present in the sample and whether it is organic unoxidized carbon (black or dark color) or oxidized carbon as in carbonate (effervesces in HCl). For some of these you may have to think about whether they can be burned. (*Beware:* Not all black substances contain carbon.) Also indicate what the sample might be used for. If needed, you can refer to uses in Table 17.3, but some you might just want to think about.

24. Write the sample numbers from Table 17.4 on your carbon cycle diagram in Figure 17.18 in the appropriate ellipses.

Humans and the Carbon Cycle

As you can see from the human sphere in your carbon cycle diagram (Fig. 17.18), humans interact in numerous ways with the carbon cycle and modify it considerably. We obtain many resources from the carbon cycle. Also, significant human influence of the carbon cycle is prevalent today in the re-lease into the atmosphere of two important greenhouse gases, carbon dioxide and methane (Fig. 17.19a and c). Greenhouse gases that are not significant parts of the carbon cycle are also on the rise (Fig. 17.19b and d). The most important greenhouse gas, carbon dioxide, has been rising since the beginning of the industrial revolution (Fig. 17.20).

25. What resources come from the carbon cycle?

_____

_____

_____

_____

26. Highlight each process (arrow) on the carbon cycle diagram that is either exclusively performed or substantially influenced by humans. Not all of these move from or to the human sphere.

27. Without the "human sphere," the natural carbon cycle produces a balance of carbon in the different carbon reservoirs. How do humans influence the carbon cycle?

_____

_____

_____

_____

28. Is that human influence significant in changing the balance? _____ If so, what carbon reservoir(s) is/are gaining carbon due to these influences? _____

_____

What reservoirs are losing carbon because of humans?

_____

29. In light of the greenhouse effect and the carbon cycle, what is a likely future result of these human activities that move carbon from one carbon reservoir to another?

_____

_____

_____

_____

Fossil fuels provide energy resources. In the carbon cycle, we saw additional resources including some biological and some mineral resources containing carbon. Next we will explore other mineral resources.

Mineral Resources

So many of the common materials and objects we use every day are made from mineral resources. Even if those everyday

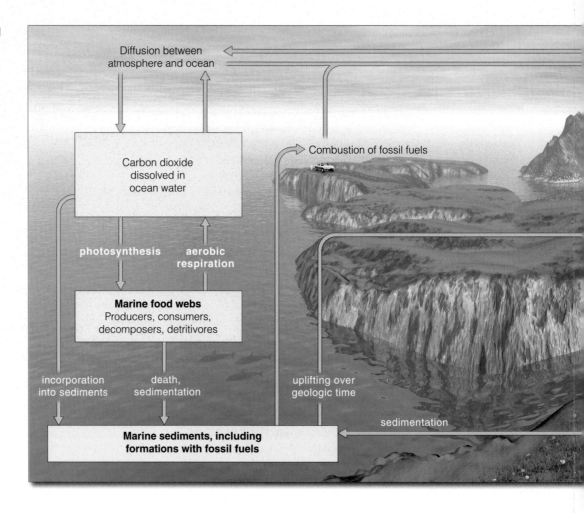

Figure 17.17 A pictorial representation of the carbon cycle. Marine systems are on the left and terrestrial (land) on the right.

Diffusion between atmosphere and ocean

Combustion of fossil fuels

Carbon dioxide dissolved in ocean water

photosynthesis   aerobic respiration

Marine food webs
Producers, consumers, decomposers, detritivores

incorporation into sediments

death, sedimentation

uplifting over geologic time

sedimentation

Marine sediments, including formations with fossil fuels

objects are not made of an Earth resource, they require mineral resources of many types to make them available to us. Consider a piece of bread. Although it has a biological source, the process of producing it and bringing it or its ingredients to you involves a myriad of mineral resources. Soil, water, and sunlight grew the grain. Iron, nickel, chromium, carbon, manganese, and silicon were used to make the stainless steel used to harvest and transport the wheat or to cut the bread. Aggregate and cement from gravel and limestone were used to make the grain elevator to store the grain or the pavement on which the trucks traveled to bring it to you. The list could go on and on. Let's look at some common objects, the material that makes them up, and the mineral resources required to produce them.

Common Materials and Their Origins

30. Around the room, a quantity of common objects have been labeled with numbers. Locate these objects; some are attached to or are part of the room, and some are loose objects. Table 17.5 lists a variety of materials that might make up common objects.
  a. Identify the composition of the objects around the room, and enter the objects in Table 17.5 in the appropriate row. Figure 17.21 may help you identify the metal objects.

  b. Fill in the rest of the table using Tables 17.6 and 17.7 where necessary.

Resource Minerals

31. From Table 17.7, list the minerals that are sources of copper. _____

_____

32. From Table 17.7, what minerals are mined for iron?

_____   _____

33. Your instructor will provide you with a set of minerals that are used as resources. Identify each of the mineral samples provided, and fill in Table 17.8 with the mineral properties. Use the mineral identification tables from Lab 3 (Tables 3.3 and 3.4). For "Matching Object(s)," indicate which object or objects from Table 17.5 would be made from this resource mineral.

Scarcity versus Abundance   Mineral resources are nonrenewable (Fig. 17.1). They have a finite supply, and when that supply is used up, the only way to get more will be to recycle what has already been used. The U.S. Geological Survey and other government agencies commonly report the abun-

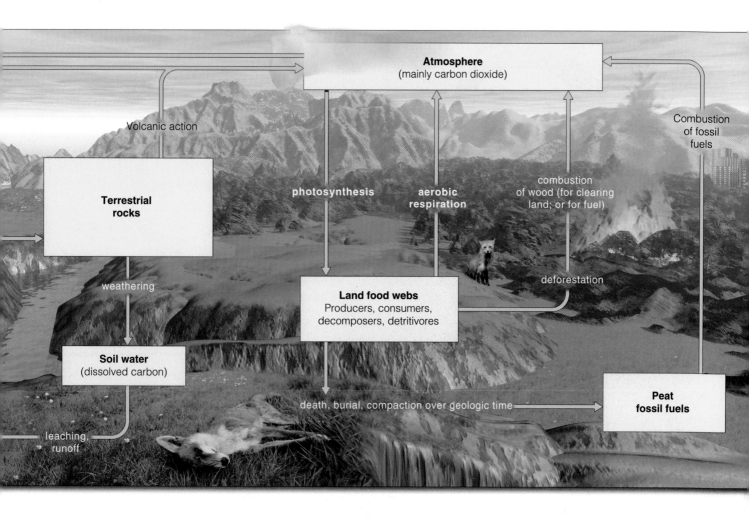

Atmosphere
(mainly carbon dioxide)

Volcanic action

Combustion
of fossil
fuels

**Terrestrial
rocks**

photosynthesis

**aerobic
respiration**

combustion
of wood (for clearing
land; or for fuel)

weathering

deforestation

Land food webs
Producers, consumers,
decomposers, detritivores

Soil water
(dissolved carbon)

death, burial, compaction over geologic time

**Peat
fossil fuels**

leaching,
runoff

dance of different elements or commodities in terms of reserves, estimated resources, and projected lifetimes. **Reserves** are the quantity of a resource that has been found and
is economically recoverable with existing technology. **Estimated resources** are an estimate of the total quantity of a
particular substance (resource) on Earth. So reserves are a
subset of estimated resources, which themselves are somewhat less than the total existing amount of the resource. **Projected lifetimes** are calculated simply by assuming that the
current rate of production of a resource will continue and
dividing that into the reserves of that resource:

$$\text{Projected lifetime} = \frac{\text{Reserves}}{\text{Production rate}}$$

34. In 1999, the worldwide reserves of iron ore were 140,000
million metric tons, with a production rate of 1000 million metric tons per year. The estimated resource of iron
ore is 800,000 million metric tons. What is the projected
lifetime of iron ore? Write out the method of calculation.
How does the number compare if you use the estimated
resource instead of the reserves in the calculation?

35. In the same year, the worldwide reserves of tin were 7.7
million metric tons, with a production rate of 0.21 million metric tons per year. What is the projected lifetime
of tin? _____. Show your calculations.

36. In a group, discuss the results of your calculations and
the following questions:
  a. Do you see any problem(s) that may result from such
  projected lifetimes?

  _____

  _____

  b. Do you think the price of these resources would influence their rate of use? _____
  If so, in what way? If not, why not?

  _____

  _____

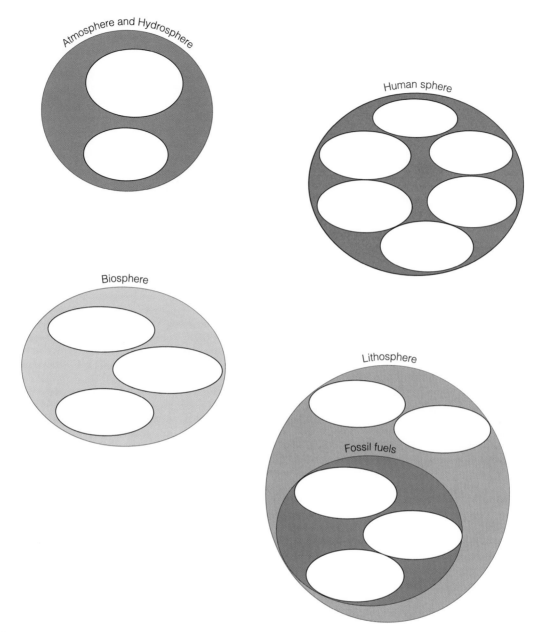

Figure 17.18 Carbon cycle. Each "sphere" is a carbon reservoir. Each ellipse within a "sphere" is a substance containing carbon. Fill in the substances from the first column of Table 17.3. Arrows (to be drawn) represent processes that move carbon from one reservoir to another. Connect the substances with arrows representing processes from column 2 in Table 17.3.

c. How would the following conditions affect the quantity of reserves? (Review the definition if necessary.)

- Value of the resource increases:

_____

_____

- New mining technique improves the efficiency of extraction of the resource:

_____

_____

d. Is the projected lifetime the actual length of time before a particular resource will run out? _____

If so, justify your answer. If not, what could cause the actual lifetime of the material to change?

_____

_____

_____

_____

Table 17.3 Substances and Processes of the Carbon Cycle

Substances That Contain Carbon—Labels for Ellipses	Processes That Move Carbon—Labels for Arrows
Carbon dioxide gas—in the atmosphere and dissolved in the oceans	**Photosynthesis**—plants take up CO_2 with the help of sunlight.
Methane gas—in the atmosphere and in the oceans	**Eating**—animals eat plants for nourishment.
Plants and their remains—part of the biosphere	**Respiration**—animals when they breathe and plants at night give off CO_2 into the oceans or atmosphere.
Animals and their remains—part of the biosphere	**Decay** and **digestion** – in the absence of oxygen, plant and animal remains decay and give off methane; some organisms such as cows and termites give off methane as part of their digestive process (including flatulence). Decay of plants during rice cultivation releases substantial methane.
Shells—made of calcium carbonate; part of the biosphere	**Burning**—of biological materials, including forests
Limestone—rocks containing calcium carbonate ($CaCO_3$)	**Lithification**—converts shells into limestone, plants into coal, and organic matter in microscopic sea critters into crude oil and natural gas
Coal—rock used to provide energy	**Degassing**—metamorphism and volcanic activity break down organic molecules and carbonates in rocks and release CO_2.
Crude oil—liquid in rock used to provide energy	**Weathering and precipitation**—CO_2 from the atmosphere combines with rain water to make a weak acid that aids in the weathering of rocks containing silicate minerals. The resulting bicarbonate and calcium ions in solution travel in rivers to the sea where they precipitate as limestone.
Natural gas—gas in rock used to provide energy (mostly methane, CH_4)	**Weathering of organic matter** in rocks or fossil fuels produces CO_2.
Organic matter in rocks (excluding fossil fuels)	**Extraction** (includes mining and pumping)— people remove material from the Earth to be used in the human sphere.
Food for humans made from plants and animals	**Tree cutting,** or **deforestation** in the extreme case—humans harvest trees for wood products and fuel.
Clothing made from plant and animal fibers or from petroleum	**Agriculture**—growing plants for human uses extracts carbon dioxide from the atmosphere through photosynthesis to grow the plants.
Refined fossil fuels—coal, refined oil, gasoline, natural gas; part of the human sphere	**Ranching** or **animal husbandry**—raising animals for human use extracts carbon from the plants to feed the animals.
Wood for construction, furnishings, and heating	**Harvesting**—plants and animals and their products for food and clothing
Limestone for lime to make cement; part of the human sphere	**Burning**—of coal, oil, gasoline (a refined oil), and wood by humans, releases CO_2
Plastics made from petroleum (or plant material)	**Making lime**—a human activity that converts limestone into lime (used to make cement) by releasing CO_2

Table 17.4 Earth Materials Containing Carbon (with one or two exceptions)

Sample Number	Sample Name	Type of Carbon, If Any	Use

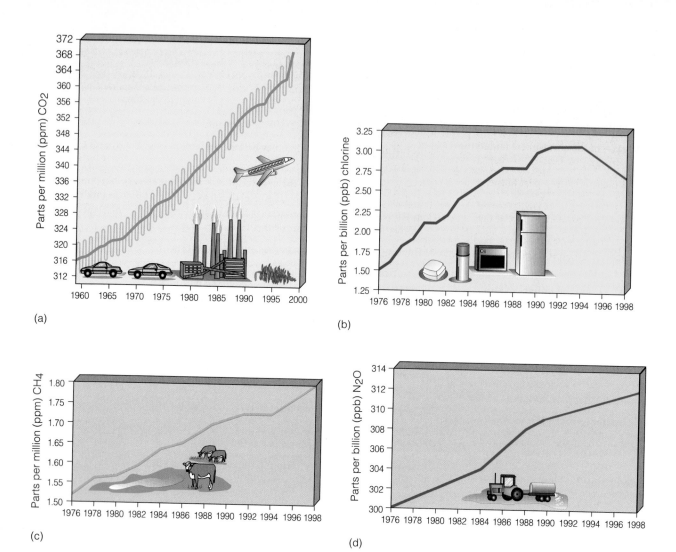

Figure 17.19 Increasing greenhouse gases in the last 30 or 40 years. (a) Carbon dioxide is the most important greenhouse gas because it is the most abundant. It is produced by burning fossil fuels (70%–75%), removing and burning plants (20%–25%). (b) Chlorofluorocarbons are a manufactured gas that was introduced in the 1940's and has been banned and is being phased out. Sources of CFCs are (or were) leaking refrigerator and air conditioner cooling units, escaping industrial solvents, foaming plastics, and propelling aerosols. CFCs are both greenhouse gases and ozone layer destroyers. (c) Methane is a strong but not abundant greenhouse gas that results from leaking gas pipelines and anaerobic decay in swamps, rice paddies, landfills, and digestive tracts of cattle, sheep and termites. (d) Nitrous oxide is a very strong greenhouse gas (200 times stronger than CO_2 per molecule). It is produced by nylon manufacturing, auto exhaust, nitrogen fertilizers, livestock waste and burning nitrogen-rich fuels such as coal and organic matter (biomass).

e. What are all the things you can think of that are likely to happen if we ran out of a resource or if its quantity dropped to a very low amount? Think about its use and what could be done about its shortage. Think about economics, price, mining. Might there be environmental consequences? You may need to select a specific resource or two as you explore the possibilities.

_____

_____

_____

_____

_____

_____

_____

_____

_____

_____

_____

The Geology and Mining of a Resource

Most mineral resources are not economically extractable unless they have been concentrated in some way by some process in or on the Earth. An **ore** is a rock containing one or more economic metal resources. Many different processes can concen-

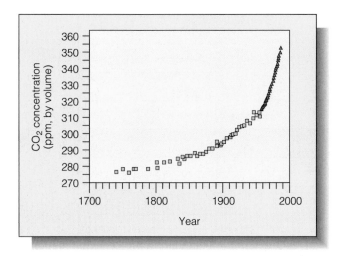

Figure 17.20 Increasing carbon dioxide. Figure 17.19a shows carbon dioxide increases from 1960, but it has been increasing since the industrial revolution. Yellow squares are measurements from ice cores, and red triangles are direct atmospheric measurements at Mauna Kea, Hawaii.

trate mineral resources. An example is the settling of heavy crystals through fluid magma to form a concentration of a mineral such as chromite (Fig. 17.22). The heat and fluids associated with the magma may help concentrate minerals in the surrounding rocks, producing hydrothermal deposits (Fig. 12.23) Aluminum is concentrated in bauxite, where tropical weathering removes almost everything but the aluminum. Streams and waves can concentrate minerals in a **placer** deposit. Erosion strips away minerals from rock upstream, and the stream transports and sorts the material (Fig. 17.23). Placers form where turbulent flowing water segregates heavier minerals from lighter ones. Water action washes away light minerals and con-

centrates heavy minerals in a way similar to the action of panning for gold. Placer minerals generally have higher than average density and must be resistant to chemical and mechanical breakdown in water; that is, they do not react chemically and they do not break apart into smaller pieces. Example placer minerals are listed in Table 17.9. Notice that many gems are concentrated in placer deposits.

Prospecting for Placer Minerals in the Stream Table Ideally we would like to prospect for gold; instead, we will observe the placer concentration of more common heavy minerals and "pan" for them. The stream table should be set up with a sand dam at the inflow end. Also provide some barriers that will deflect the flow of the stream at various places.

- Fill the lake at the inflow end with water as high as possible without overflowing.
- Quickly cut through the sand dam so that the water flows out rapidly.
- Occasionally resupply sand where the stream will erode it.
- Watch the flow of water and the movement of sand.
- Repeat the experiment, with various configurations such as stream meanders or tributaries.

37. Are there any places where the sand shows increased concentration of uncommon mineral grains? _____

If so, describe their location in terms of meanders, obstructions in the flow, the speed of water flowing, and so forth.

_____

_____

_____

_____

Table 17.5 Composition of Common Objects and Their Resources

Object Number	Object	Name of Metal or Primary Substance	Description	Made from/of	Resource Mineral(s) or Rock(s)
		Iron or steel			
		Chromium stainless steel			
		Nickel stainless steel			
		Aluminum			
		Gold			
		Brass			
		Copper			
		Concrete			
		Wallboard (or sheet rock or plasterboard)			
		Porcelain			
		Glass			
		Graphite			

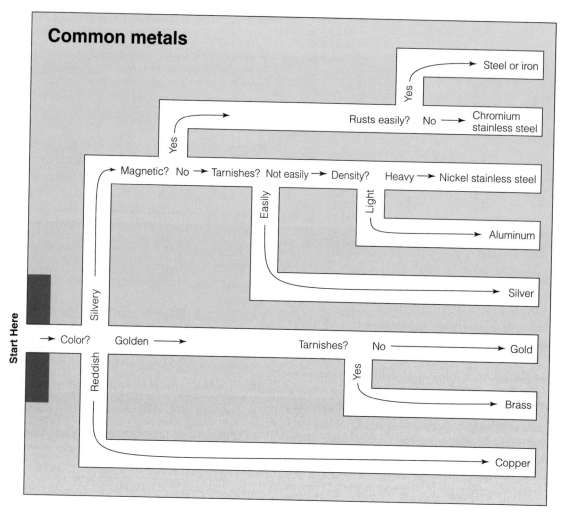

Figure 17.21   Maze for identifying some common metals

Table 17.6   Composition of Common Substances

Material or Substance	Made from/of	
Iron or steel	Iron (+carbon in steel)	
Chromium stainless steel	Iron, carbon, chromium	
Nickel stainless steel	Iron, carbon, nickel, chromium	
Brass	Copper and zinc	
Concrete	Aggregate	
	Cement	Lime (CaO), alumina (Al$_2$O$_3$), silica (SiO$_2$)
Wallboard (or sheet rock or plasterboard)	Plaster	Gypsum
	Paper	Wood
	Ceramic	Clay
Porcelain	Kaolin, feldspar, quartz	
Glass	Quartz	

38. Collect two or more samples of sand on watch glasses as follows:
    a. Average-looking sand—Label the watch glass A.
    b. Sand of a different color (placer)—Carefully scrape up just the sand from the surface. Label the watch glass B. Describe the location: _____

    _____

    c. Sand of another color, if applicable, or a second location with the same-colored sand as in **b**—Carefully scrape up the sand from the surface. Label the watch glass C. Describe the location: _____

    _____

39. Examine the sand in the watch glasses with a hand lens or binocular microscope. What colors of sand grains are present? Enter the colors in the blanks in the first row in Table 17.10 in order from most abundant to least abundant in the average sand. Describe the sand by filling in Table 17.10. For A, give a general description. Describe Sands B and C using terms relative to average Sand A. For example, Sand A might have 80% tan grains, 15% white

Table 17.7 Resource Rocks and Minerals

Resource Rock or Sediment Material	Resource Mineral Material	Used to Provide
Banded iron formation	Hematite and magnetite	Iron
Sulfide mineral deposit	Chalcopyrite	Copper
	Native copper	Copper
	Azurite	Copper
	Malachite	Copper
Sulfide mineral deposit	Sphalerite	Zinc
Layered gabbroic intrusions	Chromite	Chromium
Weathered ultramafic rocks	Garnierite	Nickel
	Pentlandite	Nickel
Placer	Cassiterite (tin oxide)	Tin
Bauxite		Aluminum
Placer	Native gold	Gold
Metamorphic rock	Graphite	Graphite
Kimberlite or placer	Diamond	Diamond
Coal		Carbon
Gravel		Aggregate
Limestone	Calcite	Lime
Clay or shale	Clay	Clay and alumina
Sandstone, quartzite, sand	Quartz	Quartz or silica
Gypsum	Gypsum	Gypsum
Kaolin clay deposit	Kaolin	Kaolin
	Feldspar	Feldspar

Table 17.8 Resource Minerals and Their Properties

Sample Number	Luster	Color	Cleavage/Fracture	Streak	Special Properties	Mineral Name	Matching Object(s)

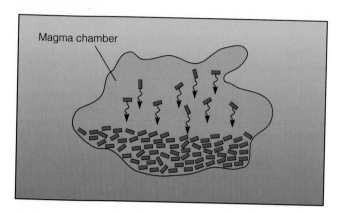

Figure 17.22 Minerals can be concentrated by crystallization of magma and gravitational settling of the denser crystals to the bottom of the magma chamber. Such a mineral deposit is known as a *cumulate* and chromite, a mineral used to extract chromium, is concentrated in this way.

grains, and 5% black grains; B might have many more black grains. You could describe these as shown in the example at the bottom of Table 17.10.

40. Explain how the placer Sands B and C became different from Sand A.

_____

_____

_____

_____

41. If you were prospecting for the mineral concentrated in Sand B or C, how would you find a rich mineral deposit?

_____

_____

_____

_____

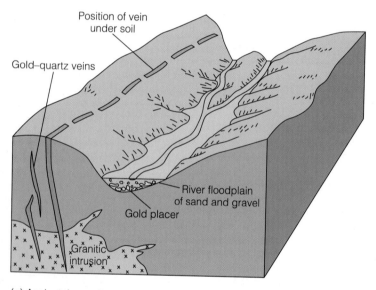

(a) Ancient river pattern

Figure 17.23 Gold may originate in hydrothermal veins associated with granitic intrusions and may be sufficiently concentrated in the veins to warrant mining, but subsequent erosion is likely to produce a higher concentration of the gold in stream beds by the action of water, known as *placer* gold. (a) First placer deposits develop from ancient streams. (b) Later placer development may have higher concentration as later streams erode through old placer terraces and process them even more.

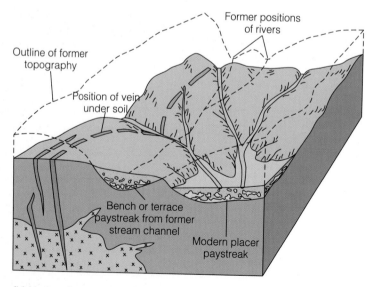

(b) Modern river pattern

Table 17.9 Placer Minerals

Placer Mineral	Density[1] (g/cm^3)	Resistance to Mechanical Breakdown due to
Gold	15–19.3	Malleable not brittle (Hardness = 2.5–3)
Diamond	3.51	Hardness = 10, resistant to abrasion
Rubies, sapphires	4.0	Hardness = 9, no cleavage
Emerald	2.7	Hardness = 8, imperfect cleavage
Garnet	4.3	Hardness = 7, no cleavage
Cassiterite (tin oxide)	7	Hardness = 6–7, imperfect cleavage
Magnetite	5.2	Hardness = 6, no cleavage
Ilmenite	4.7	Hardness = 5.5–6, no cleavage
Rutile	4.2	Hardness = 6–6.5

[1]Average density of minerals in continental crust is about 2.7 g/cm^3.

Table 17.10 Comparison of Average and Placer Sand

Sand	Grain Size (mm and compared to average sand)	Abundance of _____-Colored Sand Grains	Abundance of _____-Colored Sand Grains	Abundance of _____-Colored Sand Grains	Abundance of _____-Colored Sand Grains
A	_____ mm (same as average sand)	Most abundant			
B					
C					

Example Sand	Grain Size (mm and compared to average sand)	Abundance of *Tan*-Colored Sand Grains	Abundance of *White*-Colored Sand Grains	Abundance of *Black*-Colored Sand Grains	Abundance of _____-Colored Sand Grains
A	0.5 mm (same as average sand)	Most abundant *predominant*	*Modest amount*	*Least abundant minor*	
B	*Finer, 3/4 the size of A*	*1/2 as many*	*1/2 as many*	5 times A	

Panning for Heavy Minerals   Let's go panning for the heavy minerals in the sand. Choose a prospect site as you described in Exercise 41, and scoop up some of the sand, trying to get the highest proportion of heavy minerals you can. Put it in a watch glass. While holding the watch glass just under the water, swirl the water and sediment around. Make the swirls large enough to allow the lighter sand to swirl out of the watch glass so that you increase your concentration of heavy minerals. If after a short time you have about the same amount of sand as you started with, you need to swirl the watch glass more vigorously. If you have nothing or very little left after a short time, you need to be less vigorous; try again.

42. Describe what happens as you swirl the sand. Where are the heavy minerals in the watch glass as you swirl compared to the lighter sand?

_____

_____

_____

_____

43. a. Examine the remaining sand in your watch glass with a hand lens or under the microscope, and describe the sand. It may help to place the watch glass on a light-colored background.

_____

_____

_____

_____

b. About what percentage of the sand in the watch glass is the heavy sand? _____

44. Is gold denser or less dense than ilmenite or magnetite, two common minerals in heavy black sand (Table 17.9)?

_____ If gold were also present in your sand, assuming the other heavy minerals are about as dense as ilmenite and magnetite, would it swirl more slowly, behind the heavy minerals, or faster in front of them?

_____

Glossary

Italic terms below are also defined in this glossary.

Absolute age The numerical age or *isotopic age,* usually of a *rock,* and with some degree of error

Absolute humidity The mass of water contained as vapor in a certain volume of air, usually given in units of grams per cubic meter (g/m^3)

Abyssal plain The broad expanse of the ocean floor that is nearly flat and at a depth of about 5000 m

Accretionary wedge A wedge-shaped volume of thrusted and folded *rock* and *sediment* scraped and *thrust-faulted* off a subducting plate onto the edge of a continent. See *subduction.*

Acid mine drainage Acidic water draining from a mine

Active continental margin The edge of a continent with a current *plate boundary*

Adamantine luster A bright *mineral luster* that resembles the way diamond shines

Aftershocks Smaller earthquakes that occur after and are associated with an *earthquake*

Aggregate, mineral A group or cluster of *crystals* with their edges touching one another

Amygdaloidal Texture The *texture* that results when *vesicles* or gas cavities fill with secondary *minerals.* The secondary minerals were deposited after the *solidification* of the original *rock.*

Amygdule *vesicle* or gas cavity filled with secondary *minerals.* The secondary minerals were deposited after the *solidification* of the original *rock.*

Angular unconformity An *unconformity* where the older *rocks* below the unconformity are tilted or *folded sedimentary/volcanic rocks* and are not parallel to the unconformity surface

Anhedral An adjective describing a *grain* lacking well-formed *crystal faces*

Anticline An upward-arching *fold.* See also *syncline, upright fold,* and *plunging fold.*

Anticyclone A high-pressure weather system with dry sinking stable air and clear skies with a clockwise wind rotation in the Northern Hemisphere and a counterclockwise wind rotation in the Southern Hemisphere

Aphanitic texture A *texture* of *igneous rocks* with *crystals* so *fine-grained* that they cannot be seen with the naked eye

Aquiclude A *rock* or *sediment* adjacent to an *aquifer* that is impermeable and thus prevents the flow of water through it. Compare with *aquitard.*

Aquifer A *permeable* body of *rock* or *sediment* below the *water table* that can sustain a productive water *well*

Aquitard A *rock* or *sediment* that has lower *permeability* than an adjacent *aquifer* and retards the flow of water through it. Compare with *aquiclude.*

Artesian aquifer A *confined aquifer* that is under pressure

Ash Sand-sized to powdery *volcanic* material produced by explosive *volcanic eruptions* when a spray of magma and particles of *rock* spew out of the *volcano.* Short for *volcanic ash.*

Asthenosphere A layer of the Earth that is made of weak *plastic rock,* below the *lithosphere* and above the *mesosphere.* The asthenosphere is part of the upper *mantle.*

Aureole, Contact A zone surrounding an *igneous intrusion* where *contact metamorphism* has occurred

Axes Plural of *axis*

Axis of a fold The line around which folded layers are bent; a *fold hinge*

Axis of Earth's rotation The line around which the Earth rotates

Barometer An instrument used to measure air pressure

Barrovian zone A *metamorphic zone* encompassing an area with a distinct pelitic *mineral assemblage* in a region of progressive metamorphism of *pelitic rocks.* The Barrovian zone is named for one of the *index minerals* chlorite, biotite, garnet, staurolite, kyanite, or sillimanite.

Base level The level, or elevation, below which a *stream* cannot erode its *bed,* usually sea level, the level of a lake the stream enters, or the level of an especially *erosion* resistant *rock* the stream crosses. The latter two are temporary base levels.

Base line For the *Township and Range System,* an east-west line through the origin or reference point of the system

Basin, drainage The region drained by a *stream*

Batholith A roughly equidimensional *igneous intrusion* of large size, with an *outcrop* area greater than 40 mi^2 (or about 100 km^2).

Bathymetric map A map that depicts the shape of the bottom of the ocean. See also *topographic map.*

Bathymetry The shape of the bottom of the ocean. See also *topography.*

Bay-mouth bar An extension of a beach built out across a bay or inlet by the *longshore current.*

Beach drift *Sediment* moving parallel to the shoreline as a result of the *longshore current; longshore drift.*

Bed (a) (stratigraphy)a single layer of *sediment* or *sedimentary rock* with distinct surfaces separating it from other layers; *stratum;* (b) (hydrology) the bottom or base of any body of water, as in *stream bed*

Bedding The arrangement of *sediment* or *sedimentary rock* in parallel (or subparallel) layers formed at the time of deposition

Biochemical sediment *Sediment* that results from the actions of organisms

Body fossils A *fossil* made of the actual remains of an organism or parts of the organism

Body wave A *seismic wave* that moves through the Earth

Breakwater An artificial structure built in the water parallel to the shore to aid sand *deposition* between it and the shore

Brittle A variety of *tenacity* of a *mineral* in which the mineral shatters, breaks, or fractures rather than bends, flows, or dents when struck with sufficient force to deform it

Brittle deformation *deformation* that occurs when *rocks* break, rather than flow, after little deformation

Carbonate mineral group (carbonates) A group of *minerals* with members containing carbon and oxygen and one or more metals. Examples are calcite, dolomite, malachite, and azurite.

Carbonization A process of *fossilization* in which the original *organic* matter has been reduced to carbon

Cartography Map making

Cast A *fossil* made by filling a *mold* with *mineral* or *sediment*

Cement Material that precipitates between *sediment grains* and so holds the sediment together

Cementation A process involving water moving through *sediment* and precipitating minerals that essentially "glue" the *sediment* together. See also *cement.*

Channel A long, narrow trough occupied by the water in a *stream* or a connection between two bodies of water

Chemical sediment *Sediment* that forms by chemical precipitation of compounds out of a water solution

Chilled margin Generally *finer-grained igneous rock* along the edge of an *intrusion* produced where the *magma* comes in contact with the *country rock.* The edges are cooled more quickly than the centers, producing the finer *texture.*

Cinder A piece of *vesicular pyroclastic* material (pumice or scoria) thrown from a *volcano* that is solid when it lands

Cinder cone A small *volcano* made up of *pyroclastic* blocks, *volcanic ash,* and *cinders*

Clastic sediment *Sediment* made up of broken *rock* or *mineral* pieces; *detrital sediment*

Cleavage A physical property of a *mineral* that breaks along planes of weakness within the *crystal* structure

Cleavage, slaty The property of a *fine-grained metamorphic rock* that breaks along planes of weakness created by parallel *mineral grains*

Coarse-grained texture (a) (loosely defined) A *texture* for which the *rock* has visible *grains* (larger than about 1/16 mm). Also *granular texture.* The term *phaneritic texture* is also used for *igneous rocks* and *granoblastic texture* for unfoliated *metamorphic rocks.* This usage of the term *coarse-grained* includes medium-grained and is more commonly applied to *igneous* and *metamorphic rocks.* Coarse-grained igneous rocks include granite and gabbro, and metamorphic rocks include gneiss and marble. (b) (strictly defined) A texture for which the *rock* has grains larger than sand sized, or larger than about 2 mm. Coarse-grained *sedimentary rocks* include breccia and conglomerate.

Cold front A *front* where cold, dry air advances beneath warm, moist air. Where the colder air comes in contact with warmer air, *condensation* and *precipitation* occur. See also *warm front, occluded front,* and *stationary front.*

Compaction Bringing *grains*, especially *sediment*, closer together so they take up less space

Composite volcano A steep-sloped *volcano* made of interlayered pyroclastic deposits and lava flows; a *stratovolcano*

Compression A force of deformation that causes *shortening* or has forces moving toward each other

Conchoidal fracture A *mineral* or rock fracture where the broken surfaces are smoothly curved, often with concentric ribbing (Fig. 4.13)

Concordant intrusion A *magmatic intrusion* that intrudes parallel to existing *bedding* or *foliation*. Examples include *sills* and *laccoliths*.

Condensation The process in which water vapor molecules join together to form water droplets or ice *crystals*, thus forming clouds, fog, or moisture clinging to surfaces

Cone of depression A cone-shaped drop in the *water table* around a *well* that results when water is pumped out of the well

Confined aquifer An *aquifer* that is sandwiched between *aquicludes* or *aquitards*

Confining pressure The pressure experienced by a *rock* at depth caused by the weight of the overlying *mass* of rock

Conformal projection A map *projection* in which the shape of small areas are preserved and where different directions at any point have the same scale

Conical projection A map *projection* in which the *graticule* is projected onto a cone with an axis coincident with the *geographic axis*

Contact The boundary between *rock* units or *formations*, where one formation gives way to the next, depicted as a black line on most maps

Contact aureole A zone surrounding an *igneous intrusion* where *contact metamorphism* has occurred

Contact metamorphism *Metamorphism* resulting from the heating of *rock* near a *magmatic intrusion*; *thermal metamorphism*

Continental crust The part of the Earth's *crust* that underlies the continents and is chemically distinct (*felsic*) from *oceanic crust*

Continental rift The separation at a *divergent margin* where a continent diverges, beginning to separate the continent into smaller pieces. If divergence continues, a continental rift develops into an ocean basin with a *mid-ocean ridge spreading center*.

Continental-rise An area along a *passive continental margin*, farther toward the deep sea from the *continental slope*, where the seafloor slopes more gently

Continental shelf A gently sloping area of shallow water between a continent and the *continental slope*

Continental slope An area between the *continental shelf* and *continental rise* with a dramatically steeper slope than either adjacent area

Contour interval The difference in elevation between two adjacent *contour lines*

Contour line A line that connects points on a map representing places on the Earth's surface that have the same elevation

Convergent margin A *convergent plate boundary*

Convergent plate boundary A plate margin where two *plates* move toward each other; *convergent margin*, or destructive plate margin. Types include *continental collision*, and ocean-continent and ocean-ocean subducting plate boundaries (see *subduction*).

Coral polyp A small colonial animal that, as a colony, builds coral reefs; an individual coral animal

Core The center part of the Earth, below the *mantle*, that is chemically distinct, primarily made up of iron. The core consists of the *inner* and *outer core*.

Coriolis effect An apparent deflection, to the right in the Northern Hemisphere and to the left in the Southern Hemisphere, of freely moving objects or substances, such as ocean water or the atmosphere, as a result of the rotation of the Earth. The deflection is relative to the Earth's surface.

Correlation The process of matching the ages of *rocks* from different localities by matching stratigraphic sequences, *fossil* assemblages, and/or distinctive stratigraphic time markers such as volcanic ash layers

Correlative An adjective that applies to two or more *rocks* that have the same age

Country rock The *rock* intruded by *magma*, or the rock surrounding a magmatic intrusion

Cross bedding Inclined *bedding* where the inclined layers formed at a low angle to the major *sedimentary* bedding; cross-stratification. Typically, the top of the inclined layer has been truncated.

Cross section A side view of the Earth's interior, generally near the surface, exhibiting the arrangement and compositions of *rocks* and rock layers; a *structure section*

Cross-cutting relationships, the principle of States that a geologic feature that cuts across another feature is younger than the feature it cuts. The most common cross-cutting features are *discordant intrusions*, *faults* and *unconformities*.

Crust The surface solid layer of the Earth that is chemically distinct from other layers below. It is primarily made up of oxygen, silicon, and aluminum and is generally *mafic* to *felsic*. The crust makes up the uppermost part of the *lithosphere*. See also *oceanic crust* and *continental crust*.

Crystal faces Planar surfaces of a well-formed *crystal* that grew during *crystallization* or *recrystallization* and reflect the *mineral's* internal atomic order and arrangement

Crystal A single *grain* of a *mineral* in which the structural planes of atoms extend in the same directions throughout the grain

Crystalline Having an orderly internal arrangement of atoms in three dimensions, or having a *crystalline texture*

Crystalline texture A *texture* where *mineral grains* crystallized in place with grain boundaries touching—an interlocking texture. Most *igneous* and *metamorphic rocks* have this texture, as do *sedimentary* evaporites.

Crystallization The formation of a *crystal* with an orderly three-dimensional arrangement of atoms

Deformation The process of change in shape or form of *rocks* after they have formed—for example *folding*, *faulting*, stretching, and flattening

Density The *mass* per unit *volume* of a substance

Deposition The laying down or accumulation of *sediment* or other material

Detrital sediment *Sediment* made up of broken *rock* or *mineral* pieces; *clastic sediment*

Deuterium A form of hydrogen atom with one proton and one neutron in its nucleus, thus having an atomic weight of 2

Dew point The temperature where the *relative humidity* reaches 100% and dew, fog, or clouds form

Differential erosion The process of *erosion* in which harder *rocks* tend to erode less and stand out as hills or ridges and softer rocks erode more to become valleys or troughs

Differential stress The pressure or stress experienced by a *rock* undergoing *deformation* in which the forces on the rock are not equal in every direction; *directed pressure*

Dike A planar *igneous intrusive* body that cuts across layers, a *discordant intrusion*, or cuts through unlayered *rocks*

Dip The direction and angle of downward tilt of a geologic plane. Dip is measured from horizontal, perpendicular to the *strike*, along the steepest slope of the plane. Planes commonly measured include *bedding*, *foliation* and *fault* planes. See *dip direction* and *dip angle*.

Dip angle The angle between a *bedding*, *foliation*, *fault*, or other geologic plane and a horizontal plane.

Dip angle is measured from horizontal to the steepest slope of the plane.

Dip direction The approximate compass direction of the steepest downward slope of a geologic plane, especially a *bedding*, *foliation*, or *fault* plane, perpendicular to the *strike*

Dip-slip fault A *fault* with displacement parallel to the *dip* of the fault plane. Displacement may be up or down the dip of the fault plane.

Directed pressure The pressure, or stress, experienced by a *rock* undergoing *deformation* in which the forces on the rock are not equal in every direction; *differential stress*

Disconformity An *unconformity* where the older *rocks* below the unconformity are *sedimentary* and/or *volcanic* rocks and parallel to the unconformity surface.

Discordant intrusion A *magmatic intrusion* that intrudes across *bedding*, *foliation*, or existing *rock* masses. Examples include *dikes*, *stocks*, and *batholiths*.

Divergent margin A *divergent plate boundary*

Divergent plate boundary A *plate* edge where two plates move away from each other; a *divergent margin*, *spreading center*, or constructive plate margin

Divide, drainage The boundary separating one *drainage basin* from another

Double refraction A *special property* of a *mineral* that breaks light passing through a clear *crystal* into two different rays, causing an image viewed through the crystal to appear double

Downwelling The movement of warmer surface seawater downward. See also *upwelling*.

Drainage basin The region drained by a *stream*

Drainage divide The boundary separating one *drainage basin* from another

Ductile A type of *tenacity* of a *mineral* in which the specimen can be pulled out into an elongated shape

Ductile deformation *Deformation* that occurs when *rocks* bend, flex, and/or flow and generally when the rocks are deep and possibly warm

Dull luster A description of the surface of a *mineral* that is not shiny; *earthy luster*

Earthquake The vibration of the ground caused by natural geologic forces

Earth's axis of rotation The line around which the Earth turns

Earthy luster A description of the surface of a *mineral* that is dull, not shiny; *dull luster*. It may also be rough and/or dusty.

Eastern boundary current A cold ocean current that flows toward the equator near the eastern edge of an ocean.

Eastings In a geographic grid, the north-south trending lines, with numbers increasing eastward

Effervescence A *special property* of a *mineral*, such as calcite, in which the mineral reacts with an acid solution by bubbling due to the release of carbon dioxide gas

El Niño The occurrence of unusually warm water in the equatorial eastern Pacific Ocean, off the coast of Peru and Ecuador

Elastic A type of *tenacity* of a *mineral* where the sample bends when force is applied but resumes its previous shape when the pressure is released; also applies to behavior of other materials that snap back to their original shape after stress or pressure has been removed

Elastic rebound A snapping-back action as bent *rocks* return to their original shape during an *earthquake*

Electromagnetic spectrum A range of radiant energy of different wavelengths and energy levels that includes gamma rays, X rays, *ultraviolet rays*, visible light, *infrared rays*, microwaves, and radio waves

Epicenter The point on the land surface directly above the *focus* of an *earthquake*

Equal area projection A map *projection* where equivalent areas on the ground are preserved as equal areas on the map

Equatorial countercurrent A warm ocean current that flows to the east at or near the equator between the *equatorial currents*

Equatorial current A warm ocean current that flows to the west just north or south of the equator

Erosion The wearing away and removal of rock or sediment

Erupt To *extrude*, spew forth, or emanate *lava* as flows, fountains, or *pyroclastic* material from the Earth

Eruption, volcanic The extrusion or emanation of *lava* as flows, fountains, or *pyroclastic* material from the Earth

Estimated resource An estimate of the total quantity of a particular natural substance (*resource*) on the Earth

Euhedral An adjective describing a *mineral grain* with well-formed *crystal faces*

Evaporite A *rock* formed by *precipitation* of *minerals* from water due to evaporation

Exaggerated profile A *topographic profile* where the vertical *scale* is larger than the horizontal scale. Exaggeration accentuates the topographic features by making the highs and lows more extreme and the slopes steeper than reality.

Explanation of a map A guide to the various colors and symbols on a map; a *map key* or *map legend*.

Extratropical cyclone A low-pressure weather system with wet unstable rising air, cloudy skies, and *precipitation*, with a counterclockwise wind rotation in the Northern Hemisphere and a clockwise wind rotation in the Southern Hemisphere

Extrude To spew forth or emanate *lava* as flows, fountains, or *pyroclastic* material from the Earth; to *erupt*

Extrusive rock A type of *igneous rock* that forms at the surface of the Earth; *volcanic rock*

Extrusion The spewing forth, emanation, or *eruption* of *volcanic* material

Faces Planar surfaces of a well-formed *crystal* that grew during *crystallization* or *recrystallization*; *crystal faces*

Fault A *fracture* or zone of fractures across which displacement has taken place so the two sides remain in contact; also to form a fault

Feel A *special property* of a *mineral* that is distinctive to the touch, such as the greasy feel (not *luster*) of graphite

Feldspar mineral group (feldspars) A group of minerals, the most abundant in the crust, that have the general formula $MAl(Al,Si)Si_2O_8$ in which M is most commonly some combination of Ca, Na, and/or K. Example feldspars with different substitutions for M are alkali feldspar (Na and K), potassium feldspar (K, an alkali feldspar), orthoclase (K), plagioclase feldspar (Na, Ca).

Felsic A chemical composition term for *igneous rocks* that indicates the presence of low magnesium and iron content and high silica content (~70%). The term *felsic* is derived from feldspar (*fel*) and *silica* (*sic*).

Fibrous fracture A *mineral fracture* where the broken surface has the appearance of many threads or fibers

Fine-grained texture A *texture* of a *rock* with individual *crystals* so small that they cannot be seen with the naked eye. Fine-grained rocks include basalt, shale, and slate. See also *aphanitic texture*.

Fissile A property of a *rock* such as shale that splits or breaks into thin platy slabs, sheets, or flakes

Flexible A type of *tenacity* of a *mineral* where the specimen bends without breaking but will not spring back to its original shape when pressure is released. Compare with *elastic*.

Flood basalts Extruded *mafic magma* that has spread out and *solidified* in wide flat *lava* sheets over extensive areas

Fluorescence A *special property* of a *mineral*, such as some varieties of fluorite, by which the mineral emits light during exposure to *ultraviolet rays*

Fluvial Having to do with *streams* or *rivers*

Focus The place beneath the Earth's surface where an *earthquake* starts

Fold A bend in *rock* layers or other planar features, generally formed by *compression* parallel to the layering; also, to form such a bend

Fold axis The line around which folded layers are bent; a *fold hinge*

Fold hinge A *fold axis*

Foliation A planar *texture* or *structure* of a *metamorphic rock* that was produced by *directed pressure*. See also *slaty cleavage, schistosity*, and *gneissosity*.

Footwall The block beneath an inclined *fault* plane. Compare with *hanging wall*.

Formation (a) (cartography) A continuous or once continuous body of *rock*—igneous, sedimentary, or metamorphic—that can be easily recognized in geologic fieldwork and is the basic unit shown on geologic maps; (b) (stratigraphy) a rock body identifiable by its geologic characteristics and its position relative to the stratigraphic sequence in an area; a fundamental rock unit with distinct features

Fossil The naturally preserved remains or trace of life preserved in a *sedimentary rock* at the time the *rock* formed

Fossil fuels Energy *resources* that come from the *organic* remains of organisms preserved in *rocks*

Fossil succession, the principle of States that organisms evolve in a definite order, that species evolve and go extinct, never to reevolve, so the evolution of a species and its extinction become time markers separating time into three units: one before the organism existed, one during the existence of the organism, and one after the organism went extinct

Fossiliferous Containing abundant *fossils*

Fossilization The process of formation of a *fossil*

Fractional scale A method of expressing a map *scale* by giving the fraction or ratio of a given distance between two points on the Earth's surface to the distance between the same two points as represented on the map; *representative fraction*

Fracture (a) (mineral) A property whereby the *mineral* exhibits irregular and nonplanar surfaces when broken; (b) (structure) a crack or break in *rock*

Front A boundary or transition zone between air masses of different density. See also *cold front, warm front, occluded front*, and *stationary front*.

Geographic axis The *axis of Earth's rotation*; the line around which the Earth rotates

Geographic pole The end, either north or south, of the *geographic axis*

Geologic map A map used to show the distribution of various *rock* bodies, *formations, structures*, and age relationships

Geologic resource A naturally occurring substance that can be used and comes from the Earth but not directly from living things

Geologist A scientist who has been trained and works in the geological sciences—those sciences having to do with the study of the Earth

Geology The science that is the study of the Earth

Geophysics The branch of *geology* that studies the physics of the Earth, including *seismology*, geomagnetism, and aspects of volcanology and *oceanography*.

Glassy texture The *texture* of glass, which has *vitreous luster* and no definite *minerals; hyaline texture*

Global Positioning System (GPS) A satellite-based system of radio location. See Lab 1 for details.

Gneissic banding *gneissosity*

Gneissosity A type of *foliation* where the *rock* is medium- to coarse-grained and light and dark *minerals* are arranged in parallel bands, streaks, or layers; *gneissic banding*. Mineral segregations are at least about 2 mm wide.

Gondwana A supercontinent that existed hundreds of millions of years ago made of the southern continents assembled together; *Gondwanaland*

Gondwanaland *Gondwana*

Graded bedding: A single layer or *bed* of *sediment* or *sedimentary rock* in which the largest *grains* are concentrated at the bottom, gradually decreasing in size upward to the smallest at the top of the bed

Gradient, stream The steepness of the slope of a *stream* usually given in feet per mile or meters per kilometer

Grain A *mineral* or *rock* particle or *crystal* that has a distinct boundary separating it from surrounding grains, *matrix*, or *groundmass*

Granoblastic texture A *coarse-grained metamorphic texture* in which the *mineral grains* are randomly oriented in the *rock*

Granular texture A *texture* where the *rock* has visible grains; *coarse-grained texture*

Graphic scale A way of representing the scale of a map (or diagram) by drawing a line on the map (or diagram) representing the actual distance; a *scale bar*

Graticule The criss-crossing *lines of latitude* and *longitude* that form a reference grid on the spheroidal surface of the Earth

Greasy luster The surface shine of a *mineral* that resembles the way petroleum jelly or a greasy surface reflects light

Great circle The circle produced by a plane passing through the center of the Earth intersecting with the surface of the Earth

Greenhouse effect A process that heats the Earth's surface and lower atmosphere as follows: Visible light from the sun passes through the atmosphere and heats the Earth's surface. The warmed surface radiates *infrared rays*, which are partially absorbed by greenhouse gases in the lower atmosphere and radiated back to the Earth's surface.

Greenwich meridian The *meridian* that has a value of zero degrees and passes through Greenwich, United Kingdom

Grid A system of lines intersecting at right angles to form rectangles

Grid north The orientation of the north-south set of *grid* lines of a regional coordinate system used on a map

Groin A artificial structure built at a beach perpendicular to the shore to slow beach *erosion*

Groundmass Smaller *grains* surrounding larger grains in an *igneous rock* with *porphyritic texture*. See also the larger grains.

Groundwater Water in the ground, mostly below the water table

Guide fossil A fossil of an organism that existed for a short period of time so that its occurrence in a *rock* suggests a narrow range of age; an *index fossil*

Habit The external shape of a *mineral* if it were free to grow

Hackly fracture The fracture of a *mineral* that breaks to produce a sharp, jagged surface

Half life The length of time it takes for one-half of a *radiogenic isotope* to decay

Halide mineral group (halides) A group of *minerals* with members containing one of the halogen elements, fluorine, chlorine or bromine and one or more metals, but no other nonmetals. Examples are halite, fluorite, and sylvite.

Hanging wall The block above an inclined *fault* plane. Compare with *footwall*.

Hardness The resistance of a *mineral* to abrasion or scratching

Headland A promontory at a shoreline that sticks out into the water

High-grade metamorphism *Metamorphism* at high temperatures and pressures. Compare with *low-* and *medium-grade metamorphism*.

Hinge, fold The line around which the *beds* of the *fold* are bent; *fold axis*

Humidity The water vapor content of the air. See also *absolute humidity* and *relative humidity*.

Hyaline texture The *texture* of glass, which has *vitreous luster* and no definite *minerals; glassy texture*

Hydraulic head The difference between the height of the surface of water open to the air, or at atmospheric pressure, and water at a point in the subsurface

Hydroxide mineral group (hydroxides) a group of *minerals* with members containing hydrogen and oxygen and one or more metals, but no other nonmetals (e.g., limonite)

Igneous intrusion A body of still molten or solidified *magma* beneath the Earth's surface; a *magmatic intrusion*

Igneous rock A *rock* that has formed by the solidification of *magma*. See Lab 4 for descriptions of individual igneous rocks.

Immature sediment *Clastic sediment* that has a high proportion of easily weathered material (e.g., *rock* fragments, mafic minerals, and feldspar) and contains angular, *poorly sorted grains*

Impermeable Having no *permeability*

Index contour A *contour* that is drawn with a heavier line and is labeled with the elevation; usually every fourth or fifth line

Index fossil A *fossil* of an organism that existed for a short period of time so that its occurrence in a *rock* suggests a narrow range of age; a *guide fossil*

Index mineral A *mineral* that is indicative of a *metamorphic zone*. The first appearance of the index mineral as the zonation progresses from lower to higher temperature marks the beginning of a metamorphic zone and coincides with an *isograd*. For example, the area that includes the first appearance of biotite to the first appearance of almandine garnet corresponds to the biotite zone. Biotite is the index mineral as long as garnet and other higher temperature index minerals are not present. If garnet is also present, then the *rock* belongs to the garnet or higher temperature zone.

Indirect dating A dating method whereby the numerical age of a *sedimentary rock* can be approximately determined by using a combination of *relative* and *isotopic dating* techniques

Infrared rays Part of the *electromagnetic spectrum* that is longer in wavelength and lower in energy than visible light but shorter in wavelength and higher in energy than microwaves

Inner core The part of the Earth's *core* that is solid metal, mostly iron

Inorganic Not *organic*

Intermediate A chemical composition term for *igneous rocks* that indicates the presence of intermediate magnesium, iron and silica content (silica, ~60%)

International Date Line The line on the Earth's surface across which the date changes, located near 180° longitude

Intrusion (a) The process by which *magma* enters *rock*; (b) short for *magmatic* or *igneous intrusion*

Intrusion, magmatic A body of still molten or solidified *magma* beneath the Earth's surface

Intrusive rock An *igneous rock* that solidified beneath the Earth's surface

Irregular fracture A type of *mineral fracture* that is uneven and not *conchoidal, fibrous,* or *hackly fracture*

Isobars Lines of equal pressure on a map or graph

Isograd A boundary between *metamorphic zones*

Isotope A type of atom of an element with a specific number of neutrons and protons in the nucleus

Isotopic dating The process of determining the numerical or *absolute age* using radioactive isotopes; *radiometric dating*

Karst topography the *topography* resulting from the flow of *groundwater* through areas of soluble *rocks*. Features of karst include caves, solution valleys, *sinkholes,* karst towers, and underground drainage, which are caused by the dissolution of limestone or other soluble rocks, such as rock gypsum

Key, map A guide to the various colors and symbols on a map; a *map legend* or *explanation*

La Niña A more extreme version of the normal current pattern of cold water along the coast of Peru and Ecuador; the opposite of *El Niño*

Laccolith An *igneous* intrusive body that intrudes parallel to layers (*concordant*) and bulges upward to make a three dimensional body, doming the layers above it

Land Office Grid System A *grid* system used in much of the United States based on 30-mi.^{2} grid, called *townships,* with square-mile subdivisions called *sections*

Landslide The event or process in which *rock* or debris moves rapidly down the surface slope in response to gravity; a type of *mass wasting;* also, the landform that results from this process

Large scale The scale of a map where the ratio between an object on a map and the same object on the Earth's surface is large. For large-scale maps, a large area on the map covers a small amount of the Earth's surface, and represented objects appear large. The term can also be used for diagrams other than maps where scale is involved. Compare with *small scale.*

Latitude, line of A locational reference line that is measured by the angular distance north or south of the equator; a *parallel.* Compare with *longitude.*

Laurasia A supercontinent that existed hundreds of millions of years ago made of North America, Europe, and Asia assembled together

Lava Molten *rock* at the Earth's surface

Lava dome *Extruded magma* that is so *viscous* that it does not spread out but instead forms a dome or mound

Lava flow *Extruded magma* that has flowed laterally and solidified in a tongue shape or as a surficial sheet

Left-lateral fault A *strike-slip fault* for which movement of the side opposite the observer, viewed across the *fault* plane, is to the left along the fault

Legend of a map A guide to the various colors and symbols on a map; a *map key* or *explanation*

Line of latitude See *latitude, line of.*

Line of longitude See *longitude, line of.*

Lineation A linear *texture* or *structure* of a *metamorphic rock* that was produced by *directed pressure*

Liquefaction Quicksandlike behavior of water-saturated *sediment* when vigorously shaken by an *earthquake*

Lithification The process by which disaggregated *sediment* turns into a consolidated *sedimentary rock;* consolidation. The processes of *cementation* and *compaction,* and in some cases *crystallization,* aid the lithification process.

Lithosphere An approximately 100-km thick layer of the Earth at the surface that is made of strong brittle *rock,* underlain by the *asthenosphere.* The upper part of the lithosphere is made up of the *crust,* and the lower part coincides with the uppermost *mantle.*

Longitude, line of Any of the half *great circles* between the North and South Poles that are used as a means of specifying the east-west position of a location by the angular distance from the *prime meridian; a meridian.* Compare with *latitude.*

Longshore current A current of water moving parallel to the shoreline as a result of waves encroaching at an angle to the coastline

Longshore drift *Sediment* moving parallel to the shore as a result of the *longshore current*

Low grade metamorphism Metamorphism at low temperatures and pressures

Luster How light is reflected from a fresh surface of a mineral. See also *metallic, submetallic, nonmetallic, vitreous, resinous, waxy, pearly, fibrous, adamantine, and earthy lusters.*

Mafic A chemical composition term for *igneous rocks* that indicates the presence of high magnesium and iron content and low silica content (~50%)

Magma Molten *rock*

Magmatic intrusion A body of still molten or solidified *magma* beneath the Earth's surface; an *igneous intrusion*

Magnetic declination The angle between *magnetic north* and geographic or *true north*

Magnetic north The direction toward which a magnetic compass (or magnetized needle) points

Magnetism A *special property* of a *mineral,* such as magnetite, that is attracted to a magnet

Mantle A chemically distinct part of the Earth that is below the *crust* and above the *core.* It is primarily made up of oxygen, silicon, and magnesium and is ultramafic in composition. The mantle includes the lower part of the *lithosphere,* the *asthenosphere* and the *mesosphere.*

Map explanation A *map key* or *legend*

Map key A guide to the various colors and symbols on a map; a *map legend* or *explanation*

Map legend A *map key* or *explanation*

Mass A measure of a body's resistance to acceleration. On Earth, the mass of an object is proportional to its weight.

Massive Lacking *bedding,* parallel layered *structure,* or *foliation*

Mass wasting The downslope movement of loose surface material due to gravity, without the aid of a transporting substance such as flowing water, glaciers, or wind

Matrix (Sedimentary) finer-grained material between and/or surrounding larger *grains* in *sediment* or *sedimentary rock;* (igneous) *groundmass*

Mature *Clastic sediment* that has no easily weathered *grains* and contains well rounded and *well-sorted* material.

Meandering stream A winding *stream; a stream* with a *sinuosity* of greater than 1.5

Medium-grade metamorphism Metamorphism at moderate temperatures and pressures

Medium-grained Having visible *grains* about the size of sand. Medium-grained rocks include sandstone and schist. Sometimes, especially for *igneous* and *metamorphic rocks,* medium-grain size is loosely included as part of the term *coarse-grained.*

Mercator projection A map *projection* where the global *graticule* is projected onto a cylinder aligned with the *geographic axis* with the cylinder touching the equator

Meridian The half *great circle* between the North and South Poles that is used as a means of specifying the east-west position of a location; a *line of longitude.*

Mesosphere (a) (solid Earth) A layer of the Earth that is made of rigid *rock,* below the *asthenosphere* and above the *outer core.* The mesosphere corresponds to the lower *mantle.* (b) (atmosphere) A layer of the atmosphere (between about 50 and 80 km altitude) above the *stratosphere* and below the thermosphere, where temperature decreases with altitude. See Figure 15.1.

Metallic luster The surface shine of a *mineral* that resembles the way metals reflect light

Metamorphic facies The set of all *mineral assemblages* that may be found together in a region where the *rocks* have different chemical composition but were all *metamorphosed* at the same conditions of temperature and pressure

Metamorphic grade An approximate measure of the amount or degree of *metamorphism,* most closely tied to metamorphic temperature, but also related to metamorphic pressure. See also *low-, medium-,* and *high-grade metamorphism.*

Metamorphic rock A *rock* that has undergone *metamorphism.* See Lab 6 for descriptions of individual metamorphic rocks.

Metamorphic zone An area or region in which a distinctive *mineral assemblage* indicates a specific range of metamorphic conditions, especially temperature

Metamorphism The combination of all of the processes that change a *rock* above 200°C in the solid state as a result of changes in temperature and/or pressure. Changes primarily involve *texture* and mineralogy.

Metamorphose (verb) To undergo *metamorphism*

Meteorology The science that is the study of the atmosphere, including weather and climate

Mid-ocean ridge A long, symmetrical mountain range or broad rise in the ocean, commonly but not always at the middle, with many small, slightly offset segments. The axes of mid-ocean ridges correspond to *spreading centers*, or *divergent plate boundaries*.

Mineral A naturally occurring, usually *inorganic*, chemically homogeneous *crystalline* solid with a strictly defined chemical composition and characteristic physical properties. See Tables 3.3 and 3.4 for descriptions of individual minerals.

Mineral assemblage A group of *minerals* that grow or coexist together at the same temperature and pressure

Mineral group A set of *minerals* that have some formally defined common characteristics. The groups *silicates, carbonates, sulfates, sulfides, phosphates, oxides,* and *hydroxides* are defined by the nonmetallic part of their chemical composition. Subclasses of the silicates, such as chain or sheet silicates, are grouped by the internal atomic arrangement of *silica tetrahedra* (Table 3.1). Some other mineral groups are *native elements* and *feldspars*.

Mold A *fossil* made of the imprint of an organism or part of an organism, such as when a clam shell dissolves leaving a cavity with its clam-shell shape

Native element mineral group (native elements) A group of *minerals* with members containing only one pure element. Examples are diamond, graphite, copper, gold, silver, sulfur.

No streak Said of a *mineral* that is too hard (H > 6 to 7) to leave a powder on a porcelain *streak* plate

Nonconformity An *unconformity* where the older *rocks* below the unconformity are *plutonic igneous* or *metamorphic rocks*

Non-metallic luster The surface shine of a *mineral* that does not resemble the way metal reflects light

Nonrenewable resource A *resource* that has a limited quantity that is diminished by use

Normal fault A *dip-slip fault* with downward movement of the *hanging wall* relative to the *footwall*

Northings In a *grid*, the east-west trending lines, with numbers increasing northward

Oblique-slip fault A *fault* with displacement between the *dip* and the *strike* of the fault plane

Occluded front A *front* where two cold air masses have come in contact and wedged out a warmer air mass above. See also *cold front, warm front,* and *stationary front.*

Oceanic crust The part of the Earth's *crust* that underlies the oceans and is chemically distinct (*mafic*) from *continental crust*

Oceanic trench A deep trough in the *bathymetry* of the ocean floor. Trenches are associated with *subduction* zones.

Oceanography The science that is the study of the oceans

Ore A *rock* or *sediment* containing one or more economic metal *resources*

Organic (a) (chemistry) Pertaining to a compound containing carbon or carbon and hydrogen, but not carbon and oxygen; (b) pertaining to or derived from living organisms

Original horizontality, principle of States that *sedimentary* layers are deposited horizontally or nearly so

Outcrop An area of *rocks* exposed at the surface without much foliage, soil, *sediment,* or artificial structures covering them

Outer core The part of the Earth's *core* that is liquid

Overburden Unwanted *rock* above a valuable mineral or coal deposit

Overturned Said of *sedimentary* layers that have been tilted more than 90°

Oxide mineral group (oxides) A group of *minerals* with members containing oxygen and one or more metals, but no other nonmetals. Examples are magnetite, hematite, and corundum.

Ozone Oxygen gas with an extra oxygen atom: O_3

Ozone layer The part of the atmosphere, in the *stratosphere*, with a relatively high concentration of *ozone*

P wave The fastest type of *earthquake* wave, with the vibration direction parallel to the direction the wave travels; a compressional *seismic body wave*

Pangaea A supercontinent that existed hundreds of millions of years ago made of all of the continents assembled together. See also *Gondwana* and *Laurasia.*

Parallel A locational reference line that is measured by the angular distance north or south of the equator; a *line of latitude*

Parent rock (a) (metamorphic) The *rock* from which a *metamorphic rock* forms; the *protolith;* (b) (sedimentary) the primary source rock from which a sediment has *weathered*

Parting The property of a *mineral* that breaks along nearly planar surfaces but that is not consistent or planar enough to be considered *cleavage*

Passive continental margin The edge of a continent without a *plate* boundary

Pearly luster The surface reflection of a *mineral* that resembles the way pearls shine

Pelitic rock A *rock* with overall chemical composition similar to clay-rich shales; a *metamorphic rock* with a shale *protolith*

Permeability A measure of the ability of *rock* or *sediment* to allow liquids or gases to flow through it

Permeable Having *permeability*

Petrifaction or **Petrification** A process of *fossilization* in which original *organic* material is replaced by other *inorganic* substances

Petrology The science that is the study of *rocks* and how they form

Phaneritic texture A *texture* of an *igneous rock* where the *rock* has visible *grains;* medium- or coarse-grained *texture*

Phenocryst A large *grain* surrounded by smaller grains, or *groundmass,* in an *igneous rock.* A *rock* containing phenocrysts has *porphyritic texture.*

Phosphate mineral group (phosphates) A group of *minerals* with members containing phosphorus and oxygen and one or more metals, but no other nonmetals. The major phosphate mineral is apatite.

Photosynthesis A chemical *synthesis* performed by plants using light from the sun to form *organic* molecules and by-product oxygen from water and carbon dioxide

Placer A concentration of a mineral that forms where turbulent flowing water segregates heavier minerals from lighter ones

Planimetric map A map that depicts the location of major cultural and geographic features such as towns, rivers, roads, and railroads

Plastic An adjective describing material that can flow, stretch, bend, and/or flex without breaking. In the Earth sciences this term is usually applied to ductile *rock,* which is not rigid.

Plate A relatively rigid section of *lithosphere* that moves as a unit and relative to other plates

Plate tectonics A field of earth science that involves the study of *plates,* their boundaries and movements, and their influences on other aspects of the Earth such as *rocks, structures, topography* mountains, mountain building, and the Earth's interior

Plate tectonics, theory of The theory that the Earth's *lithosphere* is divided into a few pieces called *plates* that move relative to each other and relative to the Earth as a whole. Deformation, crustal destruction, and regeneration are concentrated around the margins of plates, and the interior of plates remain relatively rigid and do not tend to deform.

Plunge The angle of tilt down into the ground, measured from horizontal, of the *axis* of a *plunging fold*

Plunging fold A *fold* with an *axis* that is not horizontal

Plutonic rock An *igneous rock* that solidified beneath the Earth's surface

Polar easterlies Winds between the pole and 60° latitude that blow from the east

Poorly sorted A characteristic of *sediment* that has a wide range of *grain* sizes. Compare with *well sorted.*

Pore A void or space within *rock* or *sediment*

Porosity The percentage or proportion of the *volume* of *rock* or *sediment* made up of *pore* spaces

Porphyritic An adjective used to describe *igneous rocks* with *porphyritic texture* with less than 25% *phenocrysts*

Porphyritic texture The *texture* of an *igneous rock* with large *crystals* or *phenocrysts* embedded in a more finely *crystalline* or *glassy groundmass*

Porphyroblast A large *metamorphic mineral grain* or *crystal* surrounded by smaller grains in a *rock* with *porphyroblastic texture*

Porphyroblastic texture A metamorphic *texture* in which distinctly larger *grains* are surrounded by smaller grains

Porphyry The name of an *igneous rock* with *porphyritic texture* in which *phenocrysts* comprise 25% or more of its *volume*

Potentially renewable resource A *resource* that is replenished but may or may not be used faster than the rate of replenishment

Potentiometric surface The level that water in a *confined aquifer* would rise if the aquifer were punctured

Precipitation (a) (meteorology) The process in which liquid or solid water falls from clouds, forming rain, snow, sleet, hail, and freezing rain; (b) (chemistry) a chemical process in which solids form out of a solution

Prime meridian The *meridian* that has a value of 0° and passes through Greenwich, United Kingdom.

Principal meridian For the *Township and Range System,* a north-south line through the origin or reference point of the system

Projected lifetimes A crude estimate of the length of time a particular *mineral* commodity will last, calculated by assuming that the current rate of production of a *resource* will continue and dividing that into the *reserves* of that resource

Projection A geometric technique used to convert information on a three-dimensional object, such as a sphere, to two dimensions, such as a map

Protolith The *rock* from which a *metamorphic rock* forms; the *parent rock*

Psychrometer An apparatus that measures relative humidity by comparison of the cooling effect of evaporation on a wet thermometer bulb compared to the dry bulb temperature

Public Land Survey A *grid* system used in much of the United States based on 36-mi^2 grid, called *townships,* with subdivisions, called *sections,* for every square mile; the *Township and Range System*

Pyroclastic Pertaining to *pyroclastics* or having *pyroclastic texture*

Pyroclastic texture The *texture* of *rock* or loose material made of *volcanic ash* and/or larger *rock* fragments exploded from a *volcano*

Pyroclastics Fragmental *volcanic* products such as *ash, cinders,* pumice pieces, and blocks, formed by aerial ejection or explosion from a *volcanic vent*

Radioactive Possessing *radioactivity*

Radioactivity A process whereby the nucleus of unstable atom spontaneously decays by losing or gaining a particle, thereby changing into the nucleus of another element

Radiogenic isotope An (unstable) *isotope* that is naturally *radioactive*

Radiometric dating The process of determining the numerical or *absolute age* using *radioactive isotopes; isotopic dating*

Raised relief Map *Topographic map* that is drafted onto a plastic sheet or other medium that is molded or shaped to give the viewer an idea of the *relief* in the map area in three dimensions

Recrystallization The formation of new *grains* of *mineral* material already present in a *rock*. The original material of an organism may recrystallize as part of the process of fossilization. During metamorphism, some minerals recrystallize from the *parent rock*; for example, the recrystallization of quartz as a quartz sandstone *metamorphoses* into a quartzite.

Reduction A chemical reaction that removes oxygen from a substance or reduces the electrical charge on atoms to a lesser number. For example, reduction may change Fe^{3+} to Fe^{2+}.

Refraction The bending of waves of all kinds, caused by changes in wave speed. For example, light waves refract when passing through a lens; ocean waves refract as they approach the shoreline at an oblique angle.

Regional metamorphism *Metamorphism* over a wide area or region where *directed pressure* combines with a wide range of temperatures and *confining pressure* at moderate to great depths. Regional metamorphism commonly results at *convergent plate boundaries.*

Relative age The age, usually of a *rock* or *fossil*, established in comparison to other ages, using words such as *older* or *younger*, rather than numerical ages

Relative humidity The percentage of water vapor relative to the total amount of water vapor a particular *volume* of air can potentially hold.

Relict bedding *Sedimentary bedding* that has been preserved after *metamorphism* and is visible in a *metamorphic rock*

Relief The difference in elevation between the highest and lowest point in an area. For example, steep slopes and cliffs occur in areas of high relief.

Renewable resource A *resource* for which its use does not deplete its quantity

Representative fraction A method of expressing a map *scale* by giving the ratio of a given distance between two points on the Earth's surface to the distance between the same two points as represented on the map; *fractional scale*

Reserves The quantity of a *resource* that has been found and is economically recoverable with existing technology

Reservoir (a) (water) a body of water stored in a valley behind a dam; (b) (petroleum geology) a body of *permeable rock* containing oil and natural gas

Resinous luster The surface shine of a *mineral* that resembles the way plastic reflects light

Resource (a) Any naturally occurring substance that can be used by humans; See also *geologic resource.* (b) (numerical definition) The total quantity of a particular natural substance on Earth

Reverse fault A *dip-slip fault* with upward movement of the *hanging wall* relative to the *footwall*

Right-lateral fault A *strike-slip fault* for which movement of the side opposite the observer, viewed across the *fault* plane, is to the right along the fault

River A part of a *stream* system where the stream is large

Rock An aggregate of *mineral grains* and/or mineral-like matter or a homogeneous mass of mineral-like matter. By mineral-like matter, we mean material such as opal or volcanic glass that is a natural *inorganic* solid and material such as found in coal that is the solid *organic* remains of organisms that have changed in some way since the organism's death.

Rule of V's The principle that the surface trace of an inclined planar feature in a valley forms a V that points in the direction of *dip* of the feature

S wave An *earthquake* wave with the vibration direction perpendicular to the direction the wave travels; a shearing *seismic body wave*

Scale The size reduction needed to convert the actual feature into its representative on a map or diagram

Scale bar A way of representing the *scale* of a map (or any diagram) by drawing a line on the map (or diagram) representing the actual distance; *graphic scale*

Schistosity A type of *foliation* where the *rock* is medium- to coarse-grained and mica or other *minerals* are oriented parallel to each other but are fairly evenly distributed throughout the rock. The rock tends to split parallel to this foliation.

Sectile A type of *tenacity* of a *mineral* where the sample can be shaved with a sharp blade as if it were wax or soap

Section One square mile in a *township* in the *Township and Range System*

Sediment Loose material at the Earth's surface from *rock* and *mineral* particles, from organisms and their remains, and/or from chemical precipitation

Sedimentary rock A *rock* made by the *lithification* or consolidation of *sediment*

Seismic wave An *elastic* wave or vibration generated by an *earthquake* or explosion

Seismogram The record of an earthquake recorded by a *seismograph*

Seismograph An instrument that inscribes the Earth's motion during an *earthquake* on a record called a *seismogram*

Seismologist A scientist who studies *earthquakes*

Seismology The study of *earthquakes* and their waves

Shear A force of *deformation* that has a scissorlike motion causing one *rock mass* to pass by another

Shield volcano A large and gently sloping *volcano* made up of basaltic *lava flows* and very little *ash*

Silica Silicon dioxide (SiO_2), an essential constituent in *silicate minerals*. The mineral quartz is pure silica.

Silica tetrahedron The shape of the arrangement of a silicon atom surrounded by four oxygen atoms, making the basic building block of *silicate minerals*, in which the oxygen atoms are centered at the apexes of a *tetrahedron* with the silicon atom in the center of the tetrahedron; singular of *silica tetrahedra*

Silicate mineral group (silicates or silicate minerals) A group of *minerals* with members containing silicon and oxygen as basic building blocks for their internal structures. Examples are listed in Table 3.1.

Silky luster The surface shine of a *mineral* that resembles the way silk reflects light

Sill A planar *igneous* intrusive body that intrudes parallel to layers (*concordant*)

Sinkhole A hole or closed depression in *karst* regions

Sinuosity A measure of how winding or *meandering* the course of a *stream* is

Skeletal texture A *sedimentary texture* in which the entire *rock* is essentially made up of visible *fossils*

Slaty cleavage A type of *foliation* in a *fine-grained metamorphic rock* that splits into planar slabs; rock cleavage

Slickensides Smooth, slick striations on *rocks* along *faults* where the walls of the fault have slipped past each other. Slickensides show the direction of slip.

Small circle Any circle on the globe with a radius less than that of a *great circle* and therefore not centered at the Earth's center

Small scale The *scale* of a map where the ratio between the map and the Earth's surface is a minute fraction. For small-scale maps, a small area on the map covers a large amount of the Earth's surface, and represented objects appear small. The term can also be used for diagrams other than maps where scale is involved.

Smell A *special property* of a *mineral* that has a distinctive odor

Sorted, poorly A property of *sediment* that has a wide range of *grain* sizes. Compare with *well sorted*.

Sorted, well A property of *sediment* that has a narrow range of *grain* sizes. Compare with *poorly sorted*.

Special property A *mineral* property that is only possessed by a few minerals. *Effervescence* in acid and *magnetism* are two special properties. Since every mineral has *luster* of one sort or another or breaks in one way or another, luster, *cleavage*, and *fracture* are not special properties.

Specific gravity The ratio of the *mass* of a substance to the mass of an equal *volume* of water; closely related to *density*

Spit An extension of a beach built part way out across a bay or inlet by the *longshore current*

Spoil Broken and crushed waste *rock* from a mining operation

Spreading center A plate margin where two *plates* move away from each other; a divergent plate boundary or margin

Standard parallel For conical *projections*, the *line of latitude* on which the cone rests

Statement scale The scale on a map expressed verbally; *verbal scale*

Stationary front A *front* where air masses move in such a way that neither the warmer nor the cooler air displaces the other. See also *cold front, warm front,* and *occluded front.*

Stereographic projection A map *projection* onto a tangential plane where the projection of the *graticule* is made from a projection point on the Earth's surface exactly opposite the plane

Stereoscope A device with two lenses designed to aid in viewing aerial photographs so *topography* can be seen in three dimensions. This effect gives the viewer a sense of the topographic *relief.*

Stock Roughly equidimensional *igneous intrusion* of small size, with an *outcrop* area less than 100 km^2 (or 40 mi^2)

Strata Layers of *sedimentary rock* that are visually separable from other layers; *beds*; plural of *stratum*

Stratigraphic superposition, the principle of States that *sedimentary rock* layers are deposited in sequence one on top of the other, so that at the time of *deposition*, the oldest rocks are at the bottom of a sequence and the youngest rocks are on top

Stratigraphy The study of *strata*, or *sedimentary* layers

Stratosphere The part of the upper atmosphere (between about 20 and 50 km altitude), above the *troposphere* and below the atmospheric *mesosphere*, where temperature increases with altitude. See Figure 15.1.

Stratovolcano A steep-sloped *volcano* made of interlayered *pyroclastic* deposits and *lava flows; composite volcano*

Stratum Singular of *strata*

Streak The color of a mineral when powdered. Streak is tested on a small piece of porcelain, known as a streak plate. See also *no streak*.

Stream Any body of water that flows under the force of gravity in a relatively narrow *channel*

Stream gradient The steepness of the slope of a *stream*

Strike The orientation of a horizontal line on a plane, especially a *bedding, foliation,* or *fault* plane; perpendicular to the *dip*

Strike-slip fault A *fault* with displacement parallel to the *strike* of the fault plane

Strip mining A method of mining usually used for extracting shallow horizontal or nearly horizontal *mineral* or coal deposits, where the *overburden* is removed from a strip and piled up next to it, and then the coal or mineral is removed. Overburden from the next parallel strip is piled in the previously mined strip, and the process continues.

Structure The physical arrangement of a *rock* mass. Examples include *intrusive* bodies, *unconformities*, orientation of *rock* layers, and *deformational* features such as *faults* and *folds*.

Structure contour A line on a map of equal elevation of the top or bottom of a geologic feature, such as a bed folded into an *anticline* or *syncline*

Structure contour map   A map that has *structure contours*

Structure section   A side view of Earth's interior, generally near the surface, exhibiting the arrangement and compositions of *rocks* and rock layers; *cross section*

Subduction   The process of movement of a slab of *lithosphere* downward into the *asthenosphere* at an ocean-ocean or ocean-continent *convergent margin*

Subhedral   An adjective describing a *mineral grain* with some but not all well-formed *crystal faces*

Subtropical gyre   A roughly circular flow of ocean currents with a center in subtropical latitudes at about 30°, made up of an *equatorial current*, a *western boundary current*, an eastward flowing current near 45° latitude, and an *eastern boundary current*

Sulfate mineral group (sulfates)   A group of *minerals* with members containing sulfur and oxygen and one or more metals. Examples are gypsum and anhydrite.

Sulfide mineral group (sulfides)   A group of *minerals* with members containing sulfur and one or more metals but no oxygen. Examples are pyrite, chalcopyrite, galena, and sphalerite.

Surface wave   A *seismic wave* that only travels along the Earth's exterior

Syncline   A downward bowed *fold*

Synthesis   Putting atoms together to form molecules or recombining atoms in one molecular combination to make another combination

Taste   A *special property* of a *mineral* that is bitter, sour, or salty to the tongue

Tenacity   The cohesiveness of a specimen, a description of a *mineral's* resistance to mechanical *deformation* (breaking, bending, crushing, etc.)

Tension   A force of *deformation* that pulls apart or has forces moving away from each other

Tetrahedron   A regular geometric solid with four sides that are equilateral triangles of equal size; a triangular pyramid; singular of *tetrahedra*

Texture   The arrangement and size of *mineral grains*, *rock* fragments, or glass in a rock.

Thermal metamorphism   *Metamorphism* resulting from the heating of *rock* near a *magmatic intrusion*; *contact metamorphism*

Thrust fault   A low-angle *reverse fault*

Tombolo   A sandy connection between an island and the mainland or between two islands

Topographic map   A map with color, shading, or *contours* that indicate the shape of the land surface; if drawn with contours, also known as a contour map. See also *raised relief map* and *contour interval*.

Topographic profile   A side view of the *topography* along a line on a *topographic map*

Topography   The shape of the physical features of the land surface. See also *relief*.

Township   A 6-by-6-mile square in the *Township and Range System*

Township and Range System   A *grid* system used in much of the United States based on 36-mi^2 grid, called *townships*, with square-mile subdivisions, called *sections*.

Trace fossil   A *fossil* that shows signs of an organism's existence or activities, but not involving the actual remains of the organism. Examples are footprints or animal burrows.

Trace of a contact   The line made on the Earth's surface by the intersection of the surface with the contact between two *rock* bodies.

Trade winds   Easterly winds (winds that blow from the east) that blow between 30°N and 30°S latitude, including the northeast trade winds in the Northern Hemisphere and the southeast trade winds in the Southern Hemisphere

Transform fault   The type of *fault* found at a *transform-fault plate boundary*. Transform faults are also *strike-slip faults*.

Transform-fault margin   A *transform-fault plate boundary*

Transform-fault plate boundary   A plate margin where two *plates* move horizontally past each other; *transform-fault margin*

Transverse Mercator projection   A *Mercator projection* where the *projection* cylinder touches a particular *meridian* of interest

Troposphere   The part of the lower atmosphere (up to an altitude of about 20 km), below the *stratosphere*, where temperature generally decreases with rise of altitude. See Figure 15.1.

True north   The direction of *geographic north,* toward the northern *axis of Earth's rotation*

Tsunami   A *seismic sea wave*; a series of ocean (or lake) waves generated by a sudden disturbance of the seafloor (or lake bottom) such as an *earthquake,* submarine *landslide*, or *volcanic eruption*

Ultramafic   A chemical composition term for *igneous rocks* that indicates the presence of very high magnesium and iron content and very low silica content (~40%)

Ultraviolet rays   Part of the *electromagnetic spectrum* with a shorter wavelength and higher energy than visible light but longer wavelengths and less energy than X rays; ultraviolet light

Unconformity   A substantial time gap in the *rock* record where rocks either were *deposited* then eroded or were simply not deposited

Uneven fracture   A type of *mineral fracture* that is irregular and does not fit any of the other standard fracture terms: *conchoidal, fibrous,* or *hackly*

Unifying theory   A guiding principle or integrative concept for an entire field of study

Universal Time   The time at the *prime meridian*

Universal Transverse Mercator coordinates   A coordinate system commonly included on *topographic* maps. See Lab 1 for more details.

Up indicator   A sedimentary feature or characteristic that shows which way up a *rock* or sedimentary layer was deposited. Up indicators are used to help recognize *overturned beds*.

Upright fold   An *anticline* or *syncline* that has a horizontal *fold axis* and a vertical *axial plane*

Upwelling   The movement of deep cool ocean water upward to the sea surface. Compare with *downwelling*.

Verbal scale   The *scale* on a map expressed as a statement; *statement scale*

Vesicle   A small cavity in a *rock* that was originally a gas bubble in *magma*

Vesicular texture   A *volcanic texture* that refers to the presence of small cavities in a *rock,* called *vesicles,* which were originally gas bubbles in the *magma*

Viscosity   A fluid property referring to the fluid's resistance to flow

Viscous   Having high *viscosity.*

Vitreous luster   The surface shine of a *mineral* that resembles the way glass reflects light

Volcanic ash   Sand-sized to powdery *volcanic* material produced by explosive *volcanic eruptions* when a spray of magma and particles of *rock* spew out of a *volcano*

Volcanic eruption   The extrusion or emanation of *lava* as flows, fountains, or *pyroclastic* material from the Earth

Volcanic rock   A type of *igneous rock* that formed at the surface of the Earth; *extrusive rock*

Volcanic vent   The opening in the Earth's surface where *volcanic eruptions* occur

Volcano   A hill or mountain formed where material erupts frequently or repeatedly from a *volcanic vent*

Volume   Size or extent in three dimensions

V's, rule of   The principle that a V formed by the surface trace of a planar feature in a valley points in the direction the feature's *dip*

Warm front   A *front* where warm moist air wedges out cold air. Where the warmer air comes in contact with cooler air, *condensation* and *precipitation* occur. See also *cold front, occluded front,* and *stationary front.*

Water table   The top of the *zone of saturation*

Waxy luster   The surface shine of a *mineral* that resembles the way wax reflects light

Weather map   A map showing the distribution of aspects of the weather, such as storm systems, winds, *precipitation,* temperature, pressure, and warm and cold fronts

Well   A hole drilled into the *zone of saturation,* below the *water table*

Well sorted   A property of *sediment* that has a narrow range of *grain sizes.* Compare with *poorly sorted.*

Westerlies   Winds in a belt from 30° to 60° latitude blowing from the west

Western boundary current   A warm ocean current that flows away from the equator near the western edge of an ocean

Xenolith   A piece of foreign *rock* embedded in *igneous rock*

Zone   See *metamorphic zone*—an area or region in which a distinctive mineral assemblage of coexists that indicates a specific range of *metamorphic* conditions, especially temperature. See also *Barrovian zone.*

Zone of saturation   The region below ground where water fills all of the *pores;* below the *water table*

Credits

Figs. 1.4, 1.12, 12.14, 12.24, 15.17, 15.21 from *Meteorology Today,* 6th ed. by C. Donald Ahrens, © 2000 Brooks/Cole. All rights reserved.

Figs. 1.8, 1.9, 1.18, 9.2, 9.4, 9.8, 9.9, 9.11, 9.12, 9.23a, 9.24, 12.1b, 12.8a, 12.15, 12.17, 15.3, 17.20 from *Oceanography,* 3rd Ed., by T. Garrison, © 1999 Brooks/Cole. All rights reserved.

Figs. 5.2, 9.6, 13.11, 14.2, 15.1, 17.1, 17.2, 17.7, 17.8, 17.14, 17.16 from *Environmental Science,* 8th ed., by G. T. Miller, © 2001 Brooks/Cole. All rights reserved.

Figs. 2.5, 4.2, 6.8, 8.16, 10.1, 10.2, 10.6, 10.11, 10.16, 10.20, 10.22, 12.1a, 12.7a, 12.10, 13.12, 13.17, 16.15, 17.4a-b from *Physical Geology: Exploring the Earth,* 3rd Ed., by J. S. Monroe and R. Wicander, © 1998 Brooks/Cole. All rights reserved.

Figs. 1.5, 2.3 4.1, 6.2, 7.4a, 7.24, 8.1, 9.1, 9.3b, 9.5, 9.23b-e, 10.7, 12.18, 12.19, 13.1, 13.4, 13.7, 13.9, 13.18, 14.1, 14.19a-d, 15.2, 15.4, 15.5, 15.8, 15.10, 15.11, 15.12a, 16.16 from *Earth Science Today,* by B. Murphy and D. Nance, © 1999 Brooks/Cole. All rights reserved.

Figs. 10.5, 10.9, 10.13, 12.6, 13.3, 13.6a,b, 14.23, 17.23 from *Geology and the Environment,* 2nd Ed., by B. W. Pipkin and D. D. Trent, © 1997 Brooks/Cole. All rights reserved.

Figs. 5.6, 7.34, 8.4, 8.7, 8.13, 8.15, 8.21, 9.13, 12.16, 16.11, 16.17 from *Historical Geology: Evolution of Earth and Life Through Time,* 3rd Ed., by R. Wicander and J. S. Monroe, © 2000 Brooks/Cole. All rights reserved.

CHAPTER 1 Fig. 1.3, Dept. of Natural Resources/Geology, WA; **Fig. 1.13,** Reprinted by permission of the James Ford Bell Library, University of Minnesota; **Figs. 1.14, 1.15, 1.16,** NASA; **Fig. 1.21,** D. Pirie

CHAPTER 2 Figs. 2.6, 2.7a-p, 2.8a-j, D. Pirie

CHAPTER 4 Figs. 4.3, 4.4, 4.5, 4.6, 4.7 inset, 4.9, 4.10, 4.11, 4.12, 4.13, 4.14, 4.15, 4.17 inset, 4.18, 4.20, D. Pirie; **Fig. 4.16,** USGS/Donald Millineaux; **Fig. 4.17a,** PhotoDisc; **Fig. 4.19,** USGS

CHAPTER 5 Figs. 5.3, 5.4, 5.8e, f, h, 5.10, 5.11, 5.12, 5.13, 5.14, 5.15, 5.16, 5.17, 5.18a, 5.19, 5.20a, 5.21, 5.22, 5.23, 5.24, 5.25, 5.26, 5.27, D. Pirie; **Fig. 5.7,** John S. Shelton; **Fig. 5.8a,** Courtesy of Geological Survey of Canada; **Fig. 5.8b,** Smithsonian; **Fig. 5.8c,** Stephen R. Manchester; **Fig. 5.8d,** Courtesy of Los Angeles Museum of Natural History; **Fig. 5.8g,** NAGT; **Fig. 5.8i,** Reed Wicander; **Fig. 5.8k,** Stew Monroe; **Fig. 5.9** From "A Paleobotanical Interpretation of Tertiary Climates in the Northern Hemisphere," by J. A. Wold, *American Scientist,* 1978, Vol. 66, no. 6, p. 695. Reprinted with permission of Sigma Xi, The Scientific Research Society. **Fig. 5.18b,** Elsevier; **Fig. 5.20b,** Jeremy Young and Markus Geisen, Natural History Museum, London; **Fig. 5.20c,** B. A. Masters; **Fig. 5.27 inset,** Elsevier

CHAPTER 6 Figs. 6.4a-g, 6.4 h (inset), 6.6a-g, 6.7, D. Pirie

CHAPTER 7 Fig. 7.1, USGS; **Fig. 7.3,** Photo Researchers; **Fig. 7.6,** John Waldron; **Fig. 7.8a,** Photri; **Fig. 7.8b,** USGS/J. B. Stokes; **Fig. 7.9,** Sue Monroe; **Fig. 7.10,** B. Pipkin; **Fig. 7.11,** Courtesy of the Geological Survey of Canada; **Fig. 7.13,** Tom and Pat Leeson/Photo Researchers; **Fig. 7.16** Redrawn from R. Barrick/USGS; **Fig. 7.18,** Corbis; **Fig. 7.20,** Corbis; **Fig. 7.22,** USGS; **Fig. 7.23** From H. L. James, G.S.A. Bulletin, vol. 66, Plate 1, p. 1454. Reprinted with permission of Geological Society of America.

CHAPTER 8 Fig. 8.3 From *Evolution: Concepts and Consequences,* 2nd Ed., by L. S. Dillon. Reprinted by permission. **Fig. 8.5** Redrawn from Giants of Geology, by C. L. Fenton and M. A. Fenton. Used by permission of Doubleday, a division of Bantam Doubleday Dell Publishing Group, Inc. © 1998 Brooks/Cole. All rights reserved. **Fig. 8.12,** USGS; **Fig. 8.19** Redrawn from L. W. Mintz, *Historical Geology: The Science of a Dynamic Earth,* 3rd ed., p. 27. (Westerville, Ohio: Charles E. Merrill Publishing Company, 1981).

CHAPTER 9 Fig. 9.10 From *Ocean Science,* by K. Stowe. © 1983 John Wiley & Sons, Inc. Reprinted by permission. **Fig. 9.15** Dr. Peter Sloss, et al., NOAA/NESDIS/NGDC

CHAPTER 10 Fig. 10.3 Data from NOAA; **Figs. 10.4, 10.17** Redrawn from *Earthquakes,* by B. A. Bolt. © 1978, 1988 W. H. Freeman & Co. Reprinted by permission. **Fig. 10.8,** Research by Philip Liu, Seung, Seo, Sung Yoon, Civil and Environmental Engineering, Devine, Cornell Theory Center. **Fig. 10.12** From *Nuclear Explosions and Earthquakes: The Parted Veil,* by B. A. Bolt. © 1976 W. H. Freeman and Co. Reprinted by permission. **Fig. 10.24,** USGS

CHAPTER 11 Figs. 11.20, 11.21, Crystal Productions

CHAPTER 12 Fig. 12.7b, Stew Monroe; **Fig. 12.9,** USGS/S. J. Williams

CHAPTER 13 Fig. 13.2 Modified from *U. S. News and World Report* (18 March 1991): 72–73. **Fig. 13.6c,** USGS data; **Fig. 13.19,** Frank Kujawa, University of Central Florida, GeoPhoto; **Figs. 13.20, 13.21,** USGS

CHAPTER 14 Figs. 14.3, 14.4, 14.5, 14.7 14.8, 14.9, 14.13, 14.16, 14.24, USGS; **Fig. 14.25,** USDA; **Fig. 14.19e** From *Introduction to Physical Geography,* 2nd Ed., by R. C. Scott. © 1996 Brooks/Cole. All rights reserved.

CHAPTER 15 Figs. 15.12a-l, C. Donald Ahrens

CHAPTER 16 Fig. 16.1, Damian Nance; **Figs. 16.3, 16.4, 16.12,** USGS; **Fig. 16.18,** Nevada Bureau of Mines and Geology; **Fig. 16.19,** USGS/R. E. Wallace; **Fig. 16.21,** USGS

CHAPTER 17 Fig. 17.9a, J. S. Shelton; **Fig. 17.9b,** Corbis; **Fig. 17.11,** USGS; **Fig. 17.15,** Bryan Peterson; **Fig. 17.17** From *Biology: Concepts and Applications,* 4th ed., by C. Starr. © 2000 Brooks/Cole. All rights reserved. **Fig. 17.19** Data from Electric Power Research Institute. Adapted and updated from *Biology: The Unity and Diversity of Life,* 6th ed., by C. Starr and R. Taggart. © 1992 Wadsworth. All rights reserved.